Global Warming:

The Science and the Pseudoscience

James G. Clancy

August 2012

ISBN-13: 9781478373483

I.   Introduction

Global Warming is confirmed. Global Warming is "settled" science. An overwhelming consensus of experts has proven beyond a doubt that humanity has caused the planet to heat up in an unprecedented fashion never before seen in its four billion year history. Warm weather records are being shattered like never before. Icebergs are disappearing. Glaciers are melting. Sea levels are rising fast. Islands are disappearing. Manhattan will soon be flooded; Al Gore even simulates this real-time in his "Inconvenient Truth" video. Extreme weather events have dramatically increased. Cold weather is also being caused by global warming. Scientists have told us that people have caused planetary warming and we must blindly believe the experts without question. After all, scientists are never wrong, scientists are never deceitful. Scientists are superhuman. Today, we must believe whatever scientists say, just like we used to unconditionally believe God in a previous age. Anyone who dares to challenge the scientists is a corrupt, science-hating Creationist, who has been bribed by the ever-avaricious fossil fuel industry. And oh, we must never forget the doomed fate of our children and grandchildren!!

The above is just some of what everyone is constantly bombarded with in the mainstream media. However, the Climategate emails, leaked at the end of 2009, have provided significant insight into the systemic corruption that has afflicted climate science, which was, hitherto, very well hidden from the general public. To start with, let us first examine how anthropogenic influence on global climate was first "proven beyond a doubt".

http://www.ecowho.com/foia.php?file=0845217169.txt&search=Hadley+Centre

"Tim Barnett of the Scripps Institution of Oceanography, part of the University of California at San Diego, compared model estimates of natural temperature fluctuations over the past 400 years with the best evidence from the real world -- from instruments in the past century and "proxy data", such as Briffa's tree rings, from before that....

Barnett knows how easily this can happen. He was a lead author for a critical chapter in the last IPCC scientific assessment, which investigated "the detection of climate change and attribution of causes". It formulated the IPCC case that the evidence points towards a human influence on climate, but it warned repeatedly that great uncertainties remained. "We wrote a long list of caveats in that chapter," says Barnett. "We got a lot of static from within IPCC, from people who wanted to water down and delete some of those caveats. We had to work very hard to keep them all in." *Even so, when the findings were first leaked to the New York Times, it was under the headline "Scientists finally confirm human role in global warming"*

In fact, the practice of good science is the exact opposite of what the mainstream media is pleading with and trying to "educate" the public to do – to stop thinking, to stop being skeptical, to follow without question, to believe the experts blindly. Such an approach has many more similarities with the practice of religion. Many decades ago, Richard Phillips Feynman, a Nobel Prize winning American theoretical physicist known for his work in the path integral formulation of quantum mechanics, the theory of quantum electrodynamics, and the physics of the superfluidity of supercooled liquid helium, as well as in particle physics, eloquently described science as "the belief in the *ignorance* of the experts." Good science is not about conformity, it is all about *challenging* the status quo. After all, had Einstein and Bohr never challenged the three-century old Laws[1] of Newton, had Einstein and Bohr been discouraged/persecuted/disparaged/vilified by pseudoscience bandits determined to stamp out any questioning of settled science, neither the Theory of Relativity, nor Quantum Mechanics, nor the exponential technological advances of the twentieth century would ever have materialized. The reader should also note that Einstein's Relativity and Bohr's Quantum theories were and still are at loggerheads to this day. For decades, physicists have been attempting unification without concrete success so far. But this is okay, as long as scientists remain free to formulate hypotheses based on observations and facts, wherever they may lead.

Patricia Wentworth, a British crime fiction writer of the early twentieth century succinctly captured the pitfalls of an absolutist code of belief as follows: "Dogma is an impediment to the free exercise of thought. It paralyses the intelligence. Conclusions based upon *preconceived* ideas are valueless. It is only the open mind that really thinks." In summary, good science, and the advances in human knowledge that have accompanied it, stems primarily from *retaining an open mind.* The opposite, sadly also prevalent throughout much of human history, has been religious dogma, based on a fixed system of absolute beliefs which are portrayed to the public as beyond question or challenge, thereby imposing obstacles to the natural human quest for further erudition. Probably the most notable of such exercises of dogma was the imposition by the medieval Catholic Church during the Dark Ages, of beliefs dating back millennia, and the related persecution of scientists like Copernicus and Galileo, who dared to challenge these beliefs based on astronomical observations. Another example is Creationism /Intelligent Design which continues to be a widely-held belief by many today, despite the solid science of Darwin's Theory of Evolution. The concept of religious dogma is not limited to monotheistic religions alone. Communist dictatorship in the twentieth century was also a form of religion, stifling human intellectual development and advancement. The key differentiator is a closed versus open mind. Fortunately, despite the perils and setbacks resulting from absolutism, humanity has always had the tenacity to overcome dogma with pragmatism and, thus, our knowledge has advanced exponentially over the millennia. But this has not been a uniform journey. Often, it has been one step backwards and then two steps forward.

Before delving into the controversy of global warming, let us review an example of good science. Paul J. Steinhardt is the Albert Einstein Professor of Science at Princeton University and a professor of theoretical physics. He received his B.S. at the California Institute of Technology and his Ph.D. in Physics at Harvard University. He was a Junior Fellow in the Harvard Society of Fellows and Mary Amanda Wood Professor of Physics at the University of Pennsylvania before joining the faculty at Princeton University in 1998. He is currently the Director of the Princeton Center for Theoretical Science. Thirty years ago Alan H. Guth, then a struggling physics postdoc at the Stanford Linear Accelerator Center, gave a series of seminars in which he introduced "inflation" into the lexicon of cosmology. The term refers to a brief burst of hyperaccelerated expansion that, he argued, may have occurred during the first instants after the big bang. One of these seminars took place at Harvard University, where Mr. Steinhardt was a postdoc. He was immediately captivated by the idea, and he has been thinking about it almost every day since. Many of his colleagues working in astrophysics, gravitational physics and particle physics have been similarly engrossed. To this day the development and testing of the inflationary theory of the universe is one of the most active and successful areas of

---

[1] In scientific parlance, Laws are far stronger than Theories in terms of the level to which the science is "settled"

scientific investigation. Its raison d'être is to fill a gap in the original big bang theory. The basic idea of the big bang is that the universe has been slowly expanding and cooling ever since it began some 13.7 billion years ago. This process of expansion and cooling explains many of the detailed features of the universe seen today, but with a catch: the universe had to start off with certain properties. For instance, it had to be extremely uniform, with only extremely tiny variations in the distribution of matter and energy. Also, the universe had to be geometrically flat, meaning that curves and warps in the fabric of space did not bend the paths of light rays and moving objects.

Today, cosmic inflation is so widely accepted that it is often taken as established fact. The idea is that the geometry and uniformity of the cosmos were established during an intense early growth spurt. But in the three decades that ensued, new evidence has come to light that has cast doubt on the inflation theory. So what did the creators of inflation theory, including Mr. Steinhardt do? Did they start from a preconception that inflation theory must be true and try to fit and rationalize the new evidence to conform to the theory? No. They are actively considering replacing the inflation theory.

> "As the original theory has developed, cracks have appeared in its logical foundations. Highly improbable conditions are required to start inflation. Worse, inflation goes on eternally, producing infinitely many outcomes, so the theory makes no firm observational predictions. Scientists debate among (and within) themselves whether these troubles are teething pains or signs of a deeper rot. Various proposals are circulating for ways to fix inflation or replace it." [2]

The above is good science – an open mind.

By contrast, the theory of Anthropogenic Global Warming is a striking example of bad science, of faith-based religious dogma corrupting the world of science, in other words, pre-conceptualism running amok. The stipulation of human-emitted carbon dioxide causing planetary warming has first been established as a foregone conclusion not to be challenged in any way, shape or form. What has ensued since is an elaborate litany of rationalizing or simply ignoring contrarian evidence, of reinventing global warming to climate weirding, of having continual amnesia of past explanations and predictions that have not materialized, of exaggerating variations in both temperature and carbon dioxide which pale in comparison to previous disparities in the Earth's long history, of doctoring data to falsely demonstrating a causal link between global temperature and anthropogenic causes, when an objective evaluation concludes otherwise, all in an effort to keep relevant the basic commandment that human $CO2$ emissions are causing climate change. In other words, the theory of Anthropogenic Global Warming is to science what a witch-doctor is to medicine.

[2] http://www.scientificamerican.com/article.cfm?id=the-inflation-summer

Stirring up mass-hysteria by amplifying out of all proportion cherry picked warm-weather events is a key tactic repeatedly used by proponents of Anthropogenic Global Warming to manipulate and hoodwink the public into believing in AGW. A recent example was the exaggeration by NASA's GISS of the Greenland Ice Melt in July 2012. The misinformation was then replicated globally, even in far-flung places such as Iran and India. Here is the Tehran Times:

http://tehrantimes.com/science/99954-greenland-ice-sheet-melted-at-unprecedented-rate-during-july-

Greenland ice sheet melted at unprecedented rate during July
Suzanne Goldenberg
On Line: 25 July 2012 15:03
In Print: Thursday 26 July 2012

The Greenland ice sheet melted at a faster rate this month than at any other time in recorded history, with virtually the entire ice sheet showing signs of thaw.

The rapid melting over just four days was captured by three satellites. It has stunned and alarmed scientists, and deepened fears about the pace and future consequences of climate change.

In a statement posted on NASA's website on Tuesday, scientists admitted the satellite data was so striking they thought at first there had to be a mistake.

"This was so extraordinary that at first I questioned the result: was this real or was it due to a data error?" Son Nghiem of NASA's jet propulsion laboratory in Pasadena said in the release.

He consulted with several colleagues, who confirmed his findings. Dorothy Hall, who studies the surface temperature of Greenland at NASA's space flight centre in Greenbelt, Maryland, confirmed that the area experienced unusually high temperatures in mid-July, and that there was widespread melting over the surface of the ice sheet.

Climatologists Thomas Mote, at the University of Georgia, and Marco Tedesco, of the City University of New York, also confirmed the melt recorded by the satellites.

However, scientists were still coming to grips with the shocking images on Tuesday. "I think it's fair to say that this is unprecedented," Jay Zwally, a glaciologist at NASA's Goddard Space Flight Center, told the Guardian.

The set of images released by NASA on Tuesday show a rapid thaw between 8 July and 12 July. Within that four-day period, measurements from three satellites showed a swift expansion of the area of melting ice, from about 40% of the ice sheet surface to 97%.

Zwally, who has made almost yearly trips to the Greenland ice sheet for more than three decades, said he had never seen such a rapid melt.

Now, pseudoscience hysteria aside, what does the science really tell us?

http://www.theregister.co.uk/2012/08/03/greenland_ice_sheet_not_about_to_disappear/print.html

Greenland ice sheet not going anywhere in a hurry, say boffins
Danish prof: You can stop work on that ark for now
By Lewis Page
Posted in Energy, 3rd August 2012 09:16 GMT

Doom-laden predictions that the seas are set to rise by a metre or more this century due to the melting of the Greenland ice sheet are well off the mark, a team of scientists has announced in a new study of the matter.

"It turns out that the ice sheet, in relation to this point, behaves more dynamically and is able to more quickly stabilise itself in comparison to what many other models and computer calculations otherwise predict," explains Professor Kurt Kjaer of Copenhagen uni.

According to Kjaer and his colleagues, the scenarios which predict huge melting and massive resultant sea-level rises are flawed because they rely on a very limited amount of information spanning just a few recent years: the Greenland ice has only been intensively studied for a relatively short period of time. This has led scientists to assume that rapid melting seen lately will carry on uninterrupted, pouring gigatonnes of water into the world's oceans and inundating coastal areas around the planet.

But this is mistaken: it now emerges that periods of rapid melting like the one just seen have happened in the past - but then, rather than continuing, the apparently runaway melting simply stopped.

"We've used a combination of old aerial photographs from the '80s and recent satellite data. In this way we've been able to gain an overview of the thinning of the ice sheet over the last 30 years in northwestern Greenland," says Shfaqat Abbas Khan of the Technical University of Denmark (DTU), who worked on the study alongside Kjaer.

"We are the first who have been able to show that the Greenland Ice Sheet was on as dramatic a diet at the end of the '80s as it is today. On the positive side our results show that - despite a significant thinning in peripheral regions from 1985-1992 - the thinning slowed and then died out."

Kjaer for his part predicts that the ongoing rapid melting at the moment will cease within a decade, leading to a stable period like the one his team has identified from the early '90s until 2003.

"It is certain that many of the present calculations and computer models of ice sheet conditions that built upon a short range of years since 2000 must be reassessed," states the prof, uncompromisingly. "It is too early to proclaim the 'ice sheet's future doom' and subsequent contribution to serious water problems for the world."

The new study has been deemed important enough to make today's edition of premier boffinry mag Science, where it can now be read [1] by subscribers.

A related study examining old aerial photos of the Greenland ice was published in June [2], revealing that in the 1930s the glaciers there were retreating even faster than they are today: but again, the process subsequently stopped on its own. ®

Links

- http://www.sciencemag.org/content/337/6094/569.abstract?sid=822b555a-638b-4a49-a021-3dab84f17457
- http://www.theregister.co.uk/2012/06/02/1930s_greenland_glacier_retreat/

Worse, there has been a multi-year campaign to discredit anyone who dares to challenge the theory as creationist, in the pocket of Big Oil, anti-science, lunatic, etc. Nobody has been spared in this deluge of suppression, including renowned scientists such as Don J. Easterbrook, John Christy, Roy Spencer, Richard Lindzen, Hal Lewis and many others. Below is just one of many egregious examples: the Huffington Post is now comparing anyone who takes issue with the theory of Anthropogenic Global Warming as having committed a crime against humanity:

http://www.huffingtonpost.com/james-odea/accountablity-global-warming-crime_b_1723704.html

Should Accountability for Global Warming Be Linked to Crimes Against Humanity?
James O'Dea
Posted: 07/31/2012 6:21 pm

"Who will be held accountable for inaction on global warming and how will they be brought to justice?"

"In the not-too-distant future will politicians who intentionally ignore global climate change, or who obstruct action to implement conscientious policies to prevent deterioration of climate conditions, be deemed criminally negligent?" asks says this article by James O'Dea. "The scale of death and destruction resulting from global warming may potentially exceed losses due to genocides and world wars."

"Given the tools at our disposal to measure the adverse climate impact of human behavior and the overwhelming consensus of climate scientists about the causes of global warming any conscious choice to deny it and refuse to take action must be considered extremely risky behavior and, I believe, criminal negligence."

"It is becoming increasingly clear that climate disruptions are going to increase global conflicts over resources, food and water, and create climate refugees. Global warming will destroy any chances of global peace. In international law we have established an international criminal court and war crimes tribunals to try those found guilty of crimes against humanity. But who will be held accountable for inaction on global warming and how will they be brought to justice?"

By this logic, I would guess that an international warrant should be issued for my arrest, and I should then be sent to The Hague to face trial for Crimes against humanity for simply having written this book!! The parallels to the medieval Inquisition are striking in terms of stifling skepticism and the pursuit of knowledge.

The eternally insightful President Dwight D. Eisenhower warned us many decades ago about the corruption of science that we now see with Anthropogenic Global Warming:

> "Akin to, and largely responsible for the sweeping changes in our industrial-military posture, has been the technological revolution during recent decades.
>
> In this revolution, research has become central; it also becomes more formalized, complex, and costly. A steadily increasing share is conducted for, by, or at the direction of, the Federal government.
>
> Today, the solitary inventor, tinkering in his shop, has been overshadowed by task forces of scientists in laboratories and testing fields. In the same fashion, the free university, historically the fountainhead of free ideas and scientific discovery has experienced a revolution in the conduct of research. Partly because of the huge costs involved, a government contract becomes virtually a substitute for intellectual curiosity. For every old blackboard there are now hundreds of new electronic computers.
>
> The prospect of domination of the nation's scholars by Federal employment, project allocations, and the power of money is ever present - and is gravely to be regarded.
>
> Yet, in holding scientific research and discovery in respect, as we should, we must also be alert to the equal and opposite danger that public policy could itself become the captive of a scientific-technological elite.
>
> The prospect of domination of the nation's scholars by Federal employment, project allocations, and the power of money is ever present - and is gravely to be regarded.
>
> It is the task of statesmanship to mold, to balance, and to integrate these and other forces, new and old, within the principles of our democratic system - ever aiming toward the supreme goals of our free society."

On a personal note, I have no affiliation to monotheistic religion. I am a committed atheist. I believe that propositions of Creationism and Intelligent Design are as devoid of any credible scientific basis as is Anthropogenic Global Warming. And I also do not have any connection to, or love of fossil fuel companies. In fact, after the BP oil spill disaster of 2010, I was very active on Internet blogs, openly calling for the Federal government to shut down all of BP's operations in the United States of America, confiscate all of its American assets, and sell off the assets to the highest bidding competitor, in order to pay for the cleanup of the unforgivable environmental mess the company made all over the Gulf coast.

Perhaps even more notably, from the early 1990s until the middle-2000s, I used to be a strong believer in Anthropogenic Global Warming. After all, we had to trust the media's reporting of what the scientists were saying without question, correct? In this high technology age, science was on a pedestal. To a large extent, science had replaced religion as the main anchor of confidence of the public. Like many others, I had suppressed my natural human skepticism when it came to scientific pronouncements. And, in the warming years of the 1980s, 1990s and early 2000s, the simplistic concept of human CO2 emissions causing global warming was easy to believe. When in 2000, David

Viner, a senior scientist at the CRU prophesized the imminent end of snow, and the IPCC followed-up in its Third Assessment Report with the prediction of Ice Storms replacing snow, I believed them unquestioningly. When I vacationed in Greece in the early 2000s and the temperature shot up to 45°C, I was convinced that was global warming in action. Green winters and lack of cold and snow had become much more common in the late 1990s. I was mad at governments for not banishing fossil fuels right away and not prohibiting petroleum-based cars completely. When Michael Crichton came out with the book *"State of Fear"* casting doubt on AGW, I laughed it off as the work of an author who did not understand anything about science.

Then came the extremely cold winter of 2007-2008: finally, my natural human trait of skepticism started to break through the massive wall of Pavlovian trust in the media reporting of science that had been ingrained into me over decades of continual manipulation. Ah, but maybe the cold winter was just a fluke, a one-off exception. But then an even colder winter struck the year after. I became interested. I started using my brain again. I began to dig. What I found was not pretty. Over time, after a lot of painstaking research of the facts, I came to the conclusion that it was possible for the practice of science to be corrupted, just like anything else. Climate science had negatively transformed into the exact antithesis of good scientific practice. As Richard Feynman had so eloquently summarized many years ago, good science relies primarily on skepticism - to put a hypothesis through every test imaginable and try to prove it wrong, and check if it stands up despite the tests. By contrast, in the world of climate science, the exact opposite had happened. A foregone conclusion had been made that human carbon dioxide emissions were causing planetary warming and everything else flowed from that preconception. Facts supporting this hypothesis were cherry picked and exaggerated. Facts contradicting the hypothesis were ignored, minimized, explained away, or rationalized in a way to "fit" the preconception. Reinventing new explanations, based on changing weather was also permitted and encouraged: warming to climate change to weirding; amnesia was to be strictly applied to previous prognostications which were discredited. Anyone attempting to develop alternate hypotheses was belittled. The parallel to the modus operandi of monotheistic religions was truly striking. Each of the great religions starts from the presumption that God is real and everything else is explained in that context. No questioning on the existence of God is permitted. Anyone challenging the authenticity of God is assailed as a heretic. Sadly, such a faith-based method of thinking had embedded itself into, and tarnished the world of science.

Finally, a word on why I am writing this book. Yes, I do indeed have a large axe to grind. I am convinced that over seven billion human beings are doing enormous damage to the planet's environment. Rainforests are disappearing fast. Biodiversity is being destroyed; species are going extinct faster than at any point in Earth's long four billion year history. Over-agriculture is literally destroying the earth, yet we still have a growing shortage of food, and in addition, over thirty percent of food gets spoiled or otherwise wasted during transportation. There is already an acute shortage of fresh water the world over. Overfishing is depleting the oceans so fast that they are predicted to become quasi-lifeless by mid-century. By 2050, it is clear that the planet will not be able to sustain the estimated nine billion souls at current income levels, without even considering the rapid growth in prosperity and associated consumption in emerging nations such as China and India. In its 2006 report, the *Worldwatch* Institute stated that "the world's ecological capacity is simply insufficient to satisfy the ambitions of China, India, Japan, Europe and the United States as well as the aspirations of the rest of the world in a sustainable way".[3] And what has been the focus of world governments?

World leaders have been at yearly climate summits. Enormous prioritization has been placed on reducing humanity's carbon footprint to "combat" climate change where no causal link to human carbon dioxide emissions has been proven, as this book will demonstrate. In a world of limited human and financial resources, especially in times of economic adversity, this has meant that woefully little attention is being given to the *real* environmental issues which afflict this planet. Without proper contingency planning of solutions and/or mitigation to these problems, humanity will likely face an extinction-level crisis by the middle of the twenty-first century. When was the last summit of global leaders to deal with the food and water crisis? What has been done in the past decade to counter wasteful fishing practices which trap not only intended catch but also other marine species, to reduce overfishing, to stop the massacre of sharks just for shark fin soup, to reduce wastage in food distribution, to reduce the throwing away of a significant amount of food not consumed in developed countries, etc? Where is the long-term planning of world leaders to deal with the multitude of environmental time bombs ticking down fast to inevitable detonation? Well, the short answer has been: why not kick the real environmental cans down the road, when one has the distraction of "anthropogenic" climate change to pontificate about?

And where is the media when it comes to these real environmental problems? Well, it is much more sensational to prophesize about extreme weather events, I guess.

Worse, the frenzy about anthropogenic climate change has set off a frenzy of activity on "green" technology while turning a blind eye to the negative impact of these alternative forms of energy. Multinational enterprises around the world have benefited handsomely from this anti-carbon mania, especially with the mantra of being "environmentally-friendly". For example, we think of electric batteries as being "clean", but rarely are we told about the very dirty business of mining rare-earth minerals such as lithium which are required in batteries. China produces ninety-five percent of the world rare-earths, and the multitude of extremely polluted rivers in that country attest to extreme environmental damage associated with one form of "clean" energy.

The "fight" against climate change has resulted in even bigger anti-environment and anti-humanity anomalies, most notably in the area of food-ethanol. In the first decade of the twenty-first century, anti-fossil fuel mania resulted in the ethanol craze. Below is the share of feedstock used globally for biofuel (source: OECD, FAO, EIA):
- Corn: 13.1%
- Wheat: 0.5%
- Sugar: 21.4%
- Molasses: 18.4%
- Palm Oil: 2.9%
- Soybean Oil: 5.7%
- Rapeseed Oil: 40.8

[3] http://www.worldwatch.org/node/3992

The ethanol craziness became so egregious that some international organizations had to plead for a return to sanity:

http://www.wto.org/english/news_e/news11_e/igo_10jun11_report_e.pdf

Reducing policy conflicts between food and fuel

"101. Between 2000 and 2009, global output of bio-ethanol quadrupled and production of biodiesel increased tenfold; in OECD countries at least this has been largely driven by government support policies.50 Moreover, trade restrictions by favouring domestic sources of raw material for biofuels do not maximise expected environmental benefits. Biofuels overall now account for a significant part of global use of a number of crops. On average, in the 2007-09 period that share was 20% in the case of sugar cane, 9% for both oilseeds and coarse grains (although biofuel production from these crops generates by-products that are used as animal feed), and 4% for sugar beet.51 With such weights of biofuels in the supply-demand balance for the products concerned, it is not surprising that world market prices of these products (and their substitutes) are substantially higher than they would be if no biofuels were produced. Biofuels also influence products that do not play much of a role as feedstocks, for example wheat, because of the close relations between crops on both the demand side (because of substitutability in consumption) and the supply side (due to competition for land and other inputs)."

Who has benefited from the boom in food ethanol? Multinational energy companies like Shell have been given yet another opportunity to reap enormous profit. And Shell is, of course, very proud of its new, thriving ethanol business. The company regularly boasts about this new line of business through which the company is contributing significantly to "saving the planet". In a world with over seven billion people and going to nine billion by 2050, in a world where over a billion people are still starving, food is already scarce, and we're now misusing a substantial portion of the limited food for fuel? Food is for feeding people, not for feeding cars. Food is the power of life, not the power of engines.

Another industry which has benefited enormously from the anti-carbon craze has been the nuclear industry – companies such as General Electric. There has been an explosion in construction of new nuclear power plants, in order to reduce carbon emissions. Never mind Fukushima….

The much maligned Danish skeptical environmentalist, Bjørn Lomborg, expressed a similar frustration with the misalignment of priorities in a recent interview on Fareed Zakaria's GPS show[4].

"Well, there are two main solutions. One is that we have a lot of technologies that we know how to get clean drinking water. We also know how to get much of the air pollution, most of the air pollution deaths are actually caused by indoor air pollution. People cooking with bad fuels like dung or cardboard. Let's make sure they actually get access to fossil fuels. That makes a lot of people uncomfortable, but of course that's the reality that we live with. And that's why 2 million people don't have to die in the developing world each year because of unsafe cooking and -- and heating fuels. But the long-term solution for that, of course, is to make sure that people actually get richer in the third world. It's a poverty problem. And so I'm a little concerned about the fact that we talk a lot about the Kyoto Protocol. But there's another city with a protocol that we don't talk very much about, the Doha Round. The idea of free trade. That is one that most economists

---

[4] http://transcripts.cnn.com/TRANSCRIPTS/1206/17/fzgps.01.html

would estimate would give much, much better opportunities in the long run for most countries in the world to actually get rid of their old problems, both environmental, but also all the other poverty-related problems, and then start focusing on environmental problems."

It makes me angry to see how the pre-conceptual craze about Anthropogenic Global Warming has so distracted humanity from focusing on the real and much more immediate environmental issues that threaten all life on this planet. They are not some imagined threats in the distant future, but real problems which are already having a major impact on humanity, and are confirmed to have an exponentially bigger impact on my children and grandchildren. In a world of many problems, we need pragmatism to drive policy and solutions, not obsessive dogmatism and fear-mongering around an unproven hypothesis, which is not even substantiated by good science. This is my main motivation for writing this book. If the book manages to have even a small iota of influence to reorient priorities to addressing real environmental issues, this will provide me enormous satisfaction.

The truth is that our current level of science can only predict weather very accurately for a couple of days, reasonably accurately up to a month (although they go wrong from time to time), and have some vague indications a few months into the future. Any predictions beyond a few months, as is demonstrated through multiple examples in this book of wrong forecasts by proponents of AGW, is faith-based prophesy, not science. To quote some ageless wisdom by Shakespeare:

> The earth hath bubbles, as water has,
> And these are of them.
>
> Or have we eaten on the insane root
> That takes the reason prisoner?
>
> But 'tis strange:
> And oftentimes, to win us to our harm,

Or to quote Churchill: "Truth is treason in the empire of lies."

****

This book documents the science and the pseudoscience of global warming in simple terms understandable to the layperson with limited knowledge of science. There are also references to scientific sources, should the reader feel inclined to study further. Much of the information in this book is publicly available on the internet, with the exception of some supplemental calculations and statistical analyses that I have made. This book brings a good part of that fragmented information together, so that the reader is able to acquire an understanding of the bigger picture.

Some readers of this book may long have suspected something wrong from the contradictory statements and hyperbole about climate change. This book fills in some of the details. Other readers may well be strong believers of the theory of Anthropogenic Global Warming. There is nothing wrong with that. I was a believer for many years. My only recommendation would be for those readers to keep an open mind, examine the facts documented in this book and then make an informed judgment on the veracity of the theory, just as I have done after careful and long study.

II. There is no correlation between human CO2 emissions and global temperature

This section will prove conclusively that there is no correlation whatsoever between human CO2 (Carbon Dioxide) emissions and global temperature.

The wording "human CO2 emissions" has been selected very carefully, because, what the proponents of Anthropogenic Global Warming will, of course, not tell you, there have been *significant **natural** variations* of CO2 concentration since the beginning of Earth's history some four billion years ago.

CO2 does have a warming effect on the atmosphere but....

Svante August Arrhenius

The roots of the modern theory of Anthropogenic Global Warming date back to a brilliant, Nobel Prize winning 19th century Swedish scientist named Svante August Arrhenius (19 February 1859 – 2 October 1927)[5]

Arrhenius developed a theory to explain the ice ages, and in 1896 he was the first scientist to speculate that changes in the levels of carbon dioxide in the atmosphere could substantially alter the surface temperature through the greenhouse effect. He was influenced by the work of others, including Joseph Fourier.

Arrhenius equation

Arrhenius used the infrared observations of the moon by Frank Washington Very and Samuel Pierpont Langley at the Allegheny Observatory in Pittsburgh to calculate the absorption of infrared radiation by atmospheric CO2 and water vapor. Using 'Stefan's law' (better known as the Stefan Boltzmann law), he formulated his greenhouse law. In its original form, Arrhenius' greenhouse law reads as follows:

> "If the quantity of carbonic acid increases in geometric progression, the augmentation of the temperature will increase nearly in arithmetic progression."

This simplified expression is still used today:

$$\Delta F = \alpha \ln(C/C_0)$$

There is no doubt about the science, about the physics that CO2 has a greenhouse gas and has a warming effect on the atmosphere. Proponents of global warming have been promoting that CO2 acts like a global thermostat, that the level of CO2 concentration IS the determinant factor for global temperature. Examples of such promotion are documented later in the book.

But....

The global climate system is complex. As this book documents, the global climate and temperature are dependent on a multitude of factors which work in parallel and do often have opposing effects.

---

[5] More details on Arrhenius and his work can be found here: Patrick Coffey, Cathedrals of Science: *The Personalities and Rivalries That Made Modern Chemistry*, Oxford University Press, 2008

Furthermore, the long history of the Earth's climate demonstrates clearly that there has been no statistically significant correlation between CO2 concentration and global temperature, not in the present and not in the past.

Simple Illustration

A good analogy to illustrate:
- Start a fire and put a pot of water on the fire. The fire naturally has a warming effect on the water
- Now take a large bucket full of cold water and pour it on the pot.

What will be the net effect?

- One possibility is that if it is night-time and/or cloudy and/or winter-time, the fire will be overwhelmed and the pot, the water and the entire environment around it will be cool, if not cold.

- Another possibility is that if it is summertime and the sun is shining brightly, despite the fire being extinguished and brief spell of cold, the pot, the water and the entire environment around it will quickly warm up again.

Are there other possibilities?

If one thinks a little harder, there are a multitude of other possibilities that one can document.

What does this analogy illustrate?

The analogy serves to provide a small insight into the complexity of the world's climate, only the climate is many times more complex. This includes both the number of factors affecting the climate as well as the way they interact with each other.

This is why climate science is still in its infancy. Most of what is published today is subject to change, due to new research, new facts uncovered from the past and/or observed in the future.

This book attempts to document the related science in a simplified manner with the intention that the facts are understandable even to the layperson with a minimal knowledge of science, while providing references to reliable science sources, if the reader is interested in more detailed information.

As Carbon Dioxide increases, it has less warming effect

Below is another key piece of science which the proponents of Anthropogenic Global Warming do not wish to have widely disseminated in the public domain:

http://joannenova.com.au/globalwarming/graphs/log-co2/log-graph-lindzen-choi-web.gif

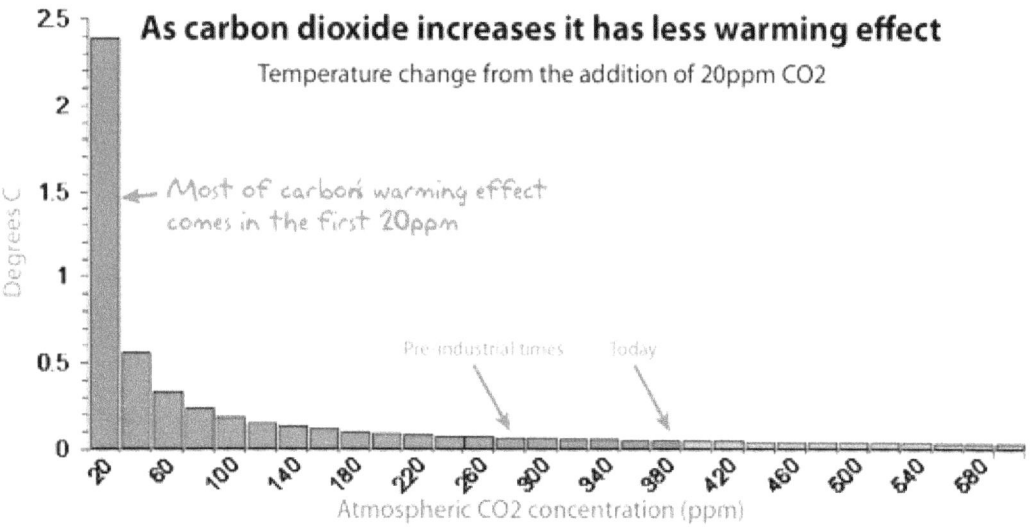

as well as the following:

http://www.junksciencearchive.com/Greenhouse/co2greenhouse-X2.png

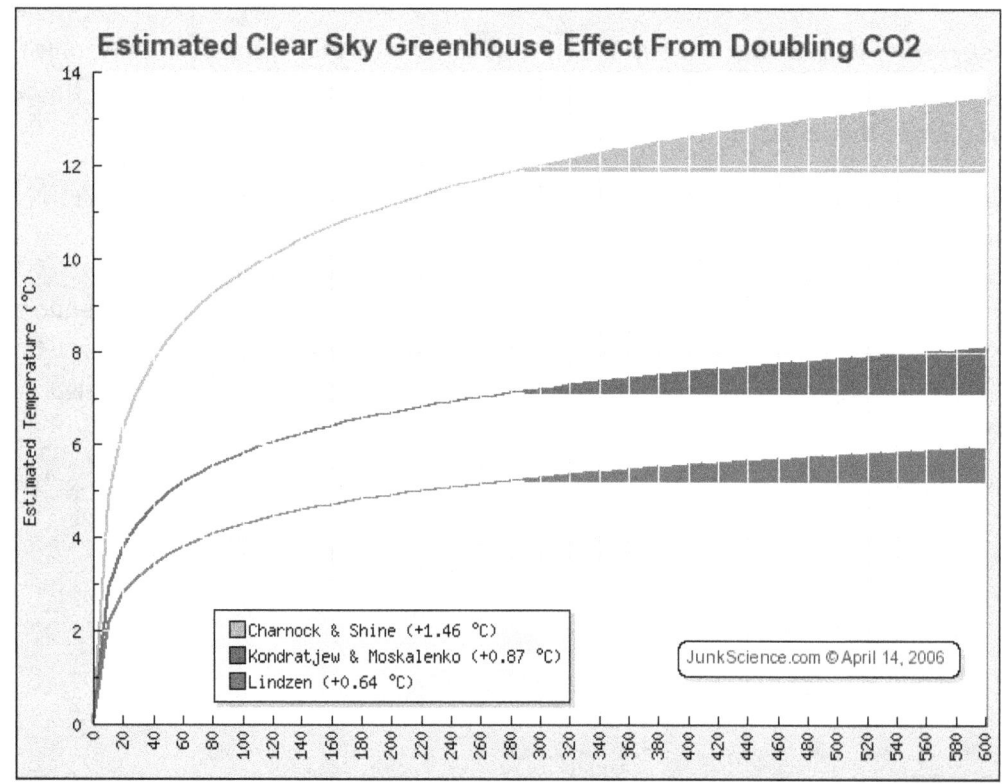

The relationship between carbon dioxide and radiative forcing is logarithmic, and thus increased concentrations have a progressively smaller warming effect. [6]

$$\Delta F = 5.35 \times \ln \frac{C}{C_0} \; \mathrm{W} \; \mathrm{m}^{-2}$$

where C is the CO2 concentration in parts per million by volume and C_0 is the reference concentration

Other Interesting Facts about Carbon Dioxide
http://www.geocraft.com/WVFossils/ice_ages.html

- Of the 186 billion tons of carbon from CO2 that enter earth's atmosphere each year from all sources, only 6 billion tons are from human activity. Approximately 90 billion tons come from biologic activity in earth's oceans and another 90 billion tons from such sources as volcanoes and decaying land plants. **6 billion out of 186 billion is 3% anthropogenic and 97% natural.**

  ⇨ Therefore, even if we suddenly stopped all human CO2 emissions, 97% of CO2 build up would still continue due to natural causes. Does it make sense to prioritize on spending trillions of dollars to convert to alternate energy, or does it make more sense to prioritize our resources to adapt to a naturally changing climate, as well as to focus on addressing real environmental issues where human activity is negatively impacting our planet?

- At 390 parts per million CO2 is a minor constituent of earth's atmosphere-- less than 4/100ths of 1% of all gases present. Compared to former geologic times, earth's current atmosphere is CO2-impoverished.

- CO2 is odorless, colorless, and tasteless. Plants absorb CO2 and emit oxygen as a waste product. Humans and animals breathe oxygen and emit CO2 as a waste product. Carbon dioxide is a nutrient, not a pollutant, and all life-- plants and animals alike-- benefit from more of it. All life on earth is carbon-based and CO2 is an essential ingredient. When plant-growers want to stimulate plant growth, they introduce more carbon dioxide. CO2 is as essential to plant life as oxygen is to animals.

- CO2 is not a poison at atmospheric levels. It takes a CO2 concentration of 110,000 parts per million (ppm), 275 times atmospheric concentration, for CO2 to have any sort of asphyxiating effect on human beings[7]

- CO2 that goes into the atmosphere does not stay there but is continually recycled by terrestrial plant life and earth's oceans-- the great retirement home for most (but not all) terrestrial carbon dioxide.

---

[6] Myhre et al., New estimates of radiative forcing due to well mixed greenhouse gases, Geophysical Research Letters, Vol 25, No. 14, pp 2715–2718, 1998

[7] Hamilton and Hardy 1974.

The Long-Term Record

Columbia University has done some excellent research in reconstructing the long-term variation in global temperature, CO2 concentration and sea level, based on data obtained from the Vostok ice cores in the Antarctic:

The graph is displayed below:

http://www.columbia.edu/~mhs119/Storms/Storms_Fig.03.gif

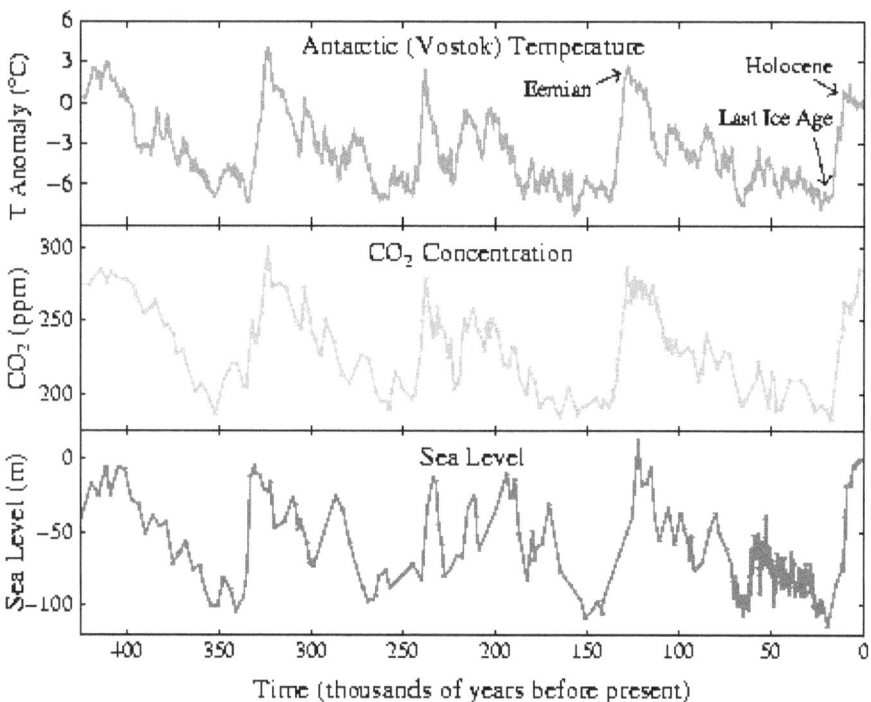

Observations

Although, a superficial review of the graphs might lead one to believe that there is a strong correlation between CO2 concentration and global temperature, a closer inspection leads to some observations which disprove such a simplistic conclusion:

- CO2 concentration has been rising naturally for the past 10,000 years, in other words starting long before the human industrial revolution
- In the same period of increasing CO2 concentration, temperatures have been falling, culminating in the little Ice Age which only ended at the beginning of the 20th century
- Sea Level versus CO2 concentration does not even exhibit a superficial similarity
- Further analysis is described in the ensuing sections

Natural CO2 variations during Earth's history

Proponents of Anthropogenic Global Warming would like you to believe that human beings with their industrial revolution and ever-increasing consumption have artificially and exponentially increased CO2 concentration in the atmosphere never before seen in Earth's natural history.

Not true. The history of Earth includes atmospheric CO2 lagging both glaciations and deglaciations, and large rapid spontaneous jumps in global temperatures without any important changes in atmospheric CO2 and without tipping Earth off into runaway catastrophes.

References:

- Aldrich, J. 1995. "Correlations Genuine and Spurious in Pearson and Yule." Statistical Science 10, 364-376.
- Adams, J. M. Maslin and E. Thomas. 1999. "Sudden Climate Transitions During the Quaternary" (Dansgaard-Oeschger events)
- " Progress in Physical Geography 23, 1-36; G. G. Bianchi and I. N. McCave. 1999"
- "Holocene Periodicity in North Atlantic Climate and Deep Ocean Flow South of Iceland." Nature 397, 515-517; M. McCaffrey, D. Anderson, B. Bauer, M. Eakin, E. Gille, et al. 2003.
- "Variability During the Last Ice Age: Dansgaard-Oeschger Events." NOAA Satellites and Information. http://www.ncdc.noaa.gov/paleo/abrupt/data_glacial2.html Last accessed on: 14 September 2007
- L. C. Gerhard. 2004. "Climate change: Conflict of Observational Science, Theory, and Politics." AAPC Bulletin 88, 1211-1220.

In geologic history, CO2 concentrations in the atmosphere were much higher, but temperatures often were lower. During the last 2 billion years the Earth's climate has alternated between a frigid "Ice House", like today's world, and a steaming "Hot House", like the world of the dinosaurs.

440 million years ago, in the Late Ordovician Period, CO2 concentrations were 10.5 times higher than they are today: 4200 parts per million (ppm). The Late Ordovician Period was an ice age.

35 million years ago, during the Eocene and Oligocene time span, the CO2 concentration of the Earth's atmosphere hovered between 1000 & 1500 ppm as compared to 390 parts per million today. At that time, life on earth was in the middle of being the most prolific, in terms of both flaura and fauna.

References:

- http://www.theresilientearth.com/?q=content/co2-temperature-during-middle-eocene-climatic-optimum
- http://www.nature.com/ngeo/journal/v3/n12/full/ngeo1014.html
- http://www.scotese.com/climate.htm
- http://www.scotese.com/mlordcli.htm
- http://www.scotese.com/oligocen.htm
- Science Magazine, 22 July 2005, volume 309 pp. 600-603.

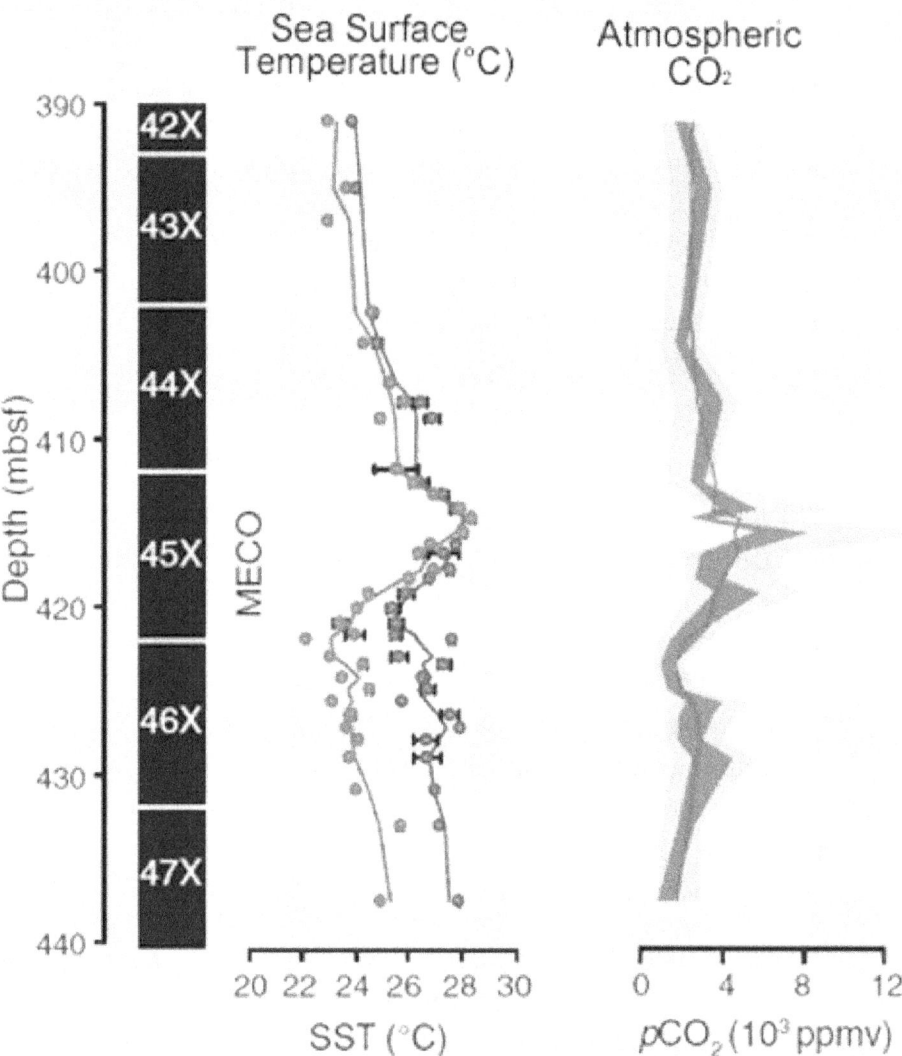

You will find more details in the later chapter: History of Global Temperature

The 800-year lag

Proponents of Anthropogenic Global Warming want you to believe that an increasing carbon dioxide levels cause global temperatures to rise. However, the *evidence* from the Vostok Ice Cores indicates the exact opposite. In other words, carbon follows temperature in the Vostok Ice Cores.

In the 1990's, the classic Vostok ice core graph showed temperature and carbon in lock step moving at the same time. It made sense to worry that carbon dioxide did influence temperature.

But by 2003 new data came in and it was clear that carbon lagged behind temperature. The link was back to front. Temperatures appear to control carbon, and while it's possible that carbon also influences temperature these ice cores don't show much evidence of that. After temperatures rise, on average it takes 800 years before carbon starts to move.

The extraordinary thing is that the lag is well accepted by climatologists, yet virtually unknown outside these circles. Unfortunately, to many people, such withholding of information will not come as a surprise: naturally, the public is fed only the information (and misinformation) which buttresses the stipulation of human CO2 emissions causing global warming.

It is impossible to see a lag of centuries on a graph that covers half a million years so the below link re-graphs the data from the original sources, but scales the graphs out so that the lag is visible to the naked eye. What follows is the complete set from 420,000 years to 5,000 years before the present.

http://joannenova.com.au/global-warming/ice-core-graph/

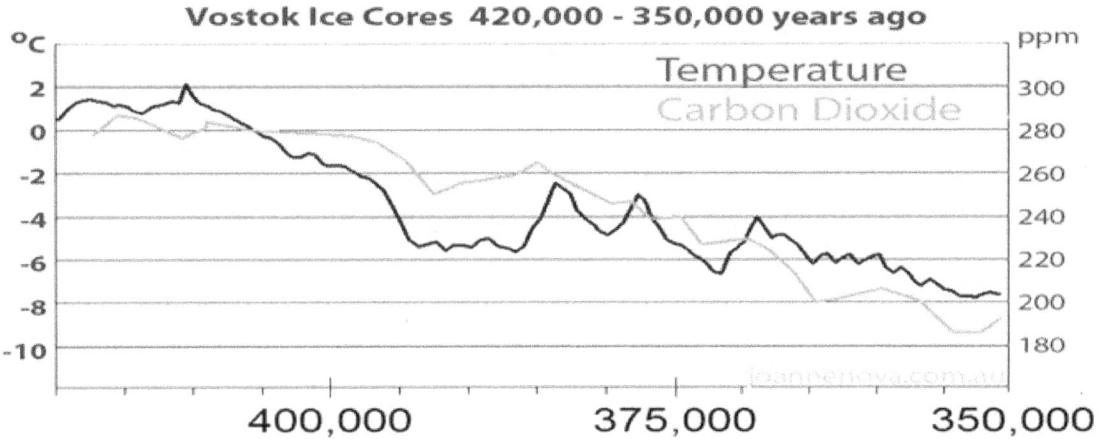

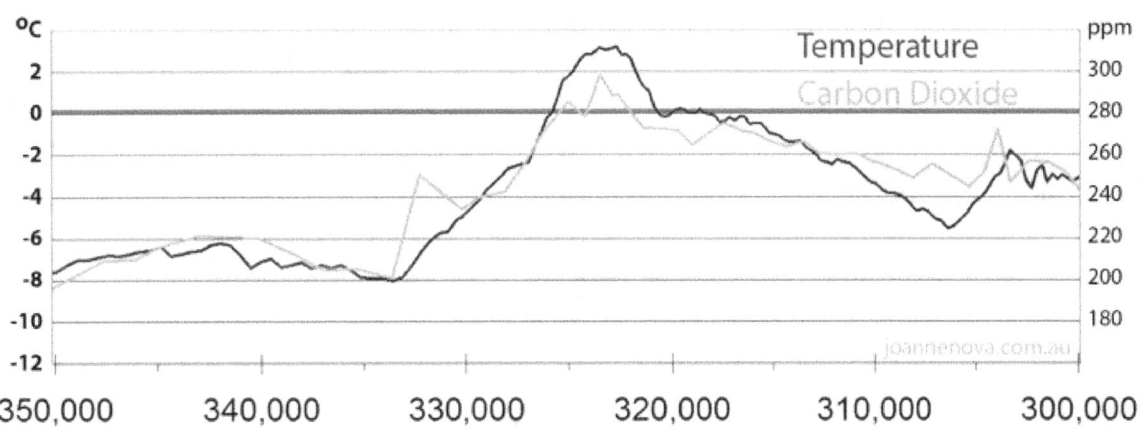

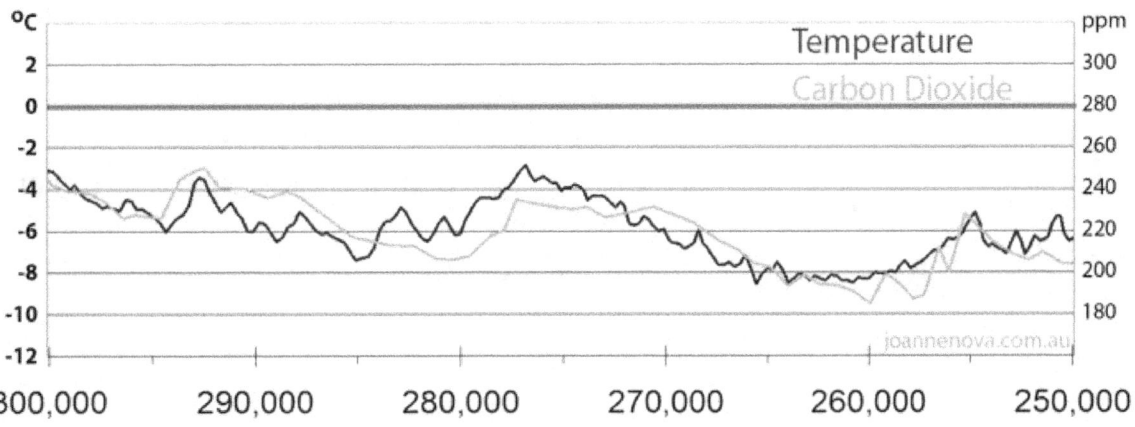

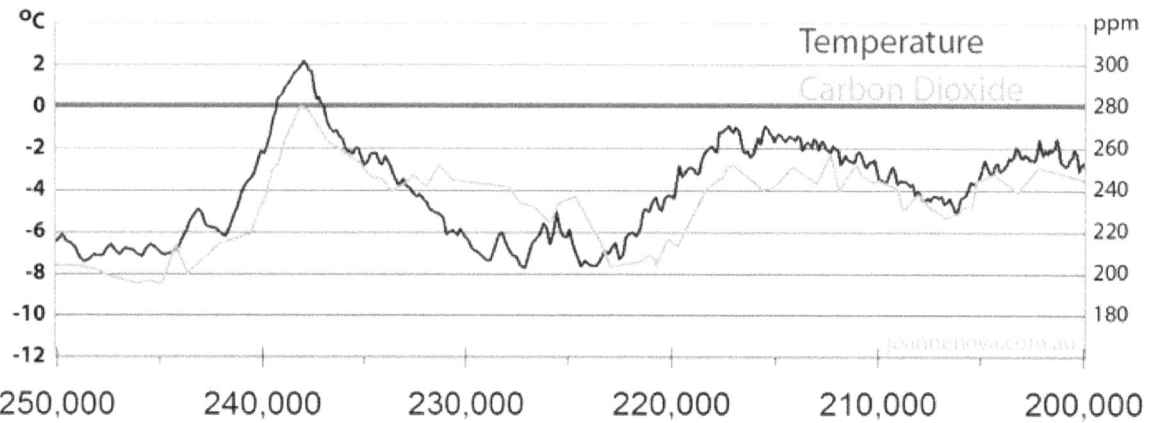

Vostok Ice Cores   250,000 – 200,000 years ago

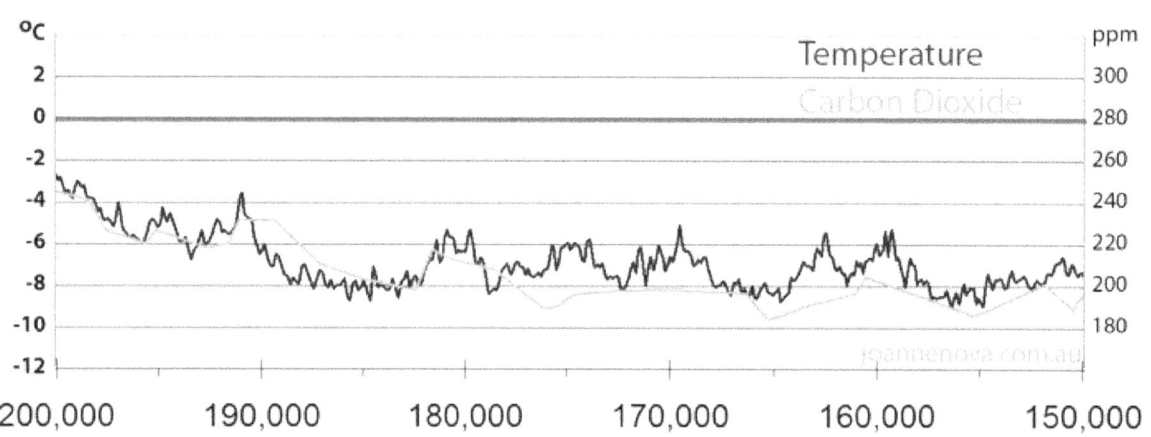

Vostok Ice Cores   200,000 – 150,000 years ago

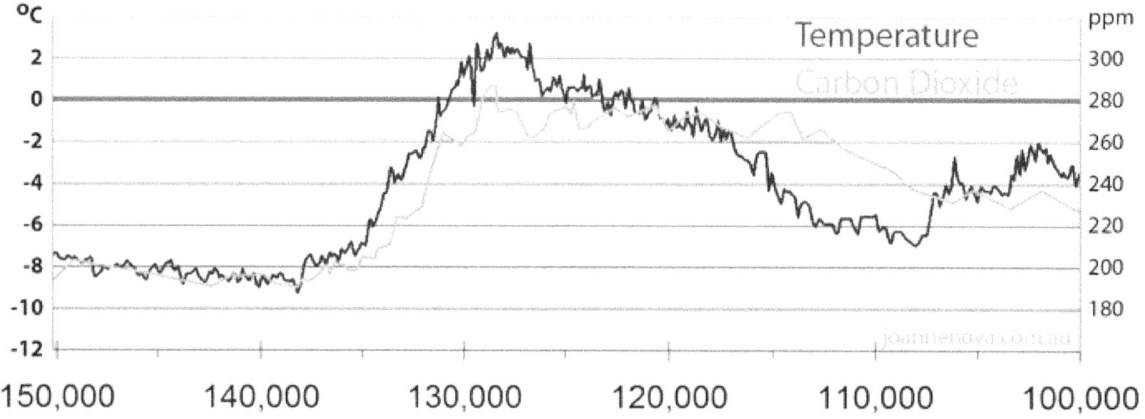

Vostok Ice Cores   150,000 – 100,000 years ago

Vostok Ice Cores   100,000 - 50,000 years ago

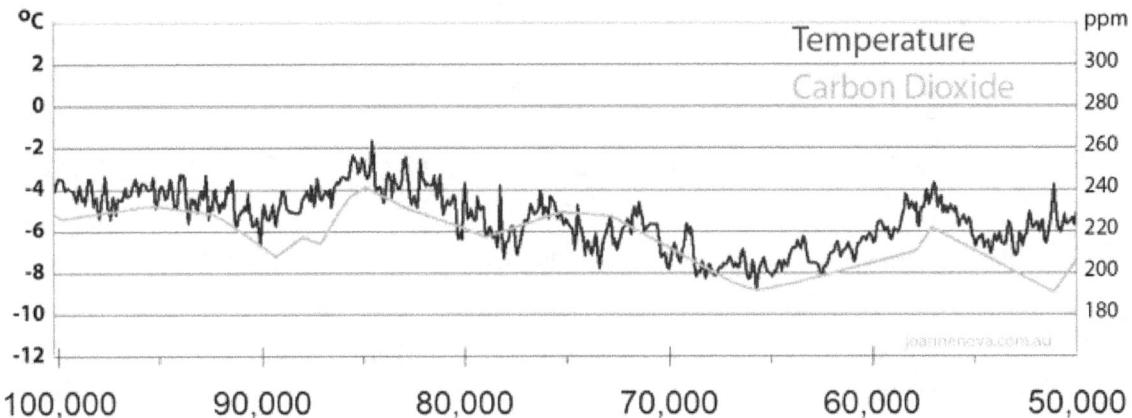

Vostok Ice Cores   50,000 - 2,500 years ago

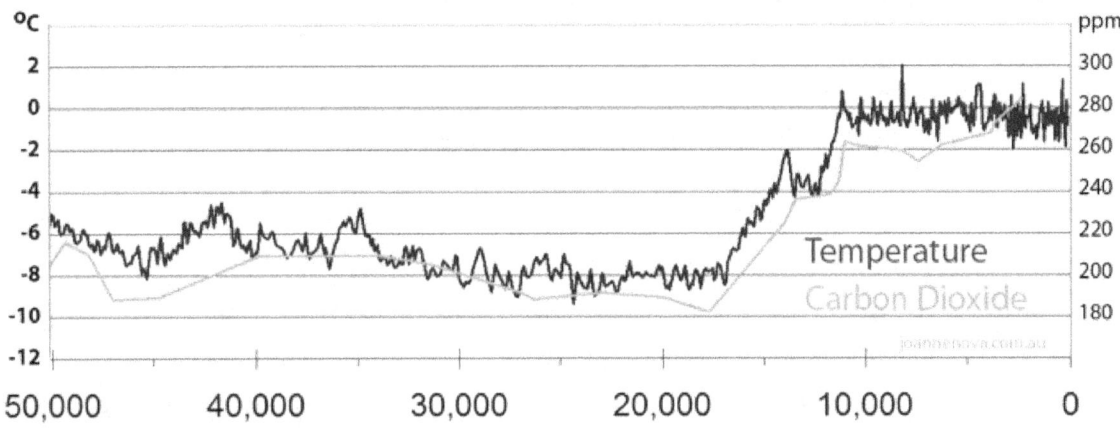

Notes:

- What really matters here are the turning points, not the absolute levels.
- The carbon data is unfortunately far less detailed than the temperature data. Beware of making conclusions about turning points or lags when only one single point may be involved.
- The graph which illustrates the lag the best, and also has the most carbon data is 150,000-100,000 years ago

The $45 Trillion Question

The IEA has advocated spending $45 Trillion until 2050 to combat anthropogenic global warming. Therefore, the $45 Trillion Question is why we are apparently seeing a significant warming due to human CO2 emissions a little two centuries after the industrial revolution really took off, when we are supposed to wait for a lag of 800 years?

References and more information

- Petit et all 1999 — analysed 420,000 years of Vostok, and found that as the world cools into an ice age, the delay before carbon falls is several thousand years. (http://www.nature.com/nature/journal/v399/n6735/abs/399429a0.html)

- Fischer et al 1999 — described a lag of 600 plus or minus 400 years as the world warms up from an ice age. (http://www.sciencemag.org/cgi/content/abstract/283/5408/1712)

- Monnin et al 2001 – looked at Dome Concordia (also in Antarctica) – and found a delay on the recent rise out of the last major ice age to be 800 ± 600 (http://www.sciencemag.org/cgi/content/abstract/291/5501/112)

- Mudelsee (2001) - Over the full 420,000 year Vostok history Co2 variations lag temperature by 1,300 years ± 1000. (http://www.manfredmudelsee.com/publ/pdf/The_phase_relations_among_atmospheric_CO2_content_temperature_and_global_ice_volume_over_the_past_420_ka.pdf)

- Caillon et al 2003 analysed the Vostok data and found a lag (where CO2 rises after temperature) of 800 ± 200 years. (http://icebubbles.ucsd.edu/Publications/CaillonTermIII.pdf)

- See Palisad for the most informative detailed graphics on what the Vostok and Dome Ice cores mean and why they strongly mathematically suggest CO2 follows temperatures and has little effect on them.This is what you need to see to understand "feedback" or the postulated "amplification". (http://www.palisad.com/co2/slides/siframes.html)

- A colorful but informative and link-filled presentation is here: http://motls.blogspot.com/2006/07/carbon-dioxide-and-temperatures-ice.html

- Excellent summary of the papers on the lag… at CO2 science. (http://www.co2science.org/articles/V6/N26/EDIT.php)

The Industrial Revolution: Carbon Dioxide versus Global Temperature

The industrial revolution began in the mid-eighteenth century and then accelerated in the western world in the nineteenth century. Before the exploitation of petroleum, unclean coal burning was the primary source of energy. Significant amounts of CO2 were discharged into the atmosphere. Anyone who has read Charles Dickens will easily recognize that the level of smog[8] in the industrialized world was significantly higher than those generally experienced by populations in the 20th and 21st centuries – higher than the legendary smog in Los Angeles, and higher even than the pollution levels in modern-day China. Many respiratory ailments in England, France and elsewhere, were attributed to the heavy smog.

---

[8] Smog is a type of air pollution; the word "smog" was coined in the mid 20th century as a portmanteau of the words smoke and fog to refer to smoky fog.] The word was then intended to refer to what was sometimes known as pea soup fog, a familiar and serious problem in London from the 19th century to the mid 20th century. This kind of smog is caused by the burning of large amounts of coal within a city; this smog contains soot particulates from smoke, sulfur dioxide and other components. Modern smog, as found for example in Los Angeles, is a type of air pollution derived from vehicular emission from internal combustion engines and industrial fumes that react in the atmosphere with sunlight to form secondary pollutants that also combine with the primary emissions to form photochemical smog. *Reference: Schwartz Cowan, Ruth (1997) - A Social History of American Technology. Oxford University Press. ISBN 978-0-19-504605-2*

Soot is a general term that refers to impure carbon particles resulting from the incomplete combustion of a hydrocarbon. It is more properly restricted to the product of the gas-phase combustion process but is commonly extended to include the residual pyrolyzed fuel particles such as coal, cenospheres, charred wood, petroleum coke, and so on, that may become airborne during pyrolysis and that are more properly identified as cokes or chars. *Reference: Rundel, Ruthann, "Polycyclic Aromatic Hydrocarbons, Phthalates, and Phenols", in Indoor Air Quality Handbook, John Spengleer, Jonathan M. Samet, John F. McCarthy (eds)*

Reliable data on CO2 concentration is available, based on Law Dome, Antarctica

http://cdiac.ornl.gov/trends/co2/graphics/lawdome.smooth75.gif

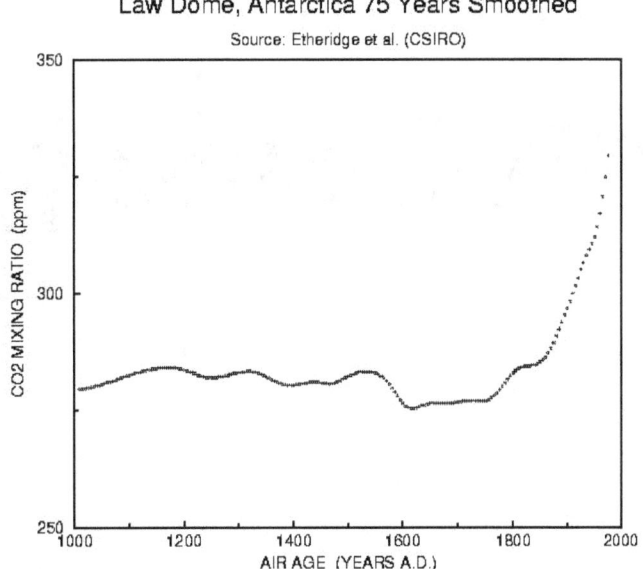

The next graph zooms in on the CO2 concentration from 1832 to 1910

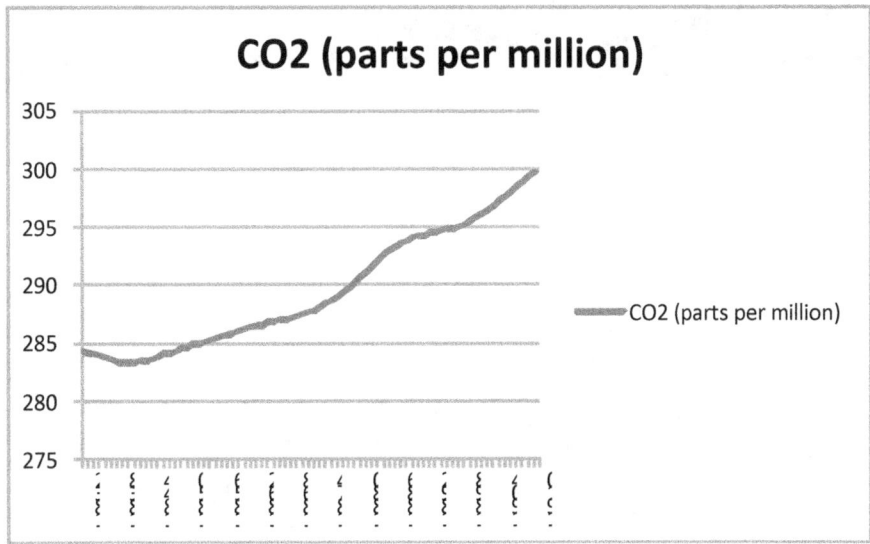

In summary, CO2 concentration increased globally from 284.3 parts per million (ppm) in 1832 to 299.7 ppm in 1910, an increase of about 5.5%.   Regional increases of CO2 concentration in rapidly industrializing parts of the world, such as Europe and North America were, obviously much higher, although exact figures are unavailable.

With a global increase in CO2 concentration of over 5%, one would have expected temperatures to rise. However, quite to the contrary, the eighteenth and nineteenth centuries were the height of the Little Ice Age, and global temperatures plummeted. Even in the industrializing parts of the world, any "urban heat island" effect was dwarfed by far more powerful natural forces dragging temperatures significantly lower all over the world.

http://www.globalwarmingclassroom.info/images/12LittleIceAge_lg.jpg

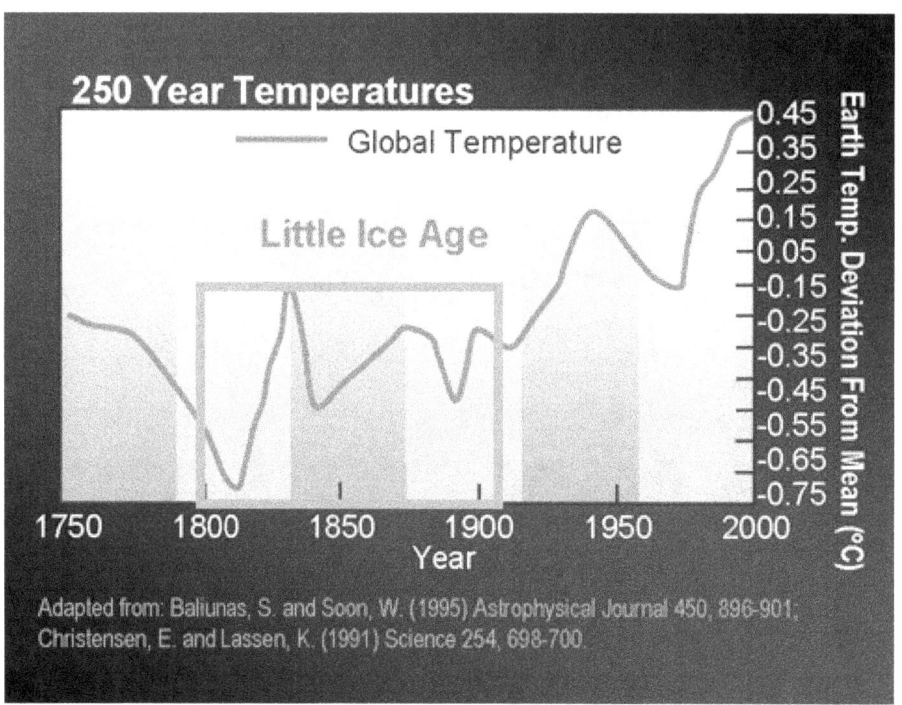

Following link provides an interactive graph to temperatures in key industrializing cities in Europe: http://climatereason.com/LittleIceAgeThermometers/Europe.html

For example:

Greenwich, United Kingdom:

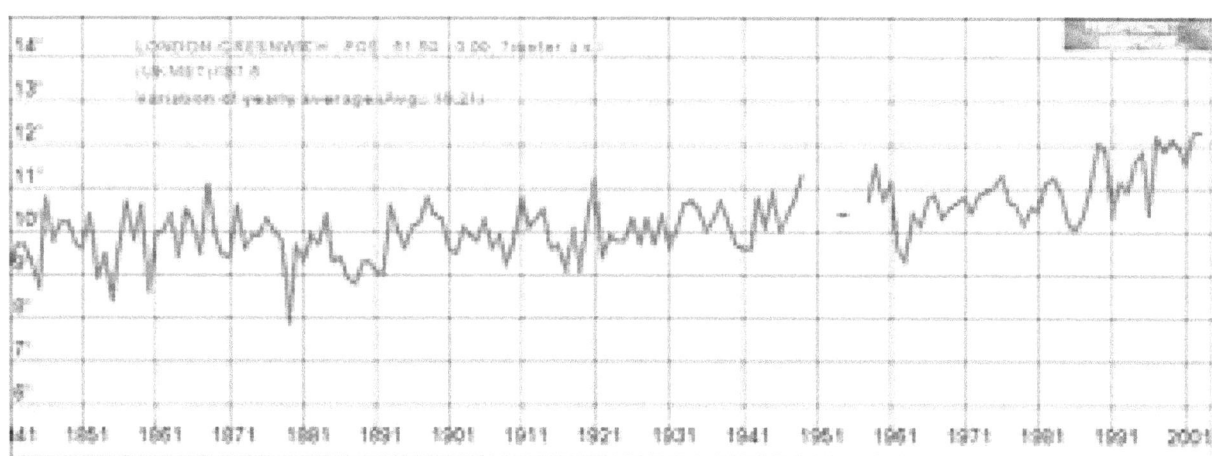

Berlin, Germany:

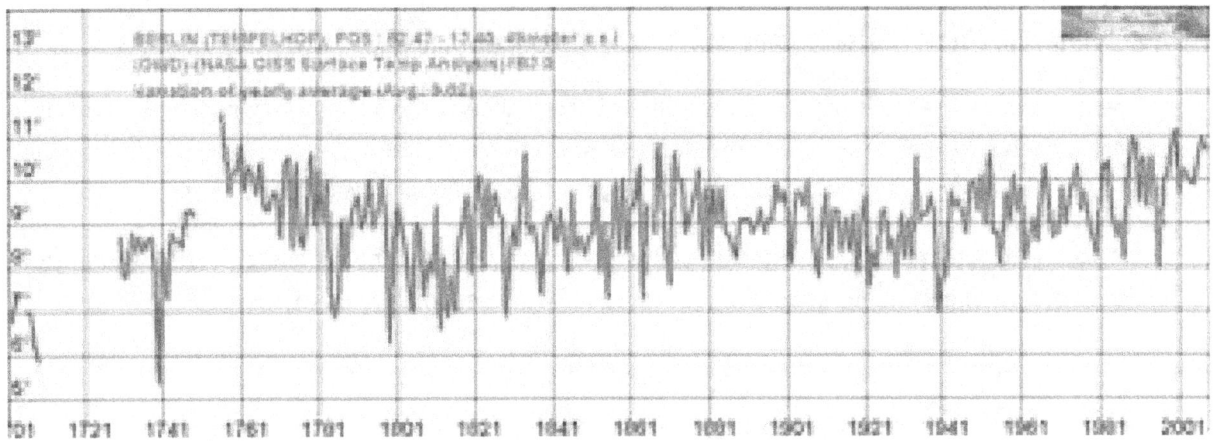

The 20th Century: Carbon Dioxide versus Global Temperature

The following graph shows the evolution of both CO2 concentration[9] and global temperature from 1880 to 2008[10].

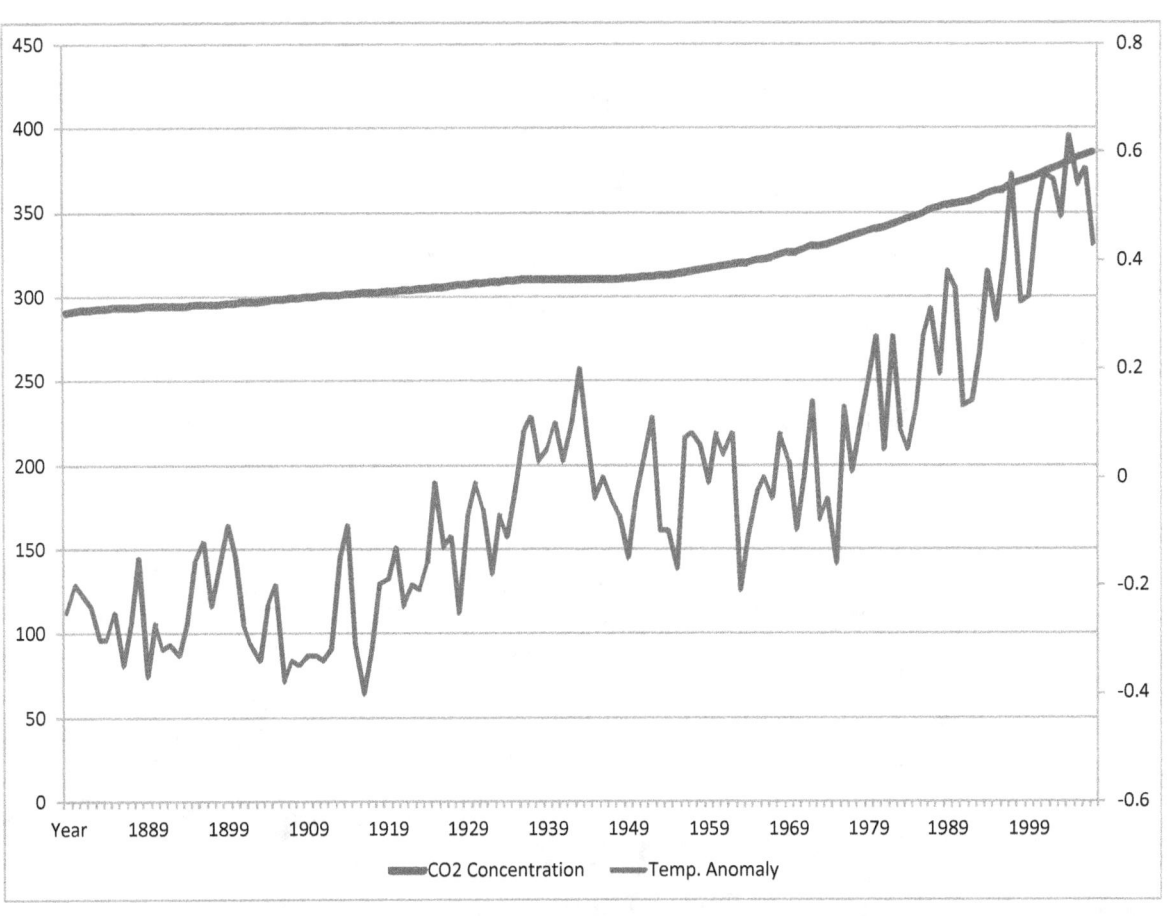

---

[9] Source for CO2 data : http://colli239.fts.educ.msu.edu/2003/12/31/parts-2003/
[10] Source for temperature data: http://www.ncdc.noaa.gov/img/climate/research/2008/ann/global-jan-nov-error-bar-pg.gif. By temperature we mean here the anomaly (difference) with the average global temperature of the entire 20th century.

To a layperson, it would *appear* that there is a direct relationship between the increase in global temperatures and the increase in CO2 concentration. This is exactly what proponents of Anthropogenic Global Warming want you to deduce by common sense.

However, common sense is often not necessarily correct when drawing conclusions from data. This is the reason why statistical methods were devised more than an century and a half ago, and enhanced ever since. What you see may not necessarily what is statistically significant or accurate.

A simple anecdotal example is illustrated below: the average grade of boys and girls is 14, therefore are the boys and girls equally good as common sense would encourage us to deduce?

The answer is no; the girls are better, because their standard deviation from the ideal score is lower than the boys, even if one boy had the highest score of 17

The parallel in the world of climate would be that even if there was an extreme cold wave, it may actually be warmer if generally temperatures have been warmer. Alternatively, even if there was an extreme heat wave, it may actually be colder, if generally temperatures have been cooler. You can now, perhaps, start to see, why there are such extensive opportunities for exaggeration and cherry picking in the domain of climatology, on either side of the Anthropogenic Global Warming debate. If the reader is interested in learning more about the proper application of statistics to analyze data, the book, *"The Basic Practice of Statistics"* by *David S. Moore* is an excellent reference.

Now, let us perform a statistical analysis of the 20th Century Carbon Dioxide and Temperature data.

1) Fitted Line Plot

The fitted line plot is the first step in simple regression analysis of continuous data

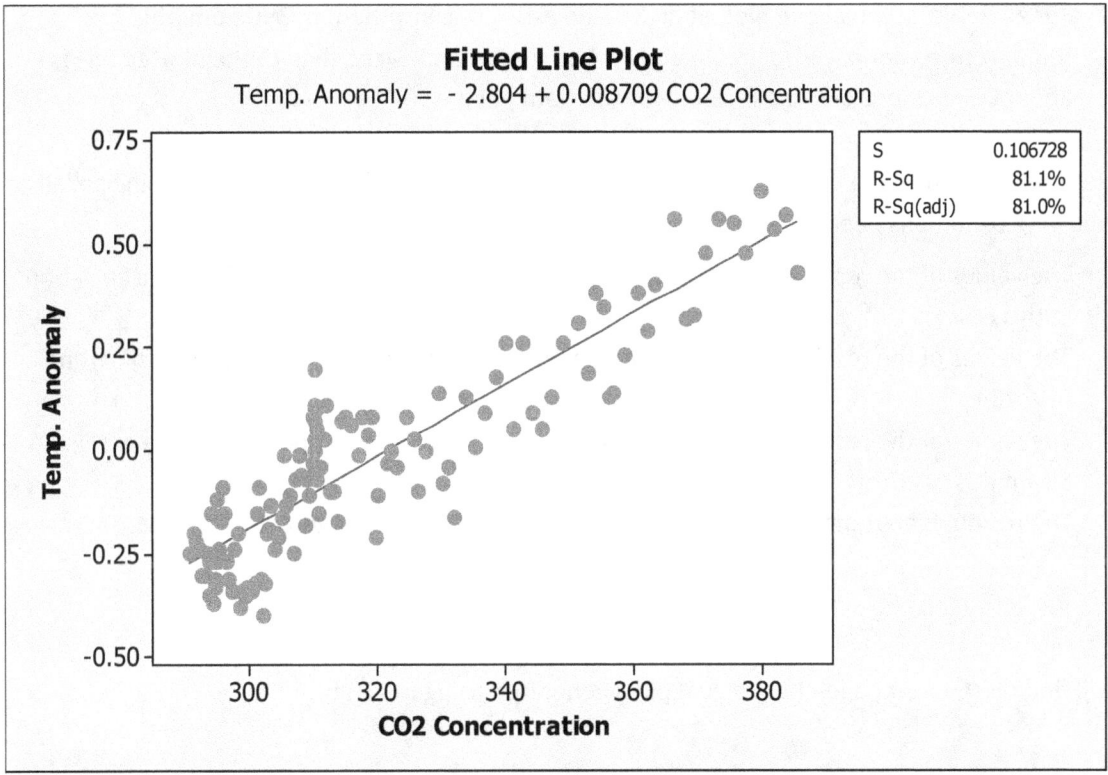

At first glance, an R-Squared value of 80% or higher would appear to indicate a strong correlation between Temperature Anomaly and CO2 Concentration. In fact, many professionals in different fields, without a strong understanding of statistics, would perform such so-called "regression" analysis in Microsoft Excel and would at this point, declare a "proven" relationship. The result is that there are many instances of incorrect conclusions being made in the world of business, and, yes, occasionally in the world of science as well, due to statistical methods being improperly and incompletely applied.

Correct application of statistical practice implies that we also have to check that the data also has to fit the plotted line, in order to confirm the relationship. More specifically, outliers (data points well outside the fitted line) can invalidate a hypothesized correlation (called the null hypothesis), no matter how strong the value of R-Squared or r are.

2) Analysis of Residuals

In order to check for outliers, one performs, what is called in statistics, a residuals analysis. You will not be able to perform residuals analysis in Microsoft Excel. You can either perform this exercise by hand or, better, by making use of a professional statistics tool such as MiniTab.

You may be scratching your head by this time. I shall try to explain as simply as possible to facilitate understanding by a layperson who may not have any notion about statistics.

First of all, what is a residual? A residual is the *observed* Y value (in this case Temperature Anomaly) MINUS the *predicted* Y value (i.e. the value according to what it should be on the hypothesized relationship line).

- For example, in the above plot, at an X value (CO_2 concentration) of 390 ppm, the temperature anomaly is approximately 0.4 while the value according to the plotted line is about 0.5 → therefore, the residual for this data point is -0.1

In order for the residuals to confirm that the data indeed fits the hypothesized relationship line, all of the following conditions need to be satisfied:

i.   The values of the residuals do not show any relationship to the X-values (CO_2 concentration in this case)

ii.   The values of the residuals are stable and independent, i.e. do not show any specific trend over time

iii. The values of the residuals are relatively constant, i.e. they do not increase as the predicted Y's (in this case Temperature Anomaly) increase.

iv.   The residuals are normally distributed with an average (mean) of zero

How do we verify the above? We use residual plots.

You will find on the next page, the residual plots generated using MiniTab.

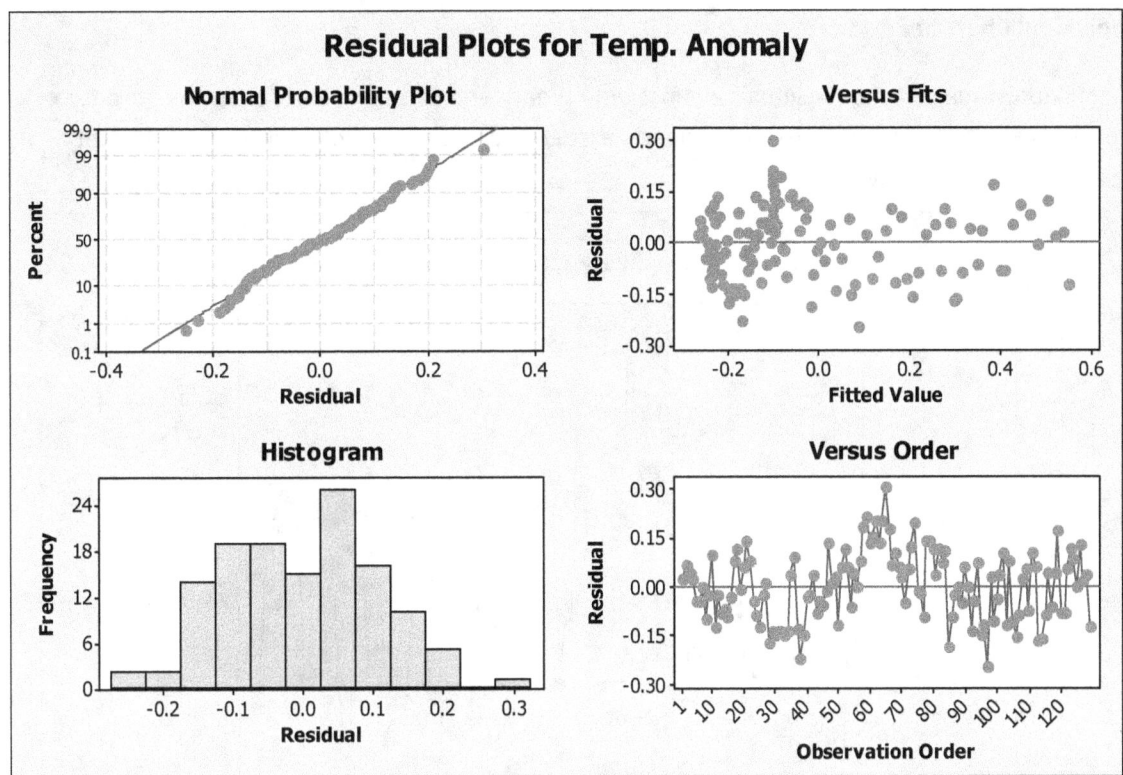

Residual Plots for Temp. Anomaly

Observations

- The Normal Probability Plot and Histogram both indicated a reasonably normal distribution, therefore, we can conclude that condition iv. Is satisfied
- The Residual versus Fits graph shows a small funnel shape but nothing significant, therefore, conditions I and iii are probably met
- The point of concern is the Residual versus Order graph, where we can clearly see some significant variation in the residuals, especially around observation order 65 and 100. So there is some doubt about condition ii (residuals may not be stable and independent)

So how do we verify (or otherwise) this doubt? We plot a statistical control chart: an Individual Chart of Residuals against time.

3) Individual Chart of Residuals

Plotting an individual chart of Residuals against time (Year), you can see clearly that there are two clear outliers – in 1945 and in 1976 – that clearly exceed the upper control limit (UCL) and lower control limit (LCL) respectively.

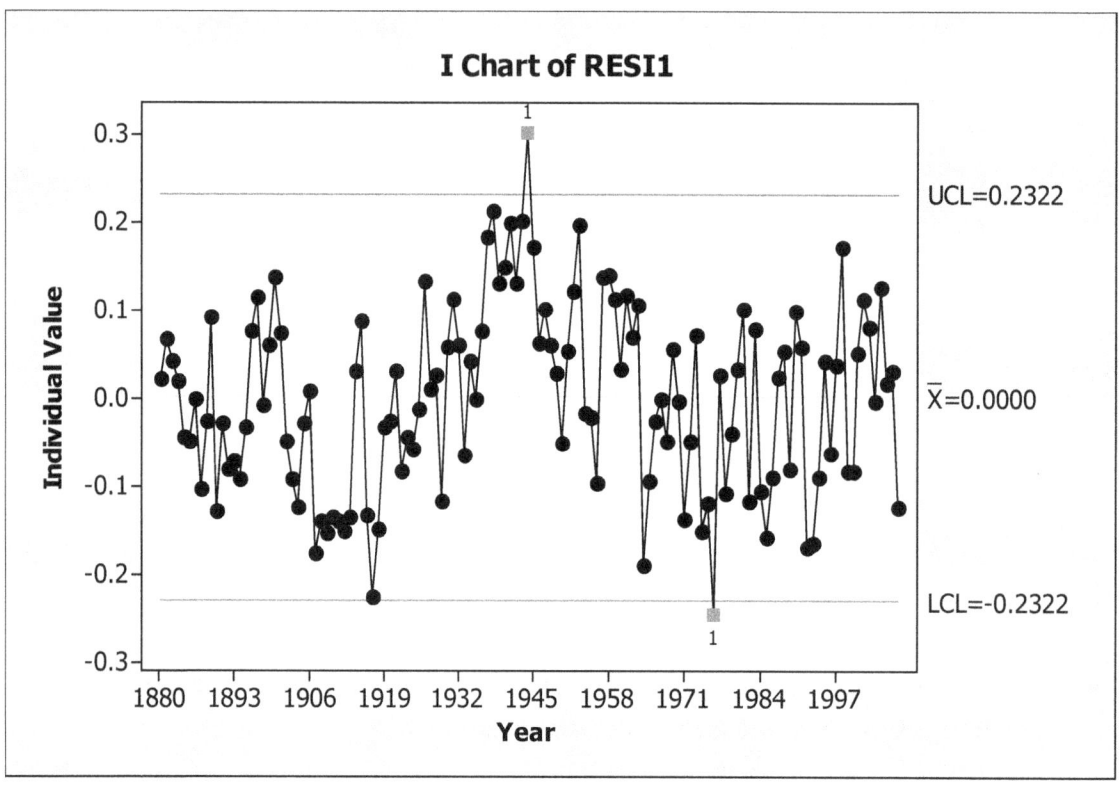

What does this mean?

- Temperature in 1945 was too high (above the upper limit with relation to the hypothesized regression equation) for the relatively low level of CO_2 concentration at that time
- Temperature in 1976 was too low (below the lower limit with relation to the hypothesized regression equation) for the higher level of CO_2 concentration at that time.

Avoiding Bias in Statistics

For conclusions to be valid using sound statistical practice, samples must be representative, meaning:

- The data you collect fairly represents all the data from the process or the population under study
- There are no systematic differences between the data you collect and the data you do not collect
- Every item stands a known and usually equal chance of being included

Statisticians refer to this as avoiding bias. Some examples of bias:

- Collecting data that are convenient for you to collect
- Using a pattern of selection that matches some structure in the data
- Some change in the environment/process means the sample is no longer representative
- Non-response
- Faulty measurement instrument or method
- Sampling plan executed improperly

Throughout this book, I shall repeatedly be referring to the above **golden rules** in the honest practice of statistics. Just like medicine, it is relatively simple to engage in statistical malpractice by misusing statistical methods to wrongly claim a conclusion, by deliberately or unintentionally ignoring the golden rules, such as introducing bias.

Outliers are not satisfactorily explained, therefore one cannot deduce any correlation between CO2 concentration and global temperature

What does a statistician do with outliers?

She/he cannot simply ignore outliers, because to do so would introduce bias favoring one's preconception of a specific hypothesis, in other words, breaking the golden rule: "collecting data that are convenient for you to collect". Introducing bias is statistical malpractice; in layperson's terms it means "cooking the books".

So what does one do with outliers? Outliers, even one or two, need to be further investigated in detailed, and, either explained very clearly and rationally in the context of the underlying process being measured, or the regression hypothesis rejected.

I have spent the past few years, desperately trying to search publicly available material from the IPCC, NASA, NOAA, CRU, etc, that provide a sound explanation of the observed outliers in 1945 and 1976. None is forthcoming. In fact, the outliers are never even mentioned. In fact, as the climategate evidence described in a later section of this book will reveal, there appears to be a pattern of ignoring inconvenient outliers in the field of climatology to-date.

More about this later, but, for now, it can be concluded, using sound statistical practice that there is no proven correlation between CO2 concentration and global temperature during the period when records have been kept, i.e. from 1880 to the present day.

Temperatures have plateaued since 1998 while CO2 concentration has continued to rise significantly

http://theclimatescepticsparty.blogspot.com.au/2012/02/g-l-o-b-l-c-o-o-l-i-n-g-wo-bleibt-die.html

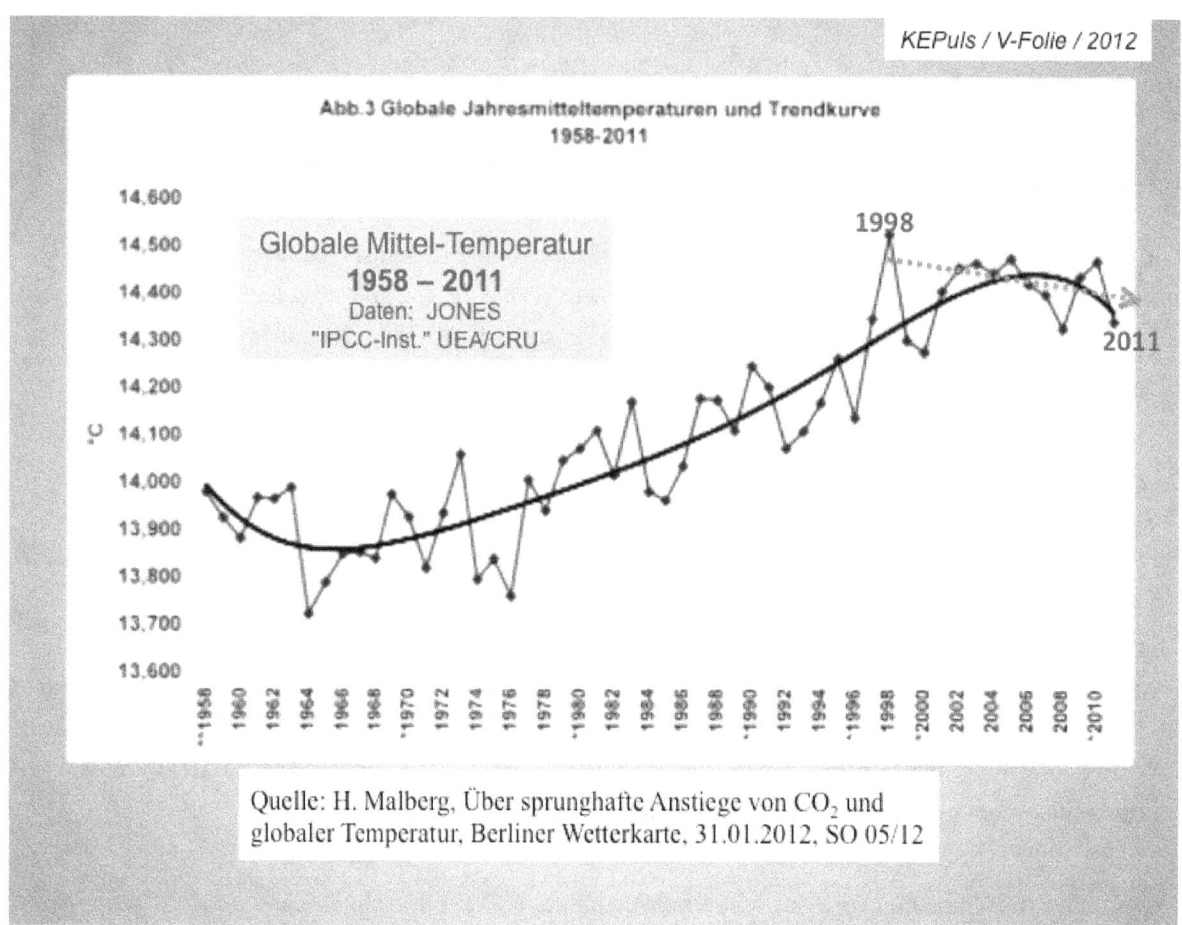

Quelle: H. Malberg, Über sprunghafte Anstiege von CO_2 und globaler Temperatur, Berliner Wetterkarte, 31.01.2012, SO 05/12

When the "theory" of Anthropogenic Global Warming (AGW) was first concocted, the authors obviously forgot to inform the public that only 3% of global carbon dioxide emissions are human-generated. Now that the public has been informed, proponents of AGW are formulating evermore creative explanations to "fit" the facts into their pre-conceptual belief of AGW. Science usually starts with facts being observed and leading to a conclusion. Although this was done initially when the theory of AGW was first described, the time of such a scientific mindset is long-over. These days, proponents of AGW usually start from the foregone conclusion that anthropogenic CO_2 is warming the planet, and then rationalize any new facts to conform to that rigid line in the sand.

Take the *New Scientist* for example:

http://www.newscientist.com/article/dn11638-climate-myths-human-co2-emissions-are-too-tiny-to-matter.html

> "Climate myths: Human CO_2 emissions are too tiny to matter
>
> So what's going on? It is true that human emissions of CO2 are small compared with natural sources. But the fact that CO2 levels have remained steady until very recently shows that natural emissions are usually balanced by natural absorptions. Now slightly more CO2 must be entering the atmosphere than is being soaked up by carbon "sinks".
>
> The consumption of terrestrial vegetation by animals and by microbes (rotting, in other words) emits about 220 gigatonnes of CO2 every year, while respiration by vegetation emits another 220 Gt. These huge amounts are balanced by the 440 Gt of carbon dioxide absorbed from the atmosphere each year as land plants photosynthesise.
>
> Similarly, parts of the oceans release about 330 Gt of CO2 per year, depending on temperature and rates of photosynthesis by phytoplankton, but other parts usually soak up just as much - and are now soaking up slightly more.
>
> Human emissions of CO2 are now estimated to be 26.4 Gt per year, up from 23.5 Gt in the 1990s, according to an Intergovernmental Panel on Climate Change report in February 2007 (pdf format). Disturbances to the land - through deforestation and agriculture, for instance - also contribute roughly 5.9 Gt per year.
>
> About 40% of the extra CO2 entering the atmosphere due to human activity is being absorbed by natural carbon sinks, mostly by the oceans. The rest is boosting levels of CO2 in the atmosphere."

Thus, in a very creative, yet clearly biased explanation to magnify the effect of human CO2 emissions, proponents of AGW have now imagined this idea of a "magical" balance that all naturally emitted CO_2 and natural absorption sinks. The scientific facts clearly show that nothing can be further from the truth. Yet this does not prevent proponents of AGW from inventing new myths to replace previous ones which have been debunked as science-fiction. As the sections on The Long-Term Record and the History of Global Temperature clearly show, there have been significant natural variations of CO2 well before the industrial revolution and well before human existence, eclipsing, by leaps and bounds, any recent variations in CO2 concentration. Equally as importantly, global temperatures have generally not correlated with CO2 concentration.

Tim Patterson, Pubs paleoclimatologist and Professor of Geology at Carleton University in Canada summarizes succinctly: "There is no meaningful correlation between CO2 levels and Earth's temperature over this [geologic] time frame. In fact, when CO2 levels were over ten times higher than they are now, about 450 million years ago, the planet was in the depths of the absolute coldest period in the last half billion years. On the basis of this evidence, how could anyone still believe that the recent relatively small increase in CO2 levels would be the major cause of the past century's modest warming?"

One will also note the following from the estimates quoted in the New Scientist article:

Total human-caused CO2 emissions per year (current):	26.4 + 5.9 = **32.3 Gigatons**
40% of human-CO2 emissions are absorbed by natural carbon sinks, therefore amount being added to the atmosphere:	**19.38 Gigatons**
Total natural CO2 emissions	220 + 330 = **550 Gigatons**
"Unabsorbed" human CO2 emissions as a percentage of the total	19.38 / (550+32.3) * 100 = **3.3%**

With an excess of 3.3% added to the atmosphere every year, one would naturally expect the CO2 concentration to double in 30 years, especially given the *New Scientist* article's stated assumption that there has historically been an "equilibrium" between natural CO2 emissions and carbon sinks absorbing an equal amount from the atmosphere.

The *rate of increase* of CO2 has remained fairly constant since 1960, according to the chart in the section, The 20th Century: Carbon Dioxide versus Global Temperature. Therefore, one wonders why the atmospheric CO2 concentration has only gone up from 316.91 parts per million (ppm) to 390 ppm in 2010 – an increase of only 23.1%, when the statement made in the New Scientist article, if factually correct, would have implied CO2 concentration to have increased by 166% in 50 years. In other words, CO2 concentration should have gone up to 526 ppm by now!

Anyone who logically digs into the detail of the material on Anthropogenic Global Warming will soon realize that these kinds of illogical fallacies litter the pseudoscience of climate change, resulting in anything but science. This book will provide numerous other examples of similar denial or misinterpretation of facts by proponents of AGW in order to conform to a preconceived conclusion.

Intelligent Design?

In fact the history of our natural system indicates rather that it is inherently unbalanced, constantly changing, and intrinsically susceptible to one-off catastrophic events. The complexity of our climate system is little understood, and there are so many variables, for instance:

- Differences in solar irradiance,
- Variation in cloud formation,
- Volcanoes discharging natural poison into the atmosphere occasionally with global ramifications,
- Large earthquakes damaging plant and animal life over a wide area,
- Mass extinction caused by asteroid impact,
- Specific animals or plants overpopulating and impacting the environment - humans obviously, but other species can and have inflicted similar damage on a smaller scale; a simple example: an overpopulated herd of elephants can do immense damage to a localized ecosystem.
- etc.

To believe in a so-called balance is environmental idealism not supported by the facts. Equilibrium can only be logical if one believes strongly in intelligent design, in other words, God orchestrating the Universe. Otherwise, the decks are strongly set against any sort of natural balance.

Misinformation on CO2 disseminated by proponents of Anthropogenic Global Warming #2

Below is *skepticalscience.com* engaging in the politics of fear by hyping an exaggerated prediction of atmospheric CO2 concentration in 2100 by the IPCC:

http://www.skepticalscience.com/exponential-increase-CO2-warming.htm

> "In short, following the 'business as usual' approach without major steps to move away from fossil fuels and limit greenhouse gas emissions, we will likely reach 850 to 950 ppmv of atmospheric CO2 by the year 2100. It will have taken approximately 200 years (from 1850 to 2050) for the first doubling of atmospheric CO2 from 280 to 560 ppmv, but it will only take another 70 years or so to double the levels again to 1120 ppmv. This will result in an accelerating rate of global warming, not a linear rate. Under Scenarios A2 and A1F1, the IPCC report projects that the global temperature in 2095 will be 2.0-6.4°C above 1990 levels (2.6-7.0°C above pre-industrial), with a best estimate of 3.4 and 4.0°C warmer (4.0 and 4.6°C above pre-industrial average surface temperatures), respectively."

First of all, according to the chart in the section, The 20th Century: Carbon Dioxide versus Global Temperature, the *rate of increase* of CO2 has remained fairly constant since 1960. The atmospheric CO2 concentration has only gone up from 316.91 parts per million (ppm) to 390 ppm in 2010; in other words an absolute increase of only 73 ppm in 50 years. It should also be noted that, within the past half-century, the technical and economic viability of renewable and/or lower carbon footprint energy – solar, wind, geothermal, electric, and other – has increased exponentially. Additionally, the increased cost of petroleum has led nations all over the world to gradually increase the usage of renewable energy. Therefore, one wonders about the assumptions used by the IPCC to predict a stratospheric increase of atmospheric CO2 concentration by 460 - 560 ppm in next 90 years, except for intentional misuse of hyperbole in yet another effort to panic the world's population. And websites such as skepticalscience.com and other pro-AGW vehicles of propaganda have broadcast this exaggeration verbatim without question or challenge. Where is the skepticism? Where is the application of common-sense? Why is the thinking mind so conspicuously absent?

Secondly, in a later section of this book,

Greenland (130,000 – 116,000 years ago), the reader will be introduced to research that proves conclusively that during the previous interglacial period preceding the last Ice Age, temperatures were 5°C higher than today while CO2 concentrations didn't exceed 300 ppm. Conversely, as documented in the chapter History of Global Temperature, during the Carboniferous period, 360 to 330 million years ago, CO2 concentration was at 800 ppm, yet global temperatures averaged 14°C, i.e. same as modern times. One wonders again what level of pre-conceptual pseudoscience has led the IPCC to be so certain that an atmospheric concentration of 850 to 950 ppm will lead to a 4°C increase in global temperature. After all, the Earth's own history empirically disproves such a notion. The fact is that the Earth's climate system is far more complex than a simplistic relationship between atmospheric CO2 concentration and global temperature. And as the IPCC's fanciful predictions indicate, we are only in the beginning stages of understanding the myriad of factors affecting the world's climate.

Misinformation on CO2 disseminated by proponents of Anthropogenic Global Warming #3

Below is *skepticalscience.com* once again, this time hinging on just one fact as a singular explanation for past periods of Earth's history when CO2 concentration was much higher, but temperatures were equivalent or much lower, while ignoring and not informing the reader of other facts which render the explanation illogical and an exercise in deliberate camouflage of the full picture:
http://www.skepticalscience.com/co2-higher-in-past.htm

> Exhibit A: "Geologists refer to ancient ice-cap formations and ice-ages as "glaciations." One such glaciation that occurred during the Late Ordovician era, some 444 million years ago has captured the attention of climate scientists and skeptics alike
>
> What about evidence for any of these short-term CO2 fluctuations? Recent research has uncovered evidence for lower ocean temperatures during the Ordovician than previously thought, creating ideal conditions for a huge spurt in marine biodiversity and correspondingly large drawdown of CO2 from the atmosphere through carbon burial in the ocean. A period of mountain-building was also underway (the so-called Taconic orogeny) increasing the amount of rock weathering taking place and subsequently lowering CO2 levels even further. The evidence is definitely there for a short-term disruption of the carbon cycle."

As documented in the chapter, History of Global Temperature, the Ordovician period lasted from 480 to 420 million years ago, in other words, a total of 60 million years, in other words, a period of time equivalent to 2.4 million times the 250 years since the start of the industrial revolution. CO2 concentration during this 60 million year period averaged 4,200 ppm (10.5 times that of today). How this can be minimized as a *"short-term"* fluctuation of CO2, while the rise of CO2 in the past two centuries from 290 to 390 ppm is considered more historically significant, is beyond logical comprehension.

> Exhibit B: "Another important factor is the sun. During the Ordovician, it would have been several percent dimmer according to established nuclear models of main sequence stars. Surprisingly, this raises the CO2 threshold for glaciation to a staggering 3000 ppmv or so. This also explains (along with the logarithmic forcing effect of CO2) why a runaway greenhouse didn't occur: with a dimmer sun, high CO2 is necessary to stop the Earth freezing over."

It is true that solar irradiance was lower 500 million years ago (about 0.95 of the current Sun). Please refer to the graph below[11]:

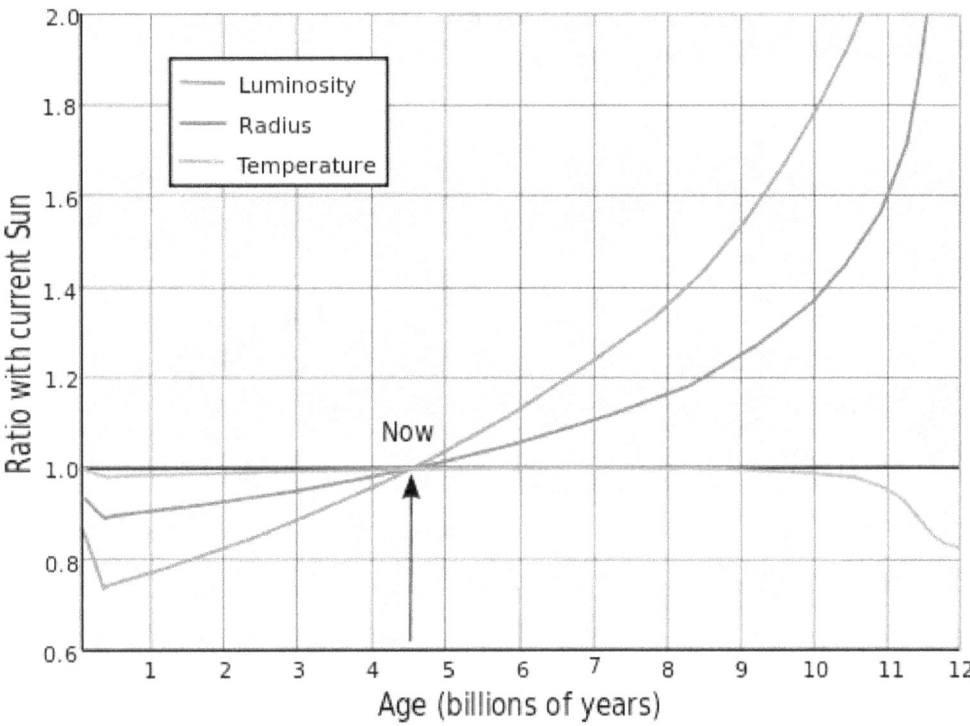

However, this limited explanation is akin to looking at a glass half-full, while concealing the half-empty part of the glass. The article fails to even mention the following:

1. The climate of the Cambrian period, 540 to 520 million years ago, pre-dating the Ordovician climate, with a lower solar irradiance was probably not very hot, nor very cold. There is no evidence of ice at the poles.[12][13]

2. During the early Ordovician climate, 480 million years ago, when solar irradiance was even lower, mild climates probably covered most of the globe. The continents were flooded by the oceans

---

[11] Ribas, Ignasi (2010), "The Sun and stars as the primary energy input in planetary atmospheres", Solar and Stellar Variability: Impact on Earth and Planets, Proceedings of the International Astronomical Union, IAU Symposium, 264, pp. 3–18, en:Bibcode 2010IAUS..264....3R

[12] http://www.scotese.com/ecambcli.htm

[13] http://www.scotese.com/mlcambcl.htm

creating warm, broad tropical seaways. [14] The Ice House World came later, some 40 million years later, when solar irradiance was higher[15]

3. During the Carboniferous period, 360 to 330 million years ago, CO2 concentration averaged 800 ppm (twice that of today) but temperatures averaged 14°C (same as modern level). For more details, please see the section Paleo Temperatures

4. Etc. Please refer to the chapter History of Global Temperature. The historical record clearly demonstrates that the mean CO2 concentration does *not* display a correlation with the mean surface temperature. In addition, the significant climatic variations within each period indicate clearly that climate can vary widely from warm to cool with the same and often much higher level of CO2 concentration than we have today

The referenced article by skepticalscience.com illustrates very well the pre-conceptual art of pseudoscience, written on the simple principle that readers will simply believe what is written without further investigation. While this may be true for some people, the authors of the article have grossly underestimated the completely natural human propensity for skepticism, in other words, to cross-check, to question, and to challenge.

---

[14] http://www.scotese.com/eordclim.htm
[15] http://www.scotese.com/mlordcli.htm

I have copied below verbatim a very interesting dissertation on the fact that water vapor is the main player in Earth's greenhouse effect, and not Carbon Dioxide as the pseudoscience myth of Anthropogenic Global Warming stipulates.

http://www.geocraft.com/WVFossils/greenhouse_data.html

Just how much of the "Greenhouse Effect" is caused by human activity?

It is about **0.28%,** if **water vapor** is taken into account-- about **5.53%,** if not.

This point is so crucial to the debate over global warming that how **water vapor is** or **isn't** factored into an analysis of Earth's greenhouse gases makes the difference between describing a *significant* human contribution to the greenhouse effect, or a *negligible* one.

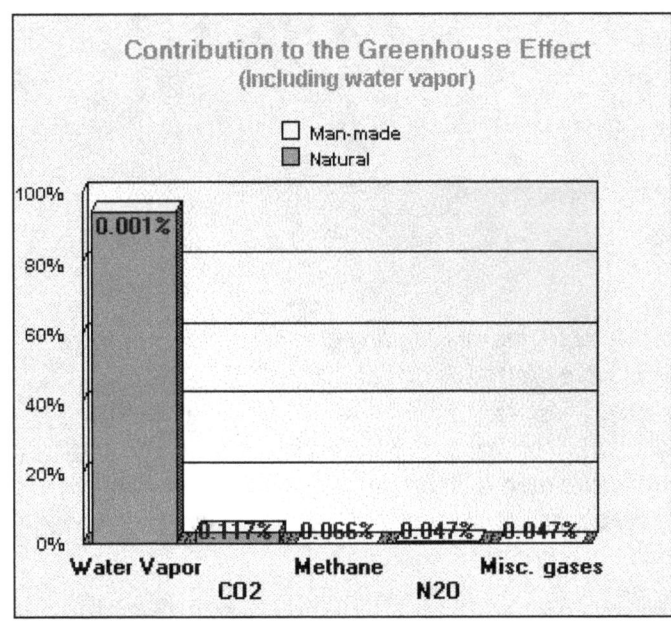

Water vapor constitutes Earth's most significant greenhouse gas, accounting for about **95% of Earth's greenhouse effect (5)**. Interestingly, many "facts and figures' regarding global warming completely ignore the powerful effects of water vapor in the greenhouse system, carelessly (perhaps, deliberately) overstating human impacts as much as 20-fold.

Water vapor is **99.999% of natural origin.** Other atmospheric greenhouse gases, **carbon dioxide (CO_2), methane (CH_4), nitrous oxide (N_2O), and** miscellaneous other gases (CFC's, etc.), are also mostly of natural origin (except for the latter, which is mostly anthropogenic).

Human activites contribute slightly to greenhouse gas concentrations through **farming, manufacturing, power generation, and transportation**. However, these emissions are so dwarfed in comparison to emissions from natural sources we can do nothing about, that even the most costly efforts to limit human emissions would have a very small-- perhaps undetectable-- effect on global climate.

For those interested in more details a series of **data sets** and **charts** have been assembled below in a 5-step statistical synopsis.

Note that the first two steps ignore **water vapor**.

1. Greenhouse gas concentrations

2. Converting concentrations to contribution

3. Factoring in water vapor

4. Distinguishing natural vs man-made greenhouse gases

5. Putting it all together

Note: Calculations are expressed to 3 significant digits to reduce rounding errors, not necessarily to indicate statistical precision of the data. All charts were plotted using Lotus 1-2-3.

Caveat: This analysis is intended to provide a simplified comparison of the various man-made and natural greenhouse gases on an equal basis with each other. It does not take into account all of the complicated interactions between atmosphere, ocean, and terrestrial systems, a feat which can only be accomplished by better computer models than are currently in use.

1. The following table was constructed from data published by the U.S. Department of Energy **(1)** summarizing concentrations of the various atmospheric greenhouse gases, and supplemented with information from other sources **(2-7)**. Because some of the concentrations are very small the numbers are stated in parts *per billion*. **DOE chose to NOT show water vapor as a greenhouse gas!**

TABLE 1.

The Important Greenhouse Gases (except water vapor)
U.S. Department of Energy, (October, 2000) (1)

(all concentrations expressed in parts per billion)	Pre-industrial baseline	Natural additions	Man-made additions	Total (ppb) Concentration	Percent of Total
Carbon Dioxide (CO2)	288,000	68,520	11,880 **(2)**	368,400	99.438%
Methane (CH4)	848	577	320	1,745	0.471%
Nitrous Oxide (N2O)	285	12	15	312	0.084%
Misc. gases (CFC's, etc.)	25	0	2	27	0.007%
Total	289,158	69,109	12,217	370,484	100.00%

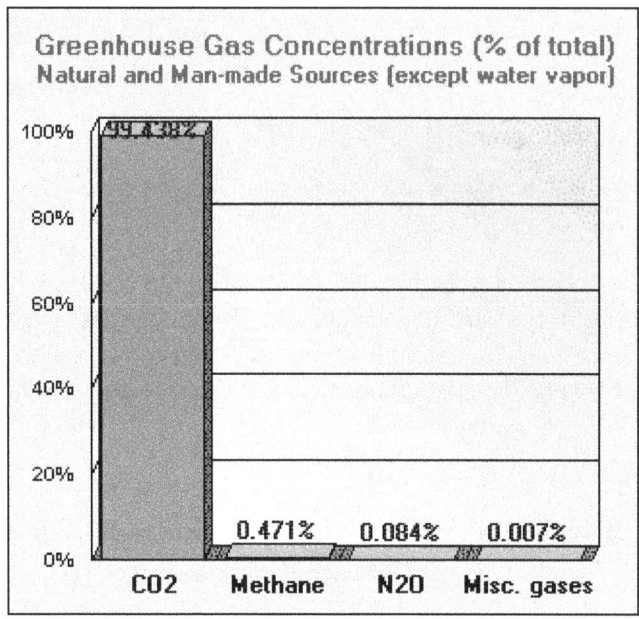

The chart above summarizes the % of greenhouse gas **concentrations** in Earth's atmosphere from **Table 1**. This is not a very meaningful view though because 1) the data has not been corrected for the actual **Global Warming Potential** (GWP) of each gas, and 2) **water vapor** is ignored.

But these are the numbers one would use if the goal is to exaggerate human greenhouse contributions:

Man-made and **natural carbon dioxide** (CO2) comprises **99.44%** of all greenhouse gas concentrations (368,400 / 370,484)--(ignoring water vapor)

Also, from **Table 1** (but not shown on graph):

- **Anthropogenic** (man-made) **CO2** additions comprise (11,880 / 370,484) or **3.207%** of all greenhouse gas concentrations, (ignoring water vapor).

- **Total combined *anthropogenic* greenhouse gases** comprise (12,217 / 370,484) or **3.298%** of all greenhouse gas concentrations, (ignoring water vapor).

The various greenhouse gases are **not equal** in their heat-retention properties though, so to remain statistically relevant *% concentrations* must be changed to *% contribution* relative to CO2. This is done in **Table 2**, below, through the use of GWP **multipliers** for each gas, derived by various researchers.

Converting greenhouse gas concentrations
to greenhouse effect contribution
(using *global warming potential*)

2. Using appropriate corrections for the **Global Warming Potential** of the respective gases provides the following more meaningful comparison of greenhouse gases, based on the conversion:

(*concentration*) **X** (the appropriate GWP **multiplier (3) (4)** of each gas relative to CO2) **=** greenhouse *contribution.*:

TABLE 2.

**Atmospheric Greenhouse Gases (except water vapor)
adjusted for heat retention characteristics, relative to CO2**

This table adjusts values in **Table 1** to compare greenhouse gases equally with respect to CO2. (#'s are unit-less)	**Multiplier (GWP)**	Pre-industrial baseline(new)	Natural additions (new)	Man-made additions (new)	Tot. Relative Contribution	Percent of Total (new)
Carbon Dioxide (CO2)	1	288,000	68,520	11,880	368,400	72.369%
Methane (CH4)	21 **(3)**	17,808	12,117	6,720	36,645	7.199%
Nitrous Oxide (N2O)	310 **(3)**	88,350	3,599	4,771	96,720	19.000%
CFC's (and other misc. gases)	see data **(4)**	2,500	0	4,791	7,291	1.432%
Total		396,658	84,236	28,162	509,056	100.000%

NOTE: GWP (Global Warming Potential) is used to contrast different greenhouse gases relative to CO2.

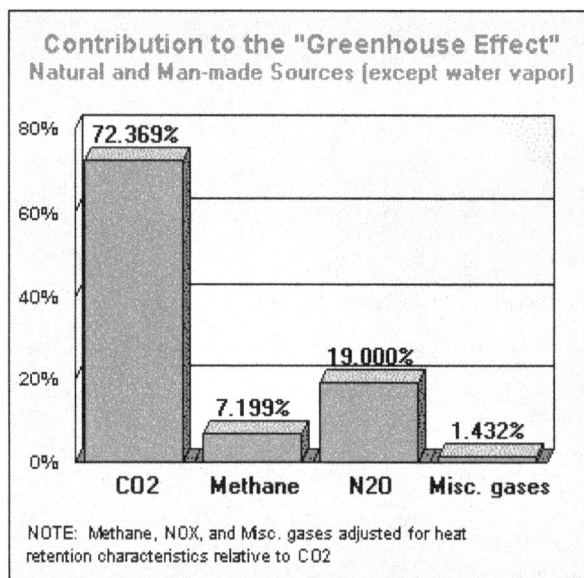

Contribution to the "Greenhouse Effect"
Natural and Man-made Sources (except water vapor)

- 72.369% — CO2
- 7.199% — Methane
- 19.000% — N2O
- 1.432% — Misc. gases

NOTE: Methane, NOX, and Misc. gases adjusted for heat retention characteristics relative to CO2

Compared to the concentration statistics in **Table 1**, the GWP comparison in **Table 2** illustrates, among other things:

Total **carbon dioxide (CO2)** contributions are reduced to **72.37%** of all greenhouse gases (368,400 / 509,056)-- (ignoring water vapor).

Also, from **Table 2** (but not shown on graph):

Anthropogenic (man-made) **CO2** contributions drop to (11,880 / 509,056) or **2.33%** of total of all greenhouse gases, (ignoring water vapor).

Total combined *anthropogenic* greenhouse gases becomes (28,162 / 509,056) or **5.53%** of all greenhouse gas contributions, (ignoring water vapor).

Relative to **carbon dioxide** the other greenhouse gases together comprise about **27.63%** of the greenhouse effect (ignoring water vapor) but only about **0.56%** of total greenhouse gas *concentrations*. Put another way, as a group methane, nitrous oxide (N2O), and CFC's and other miscellaneous gases are about **50 times more potent** than CO2 as greenhouse gases.

To properly represent the **total relative impacts** of Earth's greenhouse gases **Table 3** (below) factors in the effect of **water vapor** on the system.

Water vapor overwhelms
all other natural and man-made
greenhouse contributions.

3. **Table 3**, shows what happens when the effect of **water vapor is factored in,** and together with all other greenhouse gases expressed as a relative % of the total greenhouse effect.

TABLE 3.

**Role of Atmospheric Greenhouse Gases
(man-made and natural) as a % of Relative
Contribution to the "Greenhouse Effect"**

Based on concentrations (ppb) adjusted for heat retention characteristics	Percent of Total	Percent of Total --adjusted for **water vapor**
Water vapor	-----	**95.000%**
Carbon Dioxide (CO2)	72.369%	**3.618%**
Methane (CH4)	7.100%	**0.360%**
Nitrous oxide (N2O)	19.000%	**0.950%**
CFC's (and other misc. gases)	1.432%	**0.072%**
Total	100.000%	**100.000%**

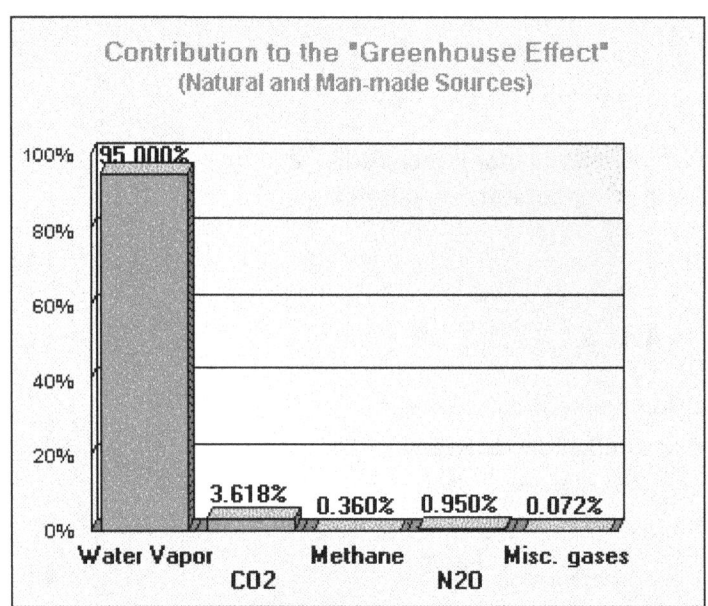

Contribution to the "Greenhouse Effect"
(Natural and Man-made Sources)

As illustrated in this chart of the data in **Table 3**, the combined **greenhouse contributions of** CO2, methane, N2O and misc. gases are small compared to **water vapor**!

Total atmospheric carbon dioxide (CO2) -- both man-made and natural-- is only about 3.62% of the overall greenhouse effect-- a big difference from the 72.37% figure in **Table 2**, which ignored water!

Water vapor, the most significant greenhouse gas, comes from natural sources and is responsible for roughly **95% of the greenhouse effect (5)**.

Among climatologists this is common knowledge but among special interests, certain governmental groups, and news reporters this fact is under-emphasized or just ignored altogether.

Conceding that it might be "a little misleading" to leave water vapor out, they nonetheless defend the practice by stating that it is "customary" to do so!

4. Of course, even among the remaining 5% of ***non-water vapor*** greenhouse gases, humans contribute only a very small part (and human contributions to water vapor are negligible). Constructed from data in **Table 1,** the charts (below) illustrate graphically how much of each greenhouse gas is **natural** vs how much is **man-made**. These allocations are used for the next and final step in this analysis-- total man-made contributions to the greenhouse effect. Units are expressed to 3 significant digits in order to reduce rounding errors for those who wish to walk through the calculations, not to imply numerical precision as there is some variation among various researchers.

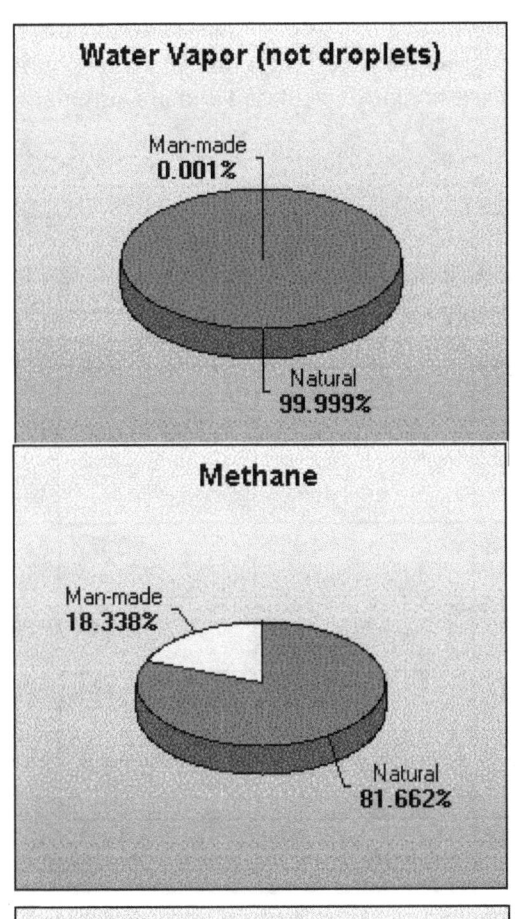

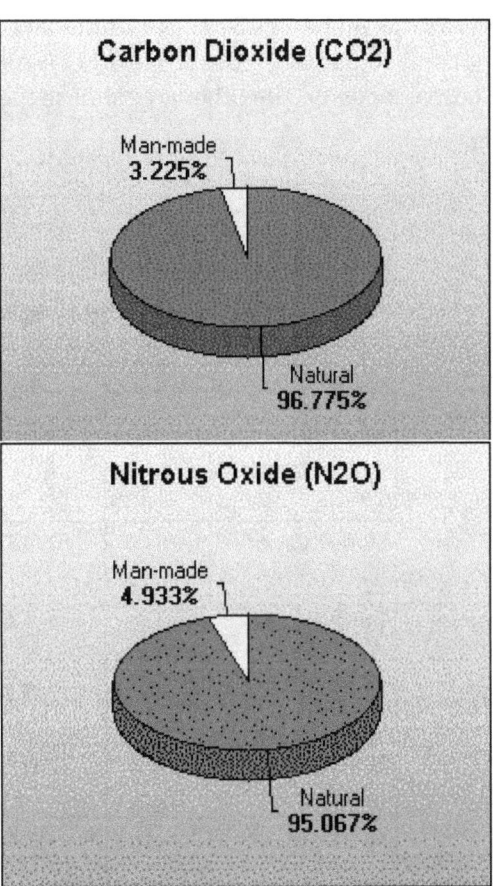

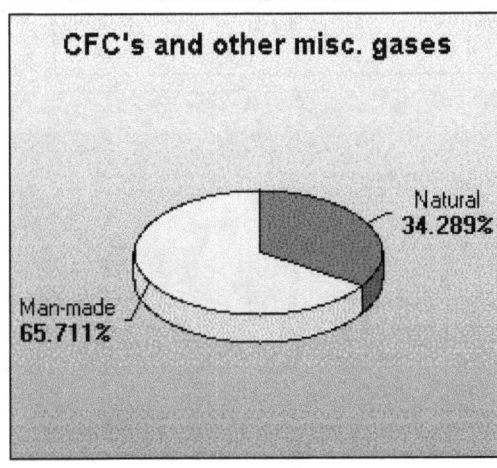

5. To finish with the math, by calculating the product of the adjusted CO2 contribution to greenhouse gases (3.618%) and % of CO2 concentration from anthropogenic (man-made) sources (3.225%), we see that only (0.03618 X 0.03225) or **0.117% of the greenhouse effect is due to atmospheric CO2 from human activity**. The other greenhouse gases are similarly calculated and are summarized below.

TABLE 4a.

Anthropogenic (man-made) Contribution to the "Greenhouse Effect," expressed as % of Total (water vapor INCLUDED)

Based on concentrations (ppb) adjusted for heat retention characteristics	% of Greenhouse Effect	% Natural	% Man-made
Water vapor	95.000%	94.999%	**0.001%**
Carbon Dioxide (CO2)	3.618%	3.502%	**0.117%**
Methane (CH4)	0.360%	0.294%	**0.066%**
Nitrous Oxide (N2O)	0.950%	0.903%	**0.047%**
Misc. gases (CFC's, etc.)	0.072%	0.025%	**0.047%**
Total	100.00%	**99.72**	**0.28%**

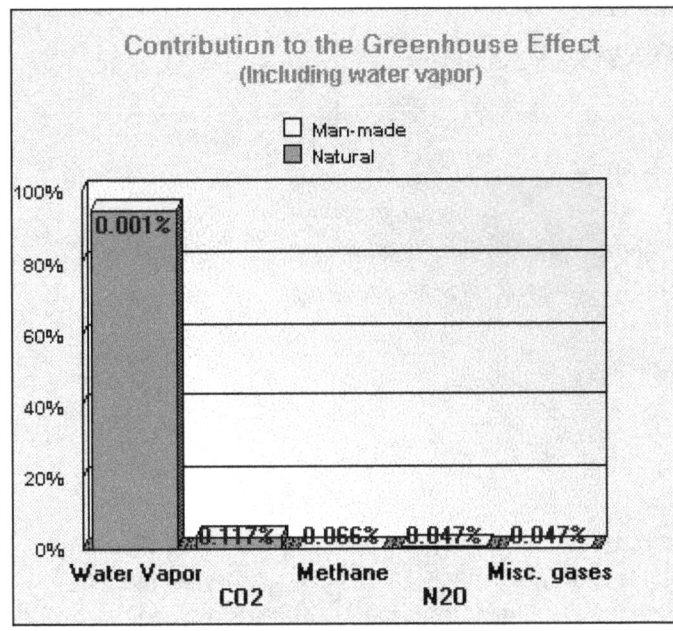

Contribution to the Greenhouse Effect (Including water vapor)

When greenhouse contributions are listed by source, the relative overwhelming component of the *natural* greenhouse effect, is readily apparent.

From **Table 4a,** both natural and man-made greenhouse contributions are illustrated in this chart, in gray and green, respectively. For clarity only the man-made (anthropogenic) contributions are labeled on the chart.

Water vapor, responsible for **95%** of Earth's greenhouse effect, is **99.999% natural** (some argue, 100%). Even if we wanted to we can do nothing to change this.

Anthropogenic (man-made) **CO2** contributions cause only about **0.117%** of Earth's greenhouse effect, (factoring in **water vapor**). This is insignificant!

Adding up all **anthropogenic** greenhouse sources, the **total human contribution to the greenhouse effect** is around **0.28%** (factoring in water vapor).

The **Kyoto Protocol** calls for mandatory carbon dioxide reductions of 30% from developed countries like the U.S. Reducing man-made CO2 emissions this much would have an undetectable effect on climate while having a devastating effect on the U.S. economy. Can you drive your car 30% less, reduce your winter heating 30%? Pay 20-50% more for everything from automobiles to zippers? And that is just a down payment, with more sacrifices to come later.

Such drastic measures, even if imposed equally on all countries around the world, would reduce total **human greenhouse contributions** from CO2 by about **0.035%**.

This is much less than the natural variability of Earth's climate system!

While the greenhouse reductions would exact a high human price, in terms of sacrifices to our standard of living, they would yield statistically negligible results in terms of measurable impacts to climate change. There is no expectation that any statistically significant global warming reductions would come from the Kyoto Protocol.

✎" There is no dispute at all about the fact that even if punctiliously observed, (the Kyoto Protocol) would have an imperceptible effect on future temperatures -- one-twentieth of a degree by 2050. "

Dr. S. Fred Singer, atmospheric physicist
Professor Emeritus of Environmental Sciences at the University of Virginia,
and former director of the US Weather Satellite Service;
in a Sept. 10, 2001 Letter to Editor, *Wall Street Journal*

Research to Watch

Scientists are increasingly recognizing the importance of water vapor in the climate system. Some, like Wallace Broecker, a geochemist at Columbia's Lamont-Doherty Earth Observatory, suggest that it is such an important factor that much of the global warming in the last 10,000 years may be due to the increasing water vapor concentrations in Earth's atmosphere.

His research indicates that **air reaching glaciers during the last Ice Age had less than half the water vapor content of today.** Such increases in atmospheric moisture during our current **interglacial period** would have played a far greater role in global warming than carbon dioxide or other minor gases.

✎" I can only see one element of the climate system capable of generating these fast, global changes, that is, changes in the tropical atmosphere leading to changes in the inventory of the earth's most powerful greenhouse gas-- water vapor. "

Dr. Wallace Broecker, a leading world authority on climate
Lamont-Doherty Earth Observatory, Columbia University,
lecture presented at R. A. Daly Lecture at the American Geophysical Union's
spring meeting in Baltimore, Md., May 1996.

Known causes of global climate change, like cyclical eccentricities in **Earth's rotation** and **orbit**, as well as variations in the **sun's energy output**, are the primary causes of climate cycles measured over the last half million years. However, secondary greenhouse effects stemming from changes in the ability of a warming atmosphere to support greater concentrations of gases like water vapor and carbon dioxide also appear to play a significant role. As demonstrated in the data above, of all Earth's greenhouse gases, water vapor is by far the dominant player.

The ability of humans to influence greenhouse water vapor is negligible. As such, individuals and groups whose agenda it is to require that human beings are the cause of global warming must discount or ignore the effects of water vapor to preserve their arguments, citing numbers similar to those in Table 4b . If political correctness and staying out of trouble aren't high priorities for you, go ahead and ask them how **water vapor** was handled in their models or statistics. Chances are, **it wasn't!**

References:

1) Current Greenhouse Gas Concentrations (updated October, 2000)
Carbon Dioxide Information Analysis Center
(the primary global-change data and information analysis center of the U.S. Department of Energy)
Oak Ridge, Tennessee

Greenhouse Gases and Climate Change (data now available only to "members")
IEA Greenhouse Gas R&D Programme,
Stoke Orchard, Cheltenham, Gloucestershire, GL52 7RZ, United Kingdom.

2) "Carbon cycle modelling and the residence time of natural and anthropogenic atmospheric CO2:on the construction of the 'Greenhouse Effect Global Warming' dogma;" Tom V. Segalstad, University of Oslo

3) Greenhouse Gases and Global Warming Potentials (updated April, 2002)
Carbon Dioxide Information and Analysis Center (CDIAC), U.S. Department of Energy
Oak Ridge, Tennessee.

4) Warming Potentials of Halocarbons and Greenhouses Gases
Chemical formulae and global warming potentials from Intergovernmental Panel on Climate Change, Climate Change 1995: The Science of Climate Change (Cambridge, UK: Cambridge University Press, 1996), pp. 119 and 121. Production and sales of CFC's and other chemicals from International Trade Commission, Synthetic Organic Chemicals: United States Production and Sales, 1994 (Washington, DC, 1995). TRI emissions from U.S. Environmental Protection Agency, 1994 Toxics Release Inventory: Public Data Release, EPA-745-R-94-001 (Washington, DC, June 1996), p. 73. Estimated 1994 U.S. emissions from U.S. Environmental Protection Agency, Inventory of U.S. Greenhouse Gas Emissions and Sinks, 1990-1994, EPA-230-R-96-006 (Washington, DC, November 1995), pp. 37-40.

5) References to 95% contribution of water vapor:

a. S.M. Freidenreich and V. Ramaswamy, "Solar Radiation Absorption by Carbon Dioxide, Overlap with Water, and a Parameterization for General Circulation Models," Journal of Geophysical Research 98 (1993):7255-7264

b. Global Deception: The Exaggeration of the Global Warming Threat
by Dr. Patrick J. Michaels, June 1998
Virginia State Climatologist and Professor of Environmental Sciences, University of Virginia

c. Greenhouse Gas Emissions, Appendix D, Greenhouse Gas Spectral Overlaps and Their Significance
Energy Information Administration; Official Energy Statistics from the U.S. Government

d. Personal Communication-- Dr. Richard S. Lindzen
Alfred P. Slone Professor of Meteorology, MIT

e. The Geologic Record and Climate Change
by Dr. Tim Patterson, January 2005
Professor of Geology-- Carleton University
Ottawa, Canada

Alternate link:
f. EPA Seeks To Have Water Vapor Classified As A Pollutant
by the ecoEnquirer, 2006

Alternate link:

g. Does CO2 Really Drive Global Warming?
by Dr. Robert Essenhigh, May 2001

Alternate link:

h. Solar Cycles, Not CO2, Determine Climate
by Zbigniew Jaworowski, M.D., Ph.D., D.Sc., 21st Century Science and Technology, Winter 2003-2004, pp. 52-65

Link:

5) **Global Climate Change Student Guide**
Department of Environmental and Geographical Sciences
Manchester Metropolitan University
Chester Street
Manchester
M1 5GD
United Kingdom

6) Global Budgets for Atmospheric Nitrous Oxide - Anthropogenic Contributions
William C. Trogler, Eric Bruner, Glenn Westwood, Barbara Sawrey, and Patrick Neill
Department of Chemistry and Biochemistry
University of California at San Diego, La Jolla, California

7) Methane record and budget
Robert Grumbine

I shall end this chapter with a little bit of humor.

The Earth's atmosphere contains 9,340 parts per million (0.9340%, i.e. nearly 1%) of Argon. This is over 200 times as much as atmospheric CO2 concentration which is 390 parts per million.

Argon does not satisfy the body's need for oxygen and is thus an asphyxiant. Argon is 25% more dense than air and is considered highly dangerous in closed areas. It is also difficult to detect because it is colorless, odorless, and tasteless. In confined spaces, it is known to result in death due to asphyxiation.

A 1994 incident in which a man was asphyxiated after entering an argon filled section of oil pipe under construction in Alaska highlights the dangers of argon tank leakage in confined spaces, and emphasizes the need for proper use, storage and handling. [16]

Thankfully, despite such a "high" concentration of Argon in the atmosphere, we are definitely not short of breath.

---

[16] Alaska FACE Investigation 94AK012 (1994-06-23): "Welder's Helper Asphyxiated in Argon-Inerted Pipe – Alaska (FACE AK-94-012)" - State of Alaska Department of Public Health.
http://www.cdc.gov/niosh/face/stateface/ak/94ak012.htm

III. Natural factors impacting global climate

Human beings today may consider themselves to be living in a technologically advanced age, and we may even delude ourselves that we have the erudition to explain every natural phenomenon. But the inconvenient truth is that we still have scant knowledge of the multitude of factors that affect global climate, and an even more flimsy conjecture on how those factors interact to change the climate over time.

This chapter examines some of the other factors that affect global climate, although this is by no means a comprehensive list. In addition, new research in the future may, and, likely will, uncover other previously undiscovered criteria which influence the evolution of climate over time

Natural Forcing of the Climate System

Below is a very good summary from the IPCC describing our current understanding of the main factors that influence global climate:

http://www.grida.no/publications/other/ipcc_tar/?src=/climate/ipcc_tar/wg1/041.htm

Houghton, J. T., Y. Ding, D. J. Griggs, M. Noguer, P. J. van der Linden, et al. Climate Change 2001: The Scientific Basis. Chapter 1. The Climate System: An Overview. Section 1.2 Natural Climate Systems, Subsection 1.2.1 Natural Forcing of the Climate System:

The Sun and the global energy balance

The ultimate source of energy that drives the climate system is radiation from the Sun. About half of the radiation is in the visible short-wave part of the electromagnetic spectrum. The other half is mostly in the near-infrared part, with some in the ultraviolet part of the spectrum. Each square metre of the Earth's spherical surface outside the atmosphere receives an average throughout the year of 342 Watts of solar radiation, 31% of which is immediately reflected back into space by clouds, by the atmosphere, and by the Earth's surface. The remaining 235 Wm^{-2} is partly absorbed by the atmosphere but most (168 Wm^{-2}) warms the Earth's surface: the land and the ocean. The Earth's surface returns that heat to the atmosphere, partly as infrared radiation, partly as sensible heat and as water vapour which releases its heat when it condenses higher up in the atmosphere. This exchange of energy between surface and atmosphere maintains under present conditions a global mean temperature near the surface of 14°C, decreasing rapidly with height and reaching a mean temperature of -58°C at the top of the troposphere.

For a stable climate, a balance is required between incoming solar radiation and the outgoing radiation emitted by the climate system. Therefore the climate system itself must radiate on average 235 Wm^{-2} back into space. Details of this energy balance can be seen in Figure 1.2, which shows on the left hand side what happens with the incoming solar radiation, and on the right hand side how the atmosphere emits the outgoing infrared radiation. Any physical object radiates energy of an amount and at wavelengths typical for the temperature of the object: at higher temperatures more energy is radiated at shorter wavelengths. For the Earth to radiate 235 Wm^{-2}, it should radiate at an effective emission temperature of -19°C with typical wavelengths in the infrared part of the spectrum. This is 33°C lower than the average temperature of 14°C at the Earth's surface. To understand why this is so, one must take into account the radiative properties of the atmosphere in the infrared part of the spectrum.

The natural greenhouse effect

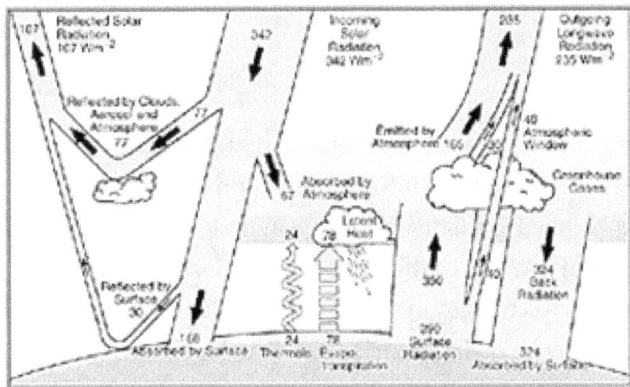

Figure 1.2: The Earth's annual and global mean energy balance. Of the incoming solar radiation, 49% (168 Wm-2) is absorbed by the surface. That heat is returned to the atmosphere as sensible heat, as evapotranspiration (latent heat) and as thermal infrared radiation. Most of this radiation is absorbed by the atmosphere, which in turn emits radiation both up and down. The radiation lost to space comes from cloud tops and atmospheric regions much colder than the surface. This causes a greenhouse effect. Source: Kiehl and Trenberth, 1997: Earth's Annual Global Mean Energy Budget, Bull. Am. Met. Soc. 78, 197-208.

The atmosphere contains several trace gases which absorb and emit infra red radiation. These so-called greenhouse gases absorb infrared radiation, emitted by the Earth's surface, the atmosphere and clouds, except in a transparent part of the spectrum called the "atmospheric window", as shown in Figure 1.2. They emit in turn infrared radiation in all directions including downward to the Earth's surface. Thus greenhouse gases trap heat within the atmosphere. This mechanism is called the natural greenhouse effect. The net result is an upward transfer of infrared radiation from warmer levels near the Earth's surface to colder levels at higher altitudes. The infrared radiation is effectively radiated back into space from an altitude with a temperature of, on average, -19°C, in balance with the incoming radiation, whereas the Earth's surface is kept at a much higher temperature of on average 14°C. This effective emission temperature of -19°C corresponds in mid-latitudes with a height of approximately 5 km. Note that it is essential for the greenhouse effect that the temperature of the lower atmosphere is not constant (isothermal) but decreases with height. The natural greenhouse effect is part of the energy balance of the Earth, as can be seen schematically in Figure 1.2.

Clouds also play an important role in the Earth's energy balance and in particular in the natural greenhouse effect. Clouds absorb and emit infrared radiation and thus contribute to warming the Earth's surface, just like the greenhouse gases. On the other hand, most clouds are bright reflectors of solar radiation and tend to cool the climate system. The net average effect of the Earth's cloud cover in the present climate is a slight cooling: the reflection of radiation more than compensates for the greenhouse effect of clouds. However this effect is highly variable, depending on height, type and optical properties of clouds.

This introduction to the global energy balance and the natural greenhouse effect is entirely in terms of the global mean and in radiative terms. However, for a full understanding of the greenhouse effect and of its impact on the climate system, dynamical feedbacks and energy transfer processes should also be taken into account..

Radiative forcing and forcing variability

In an equilibrium climate state the average net radiation at the top of the atmosphere is zero. A change in either the solar radiation or the infrared radiation changes the net radiation. The corresponding imbalance is called "radiative forcing". In practice, for this purpose, the top of the troposphere (the tropopause) is taken as the top of the atmosphere, because the stratosphere adjusts in a matter of months to changes in the radiative balance, whereas the surface-troposphere system adjusts much more slowly, owing principally to the large thermal inertia of the oceans. The radiative forcing of the surface troposphere system is then the change in net irradiance at the tropopause after allowing for stratospheric temperatures to re-adjust to radiative equilibrium, but with surface and tropospheric temperatures and state held fixed at the unperturbed values.

External forcings, such as the solar radiation or the large amounts of aerosols ejected by volcanic eruption into the atmosphere, may vary on widely different time-scales, causing natural variations in the radiative forcing. These variations may be negative or positive. In either case the climate system must react to restore the balance. A positive radiative forcing tends to warm the surface on average, whereas a negative radiative forcing tends to cool it. Internal climate processes and feedbacks may also cause variations in the radiative balance by their impact on the reflected solar radiation or emitted infrared radiation, but such variations are not considered part of radiative forcing.

The Sun

It does not take a rocket scientist to deduce that the Sun has a significant impact on global climate. What is less clear, and what the contradictory hypotheses enumerated in this section will show, is how the Sun will behave in the coming decades. We still have very limited understanding of whether the Sun will get stronger or weaker, and the evolution of global climate will depend very much on this significant unknown of what the Sun decides to do.

Strong correlation between solar activity and global temperature

The source of the following graphs is from the extensive research done by Don J. Easterbrook, Professor Emeritus at Western Washington University, Bellingham, WA. You will find a complete list of Easterbrook's publications at the following link:

http://myweb.wwu.edu/dbunny/pdfs/dje_publications.pdf

The first graph demonstrates the strong and consistent correlation between solar irradiance and global temperature between 1880 and 2000. On the right hand side, you will also find the plot of CO2 versus temperature (already described in the previous chapter) for comparison.

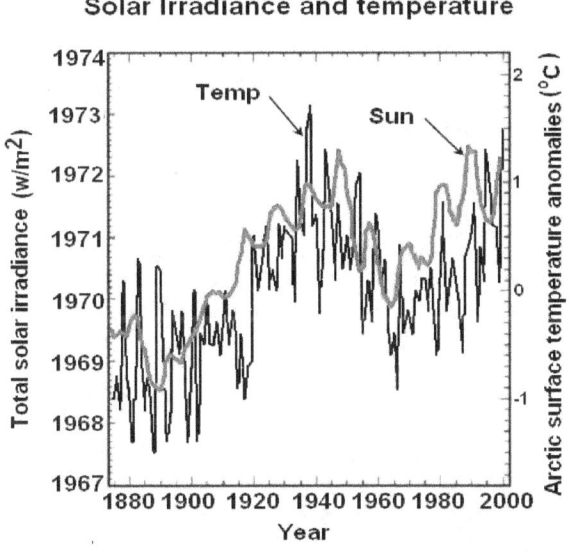
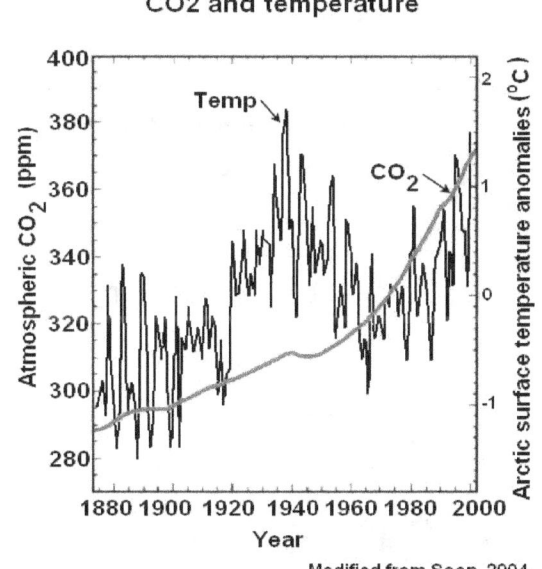

Modified from Soon, 2004

The next graph displays the strong correlation between Solar cycle length and temperature between 1750 and 2000

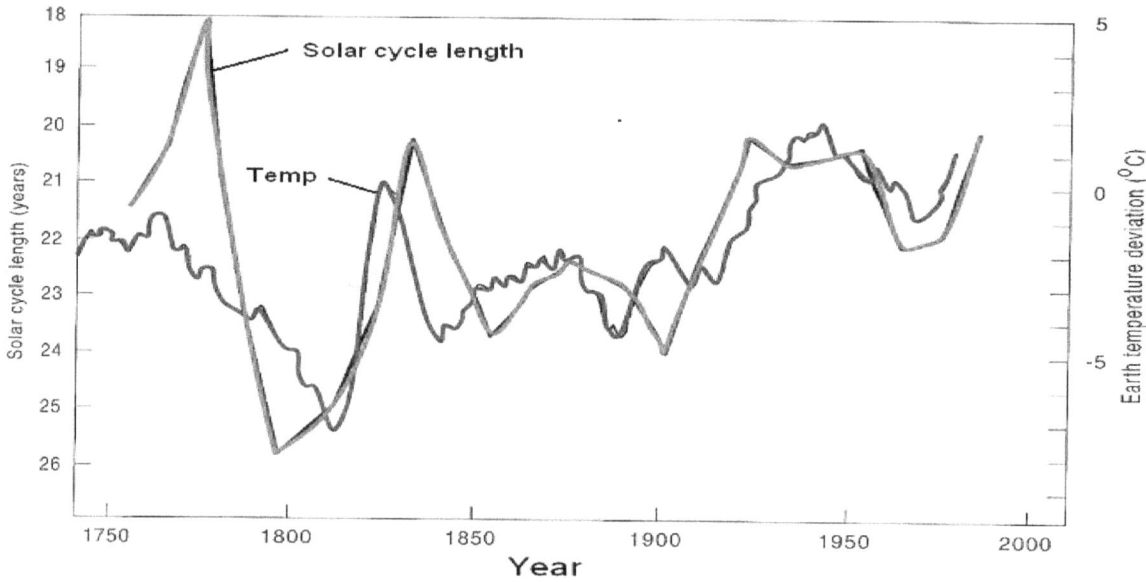

Moving 11-year average of Northern Hemisphere temperature as deviations from 1951–1970 mean versus solar magnetic cycle length. The shorter the cycle, the more active and brighter the sun (Baliunas and Soon. 1996).

The final graph from Easterbrook demonstrates that, in the past half-millennium, the climate has cooled considerably during times of few sunspots and low solar irradiance. For example, there were very few sunspots during the Little Ice Age

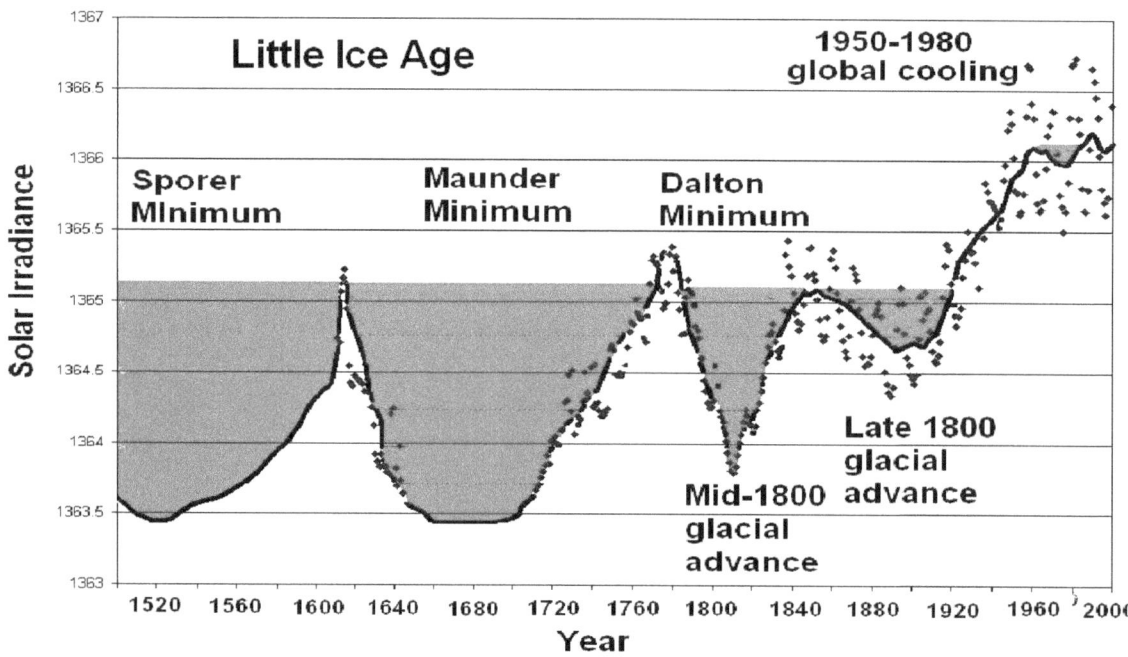

Let us now cross-check the strong correlation between solar irradiance and global temperature using another source: http://www.biocab.org/Solar_Irradiance_Climate_Change.html

1610-2000 AD

http://www.biocab.org/ISI_Lean___Loehle.jpg

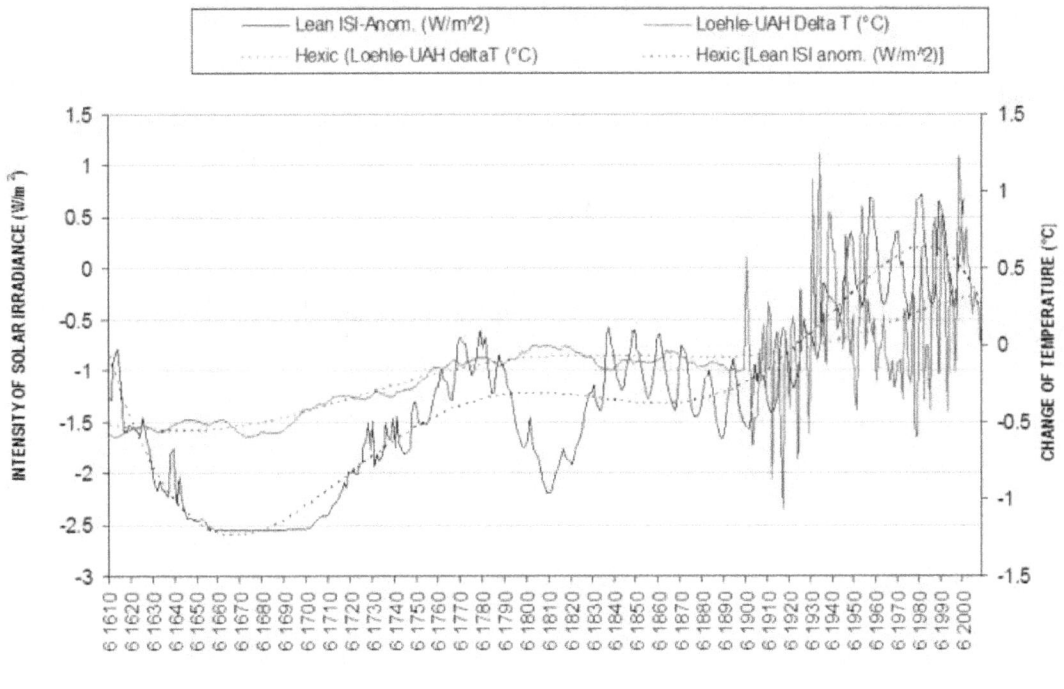

20th Century

http://www.biocab.org/SI_Lean-T_UAH_1900-2007.jpg

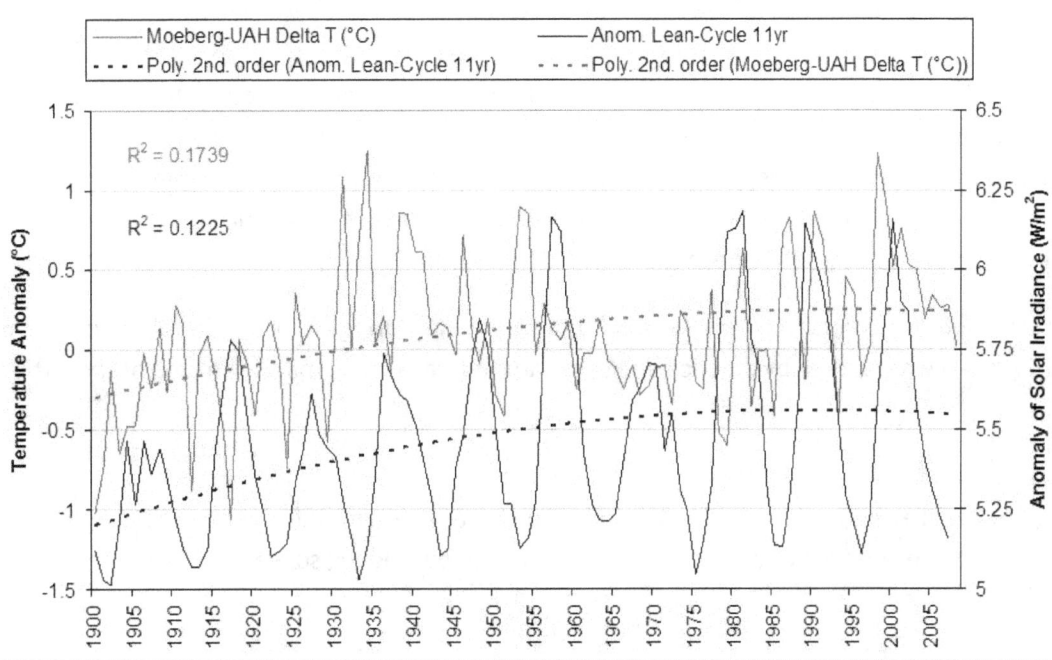

http://www.biocab.org/Preminger_TSI-UAH_T.jpg

Solar Irradiance Anomalies and Change of Temperature

| Preminger Anom SI (Z-1360 5 W/m^2) | Temp Var Möeberg/UAH (°C) |
| Hexic Anom SI (Z-1360 5 W/m^2) | Hexic (Temp Var UAH-°C) |

There is significant guesswork in the scientific community on how the Sun will behave in the next few decades. The next few sections will describe some of these sometimes contradictory hypotheses.

Hypothesis #1: Sunspots are disappearing (Don J. Easterbrook, H. Svensmark and others)

The three studies released by NSO's Solar Synoptic Network this week, predicting the virtual vanishing of sunspots for the next several decades and the possibility of a solar minimum similar to the Maunder Minimum, came as stunning news. According to Frank Hill "the fact that three completely different views of the Sun point in the same direction is a powerful indicator that the sunspot cycle may be going into hibernation."

The last time sunspots vanished from the sun for decades was during the Maunder Minimum from 1645 to 1700 AD was marked by drastic cooling of the climate and the maximum cold of the Little Ice Age.

What happened the last time sunspots disappeared?

Abundant physical evidence from the geologic past provides a record of former periods of global cooling. Geologic records provide clear evidence of past global cooling so we can use them to project global climate into the future—the past is the key to the future. So what can we learn from past sunspot history and climate change?

Galileo's perfection of the telescope in 1609 allowed scientists to see sunspots for the first time. From 1610 A.D. to 1645 A.D., very few sunspots were seen, despite the fact that many scientists with telescopes were looking for them, and from 1645 to 1700 AD sunspots virtually disappeared from the sun (Fig. 1). During this interval of greatly reduced sunspot activity, known as the Maunder Minimum, global climates turned bitterly cold (the Little Ice Age), demonstrating a clear correspondence between sunspots and cool climate. After 1700 A.D., the number of observed sunspots increased sharply from nearly zero to more than 50 and the global climate warmed.

The Maunder Minimum was not the beginning of The Little Ice Age—it actually began about 1300 AD—but it marked perhaps the bitterest part of the cooling. Temperatures dropped ~4º C (~7 º F) in ~20 years in mid-to high latitudes. The colder climate that ensued for several centuries was devastating. The population of Europe had become dependent on cereal grains as their main food supply during the Medieval Warm Period and when the colder climate, early snows, violent storms, and recurrent flooding swept Europe, massive crop failures occurred. Winters in Europe were bitterly cold, and summers were rainy and too cool for growing cereal crops, resulting in widespread famine and disease. About a third of the population of Europe perished.

Glaciers all over the world advanced and pack ice extended southward in the North Atlantic. Glaciers in the Alps advanced and overran farms and buried entire villages. The Thames River and canals and rivers of the Netherlands frequently froze over during the winter. New York Harbor froze in the winter of 1780 and people could walk from Manhattan to Staten Island. Sea ice surrounding Iceland extended for miles in every direction, closing many harbors. The population of Iceland decreased by half and the Viking colonies in Greenland died out in the 1400s because they could no longer grow enough food there. In parts of China, warm weather crops that had been grown for centuries were abandoned. In North America, early European settlers experienced exceptionally severe winters.

So what can we learn from the Maunder? Perhaps most important is that the Earth's climate is related to sunspots. The cause of this relationship is not understood, but it definitely exists. The second thing is that cooling of the climate during sunspot minima imposes great suffering on humans—global cooling is much more damaging than global warming.

Global cooling during other sunspot minima
The global cooling that occurred during the Maunder Minimum was neither the first nor the only such event. The Maunder was preceded by the Sporer Minimum (~1410–1540 A.D.) and the Wolf Minimum (~1290–1320 A.D.) and succeeded by the Daltong Minimum (1790–1830), the unnamed 1880–1915 minima, and the unnamed 1945–1977 Minima (Fig. 2). Each of these periods is characterized by low numbers of sunspots, cooler global climates, and changes in the rate of production of 14C and 10Be in the upper atmosphere. As shown in Fig. 2, each minimum was a time of global cooling, recorded in the advance of alpine glaciers.

The same relationship between sunspots and temperature is also seen between sunspot numbers and temperatures in Greenland and Antarctica (Fig. 3). Each of the four minima in sunspot numbers seen in Fig. 3 also occurs in Fig. 2. All of them correspond to advances of alpine glaciers during each of the cool periods.

In 1999, the year after the high temperatures of the 1998 El Nino, the scientists referenced below became convinced that geologic data of recurring climatic cycles (ice core isotopes, glacial advances and retreats, and sun spot minima) showed conclusively that we were headed for several decades of global cooling and presented a paper to that effect (Fig. 5). The evidence for this conclusion was presented in a series of papers from 2000 to 2011 (The data are available in several GSA papers, Easterbrook's website, a 2010 paper, and in a paper scheduled to be published in Sept 2011). The evidence consisted of temperature data from isotope analyses in the Greenland ice cores, the past history of the PDO, alpine glacial fluctuations, and the abrupt Pacific SST flips from cool to warm in 1977 and from warm to cool in 1999. Projection of the PDO to 2040 forms an important part of this cooling prediction.

So far, the cooling prediction seems to be coming to pass, with no global warming above the 1998 temperatures and a gradually deepening cooling since then. However, until now, I have suggested that it was too early to tell which of these possible cooling scenarios were most likely. If we are indeed headed toward a disappearance of sunspots similar to the Maunder Minimum during the Little Ice Age then perhaps the most dire prediction may come to pass. As the scientists referenced have said many times over the past 10 years, time will tell whether the prediction is correct or not. The announcement that sun spots may disappear totally for several decades is very disturbing because it could mean that we are headed for another Little Ice Age during a time when world population is predicted to increase by 50% with sharply increasing demands for energy, food production, and other human needs. Hardest hit will be poor countries that already have low food production, but everyone would feel the effect of such cooling. The clock is ticking. Time will tell!

References

- D'Aleo, J., Easterbrook, D.J., 2010. Multidecadal tendencies in Enso and global temperatures related to multidecadal oscillations: Energy & Environment, vol. 21 (5), p. 436–460.

- Easterbrook, D.J., 2000, Cyclical oscillations of Mt. Baker glaciers in response to climatic changes and their correlation with periodic oceanographic changes in the Northeast Pacific Ocean: Geological Society of America, Abstracts with Programs, vol. 32, p.17.

- Easterbrook, D.J., 2001, The next 25 years; global warming or global cooling? Geologic and oceanographic evidence for cyclical climatic oscillations: Geological Society of America, Abstracts with Programs, vol. 33, p.253.

- Easterbrook, D.J., 2005, Causes and effects of late Pleistocene, abrupt, global, climate changes and global warming: Geological Society of America, Abstracts with Programs, vol. 37, p.41.

- Easterbrook, D.J., 2006, Causes of abrupt global climate changes and global warming; predictions for the coming century: Geological Society of America, Abstracts with Programs, vol. 38, p. 77.

- Easterbrook, D.J., 2006, The cause of global warming and predictions for the coming century: Geological Society of America, Abstracts with Programs, vol. 38, p.235-236.

- Easterbrook, D.J., 2007, Geologic evidence of recurring climate cycles and their implications for the cause of global warming and climate changes in the coming century: Geological Society of America Abstracts with Programs, vol. 39, p. 507.

- Easterbrook, D.J., 2007, Late Pleistocene and Holocene glacial fluctuations; implications for the cause of abrupt global climate changes: Geological Society of America, Abstracts with Programs, vol. 39, p.594

- Easterbrook, D.J., 2007, Younger Dryas to Little Ice Age glacier fluctuations in the Fraser Lowland and on Mt. Baker, Washington: Geological Society of America, Abstracts with Programs, vol. 39, p.11.

- Easterbrook, D.J., 2007, Historic Mt. Baker glacier fluctuations—geologic evidence of the cause of global warming: Geological Society of America, Abstracts with Programs, vol. 39, p. 13.

- Easterbrook, D.J., 2008, Solar influence on recurring global, decadal, climate cycles recorded by glacial fluctuations, ice cores, sea surface temperatures, and historic measurements over the past millennium: Abstracts of American Geophysical Union Annual Meeting, San Francisco.

- Easterbrook, D.J., 2008, Implications of glacial fluctuations, PDO, NAO, and sun spot cycles for global climate in the coming decades: Geological Society of America, Abstracts with Programs, vol. 40, p. 428.

- Easterbrook, D.J., 2008, Correlation of climatic and solar variations over the past 500 years and predicting global climate changes from recurring climate cycles: Abstracts of 33rd International Geological Congress, Oslo, Norway.

- Easterbrook, D.J., 2009, The role of the oceans and the Sun in late Pleistocene and historic glacial and climatic fluctuations: Geological Society of America, Abstracts with Programs, vol. 41, p. 33.

- Eddy, J.A., 1976, The Maunder Minimum: Science, vol. 192, p. 1189–1202.

- Hoyt, D.V. and Schatten, K.H., 1997, The Role of the sun in climate change: Oxford University, 279 p.

- Svensmark, H. and Calder, N., 2007, The chilling stars: A new theory of climate change: Icon Books, Allen and Unwin Pty Ltd, 246 p.

- Svensmark, H. and Friis-Christensen, E., 1997, Variation of cosmic ray flux and global cloud coverda missing link in solar–climate relationships: Journal of Atmospheric and SolareTerrestrial Physics, vol. 59, p. 1125–1132.

- Svensmark, H., Pedersen, J.O., Marsh, N.D., Enghoff, M.B., and Uggerhøj, U.I., 2007, Experimental evidence for the role of ions in particle nucleation under atmospheric conditions: Proceedings of the Royal Society, vol. 463, p. 385–396.

- Usoskin, I.G., Mursula, K., Solanki, S.K., Schussler, M., and Alanko, K., 2004, Reconstruction of solar activity for the last millenium using 10Be data: Astronomy and Astrophysics, vol. 413, p. 745–751.

Washington Post Article:

By Brian Vastag

Tuesday, Jun 21, 2011

The sun is waking up.

And on June 7, it woke up Michael Hesse. At 5:49 a.m., the solar scientist received an alert on his smartphone. NASA spacecraft had seen a burst of X-rays spinning out from a sunspot. The burst was a solar flare - and a "notably large one" at that, Hesse said later.

The sun has been quiet for years, at the nadir of its activity cycle. But since February, our star has been spitting out flares and plasma like an angry dragon. It's Hesse's job to watch these eruptions.

If a big one were headed our way, Hesse needed to know, and fast, so he could alert the electric power industry to brace for a geomagnetic storm that could knock some of the North American power grid offline.

Hesse gathered his team at the Goddard Space Flight Center in Greenbelt, where he is chief of the Space Weather Laboratory, and fed the latest data from four sun-staring satellites into powerful computers.

At 7:49 Hesse got his answer. An animated chart traced the predicted path of a huge arc of plasma - hot gas - hurtling through the inner solar system. But only the tail of the plume would lick Earth, arriving June 9 and driving a dazzling display of the northern lights from Alaska through Maine.

While a video of the eruption captured by NASA's Solar Dynamics Observatory showed an enormous plume spraying from the sun, this solar tantrum would not be the big one - it would not be the 1859 event all over again.

Sept. 1 of that year saw the largest solar flare on record, witnessed by British astronomer Richard Carrington. While tracing features of the sun's surface, which Carrington had projected via telescope onto paper, he saw a sudden flash emerge from a dark spot. Although such sunspots had sparked curiosity for centuries - Galileo famously drew them, too, in the early 1600s - Carrington had no idea what the flash could mean.

Within hours, telegraph operators found out. Their long strands of wire acted as antennas for this huge wave of solar energy. As this tsunami sped by, transmitters heated up, and several burst into flames. Observers in Miami and Havana gaped skyward at eerie green and yellow displays, the northern lights pushed far south.

A knockout punch

Such a "Carrington event" will happen again someday, but our wired civilization will suffer losses far greater than a few telegraph shacks.

Communications satellites will be knocked offline. Financial transactions, timed and transmitted via those satellite, will fail, causing millions or billions in losses. The GPS system will go wonky. Astronauts on the space station will huddle in a shielded module, as they have done three times in the past decade due to "space weather," the scientific term for all of the sun's freaky activity. Flights between North America and Asia, over the North Pole, will have to be rerouted, as they were in April during a weak solar storm at a cost to the airlines of $100,000 a flight. And oil pipelines, particularly in Alaska and Canada, will suffer corrosion as they, like power lines, conduct electricity from the solar storm.

But the biggest impact will be on the modern marvel known as the power grid. And experts warn that the grid is not ready. In 2008, the National Academy of Sciences stated that an 1859-level storm could knock out power in parts of the northeastern and northwestern United States for months, even years. Report co-author John Kappenmann estimated that about 135 million Americans would be forced to revert to a pre-electric lifestyle or relocate. Water systems would fail. Food would spoil. Thousands could die. The financial cost: Up to $2 trillion, one-seventh the annual U.S. gross domestic product.

Utilities say they're studying the issue, with an eye toward understanding how to protect the grid by powering down sections of it during an hours-long solar storm.

Their efforts are motivated, in part, by the sun's increasingly frequent outbursts. Every 11 to 12 years, solar activity ramps up. After a quiet season, the sun is now spitting out flares again, with activity expected to peak in 2013 and 2014, said Dean Pesnell, a solar scientist at Goddard.

"The sun is not partisan, it doesn't listen to diplomacy, and sanctions don't work," said Peter Huessy, president of GeoStrategic Analysis. Huessy wants Congress to enact rules that would force power companies to better protect the power grid. "The sun has its own clock. And we don't know what that clock is, except for once every hundred years or so, it has a coronary."

Running out of time

Predicting flares is still a nascent science. They typically spring from sunspots, which appear when tangled magnetic fields well up from deep within the sun. A burst of X-rays, flares travel at the speed of light, reaching Earth in about eight minutes. While they can interfere with the electronics in satellites, they pose no direct threat to people on the ground because Earth's magnetic field acts as a shield against this type of solar weather. This shield is weakest at the North Pole and South Pole, which is why space weather affects high latitudes the most.

Within hours of a flare, the sun often tosses in an encore: a huge plume of plasma known as a coronal mass ejection. During each solar cycle, the sun throws off hundreds of these. But only a few are large. The fastest, most damaging of these waves of charged particles can reach the Earth in about 20 hours. On arrival, these storms deform the Earth's magnetic field, charge the atmosphere and induce electric currents in power lines for several hours.

But estimating the arrival time and damage potential of such storms is tricky business. The simulations that Hesse runs at Goddard provide only a rough estimate, bracketing the arrival time of a solar storm in a 12- to 14-hour window.

More-precise alerts are sent to power companies just 20 to 30 minutes before a solar storm hits Earth. In May, 29 such alerts went out, triggered by a NASA satellite called the Advanced Composition Explorer, or ACE.

But if ACE fails, the space weather warning system will be crippled, said Tom Bogdan, who heads the Space Weather Prediction Center at the National Oceanic and Atmospheric Administration. Bogdan wants Congress to fund other satellites to replace ACE before it runs out of fuel in 2021.

An uncertain fate

One possible replacement, a satellite called DSCOVR, sits nearly finished in a hangar at Goddard, where it has languished since 2001. The vision of Al Gore, sidelined when congressional Republicans defunded its launch vehicle, DSCOVR would provide longer warning times for solar storms. Its fate is uncertain, although President Obama is expected to ask for funds to launch the probe in his 2012 budget. Bogdan said the earliest it could get off the ground is 2014.

Representatives of the power industry take issue with the worst-case scenarios.

Leaders do acknowledge that huge solar flares are a serious issue, one the industry is addressing. But "the idea of 130 million people out of power for 10 years is an overstatement," said Gerry Cauley, president of the North American Electric Reliability Corp., or NERC.

In 2007, Congress gave NERC the power to make rules for electric utilities to prevent blackouts like the one that left an estimated 50 million people in the Midwest, the Northeast and Ontario without power for up to four days in August 2003. (That outage was caused not by a solar flare but by high demand and a tree that fell on a power line; a cascading failure knocked some 100 power plants offline.) "The potential is there for damage to equipment and possibly even outages," Cauley added. "But the grid itself is very resilient."

The grid's weak spots

In 1989, the grid got its most severe solar test, and sections did not fare well. A solar storm one-tenth the strength of the 1859 event triggered a cascade of failures in Quebec in just 90 seconds. Several million people went without power for nine to 12 hours, causing hundreds of millions of dollars in damage. In South Africa, the storm destroyed huge transformers.

Each the size of a house and costing several million dollars, transformers are the grid's weak spots. They boost the voltage of electricity for transmission along high-voltage lines, but they also absorb extra loads coming down those lines. During the 1989 event, two of South Africa's transformers overheated and fried during the storm, while nine more failed within a year, said Mark Lauby, a vice president at NERC.

Legislation under consideration in the House would force utility companies to protect 350 critical transformers from a massive solar storm. Under the bill, called the SHIELD Act, the one-time cost of $100 million to $300 million would be passed on to customers. Last year the bill passed in the House unanimously, only to stall in the Senate.

But the SHIELD Act is not dead. In May, the subcommittee on energy and power held a hearing on the bill, where military officials and government regulators warned of the dangers of space weather. Advocates expect the legislation to be reintroduced.

In the meantime, Bogdan will be losing sleep over losing ACE, the sun storm sentinel.

"It's the extreme solar events I'm worried about," he said. "It might not happen this solar cycle. But sometime in my lifetime or my children's, that storm will be here. The question is `Will we be prepared for it?' "

vastagb@washpost.com

Hypothesis #3: Cycle 25

A recent theory claims that there are magnetic instabilities in the core of the Sun that cause fluctuations with periods of either 41,000 or 100,000 years. These could provide a better explanation of the ice ages than the Milankovitch cycles

1) http://www.dailymail.co.uk/sciencetech/article-2093264/Forget-global-warming--Cycle-25-need-worry-NASA-scientists-right-Thames-freezing-again.html

The supposed 'consensus' on man-made global warming is facing an inconvenient challenge after the release of new temperature data showing the planet has not warmed for the past 15 years.

The figures suggest that we could even be heading for a mini ice age to rival the 70-year temperature drop that saw frost fairs held on the Thames in the 17th Century.

Based on readings from more than 30,000 measuring stations, the data was issued last week without fanfare by the Met Office and the University of East Anglia Climatic Research Unit. It confirms that the rising trend in world temperatures ended in 1997

Meanwhile, leading climate scientists yesterday told The Mail on Sunday that, after emitting unusually high levels of energy throughout the 20th Century, the sun is now heading towards a 'grand minimum' in its output, threatening cold summers, bitter winters and a shortening of the season available for growing food.

Solar output goes through 11-year cycles, with high numbers of sunspots seen at their peak.

We are now at what should be the peak of what scientists call 'Cycle 24' – which is why last week's solar storm resulted in sightings of the aurora borealis further south than usual. But sunspot numbers are running at less than half those seen during cycle peaks in the 20th Century.

Analysis by experts at NASA and the University of Arizona – derived from magnetic-field measurements 120,000 miles beneath the sun's surface – suggest that

2) http://sc25.com/

Abstract #1

Temporal changes in the power of the longwave radiation of the system Earth-atmosphere emitted to space always lag behind changes in the power of absorbed solar radiation due to slow change of its enthalpy.

That is why the debit and credit parts of the average annual energy budget of the terrestrial globe with its air and water envelope are practically always in an unbalanced state. Average annual balance of the thermal budget of the system Earth-atmosphere during long time period will reliably determine the course and value of both an energy excess accumulated by the Earth or the energy deficit in the thermal budget which, with account for data of the TSI forecast, can define and predict well in advance the direction and amplitude of the forthcoming climate changes.

From early 90s we observe bicentennial decrease in both the TSI and the portion of its energy absorbed by the Earth. The Earth as a planet will henceforward have negative balance in the energy budget which will result in the temperature drop in approximately 2014.

We can expect the onset of a deep bicentennial minimum of total solar irradiance (TSI) in approximately 2042±11 and the 19th deep minimum of global temperature in the past 7500 years – in 2055±11. After the maximum of solar cycle 24, from approximately 2014 we can expect the start of deep cooling with a Little Ice Age in 2055±11." --Habibullo I. Abdussamatov, Russian Academy of Science, 1 February 2012

Abstract #2
Temporal changes in the power of the longwave radiation of the system Earth-atmosphere emitted to space always lag behind changes in the power of absorbed solar radiation due to slow change of its enthalpy.

That is why the debit and credit parts of the average annual energy budget of the terrestrial globe with its air and water envelope are practically always in an unbalanced state. Average annual balance of the thermal budget of the system Earth-atmosphere during long time period will reliably determine the course and value of both an energy excess accumulated by the Earth or the energy deficit in the thermal budget which, with account for data of the TSI forecast, can define and predict well in advance the direction and amplitude of the forthcoming climate changes.

From early 90s we observe bicentennial decrease in both the TSI and the portion of its energy absorbed by the Earth. The Earth as a planet will henceforward have negative balance in the energy budget which will result in the temperature drop in approximately 2014.

References
- Ehrlich, R. (2007). "Solar Resonant Diffusion Waves as a Driver of Terrestrial Climate Change". Journal of Atmospheric and Solar-Terrestrial Physics 69 (7): 759. arXiv:astro-ph/0701117. Bibcode 2007JASTP..69..759E. doi:10.1016/j.jastp.2007.01.005.
- Clark, S. (2007). "Sun's fickle heart may leave us cold". New Scientist 193 (2588): 12. doi:10.1016/S0262-4079(07)60196-1. http://environment.newscientist.com/channel/earth/mg19325884.500-suns-fickle-heart-may-leave-us-cold.html.
- "We can expect the onset of a deep bicentennial minimum of total solar irradiance (TSI) in approximately 2042±11 and the 19th deep minimum of global temperature in the past 7500 years – in 2055±11. After the maximum of solar cycle 24, from approximately 2014 we can expect the start of deep cooling with a Little Ice Age in 2055±11." --Habibullo I. Abdussamatov, Russian Academy of Science, 1 February 2012

Hypothesis #4: Earth may be heading for a little ice age
http://www.newsmax.com/Newsfront/littleiceage-sunspots-sun-earth/2011/06/15/id/400199

The Earth could be heading straight for a little ice age — and not a global warming phase as dictated by Al Gore and others — posits the U.S. National Solar Observatory (NSO) and the U.S. Air force Research Laboratory.

The news comes from the results of three separate analyses according to The Register. Low sun activity is cited as the reason. In recent measurements, sunspot activity, intense solar eruptions caused by magnetic activity, has been astonishingly lower than predicted, which translates into a period of cooling — what scientists dub a "little ice age."

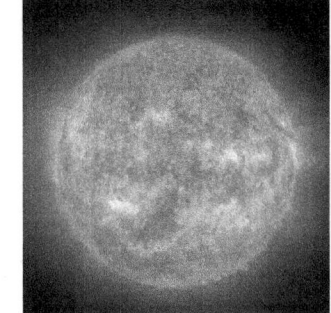

Scientists expected to find sun activity in much greater strength, meaning increased sunspot activity. They have seen this type of slowdown before, during a 70-year period, from 1645-1715. Scientists called this period the "Maunder Minimum."

During this period, researchers reported, "rivers that are normally ice-free froze and snow fields remained year-round at lower altitudes" — the culmination of a little ice age. Many rivers that do not freeze over today, such as the Thames, in fact did freeze over, at times allowing people to walk on the Thames and even armies to cross.

Solar activity (AP)

The trends that led to that Maunder Minimum seem to be repeating in the current cycle, leading scientists to predict another little ice age. Dr. Frank Hill of the NSO told The Register, "Three different views of the Sun point in the same direction . . . a powerful indicator that the sunspot cycle may be going into hibernation." In short, less sunspot activity, more ice skating.

Hill's research matched 13 years of sunspot data from the McMath-Pierce Telescope at Kitt Peak in Arizona, analyzed by physicists Matt Penn and William Livingston. Richard Altrock, an astrophysicist studying Air Force data, found a 40-year decline in magnetic activity in the corona of the Sun — further evidence of a cooling period. "Changes we see in the corona reflect changes deep in the Sun," he reported to The Register.

If proven, the upcoming little ice age could impact legislation being drawn up to deal with CO_2 emissions — including cap and trade. Not all is rosy with such a cooling period, however. Future generations may find that dealing with significantly colder temperatures is no ice skate in the park. Bundle up, Al Gore.

Volcanoes

Effects of Volcanoes[17]

There are many different types of volcanic eruptions and associated activity: phreatic eruptions (steam-generated eruptions), explosive eruption of high-silica lava (e.g., rhyolite), effusive eruption of low-silica lava (e.g., basalt), pyroclastic flows, lahars (debris flow) and carbon dioxide emission. All of these activities can pose a hazard to humans.

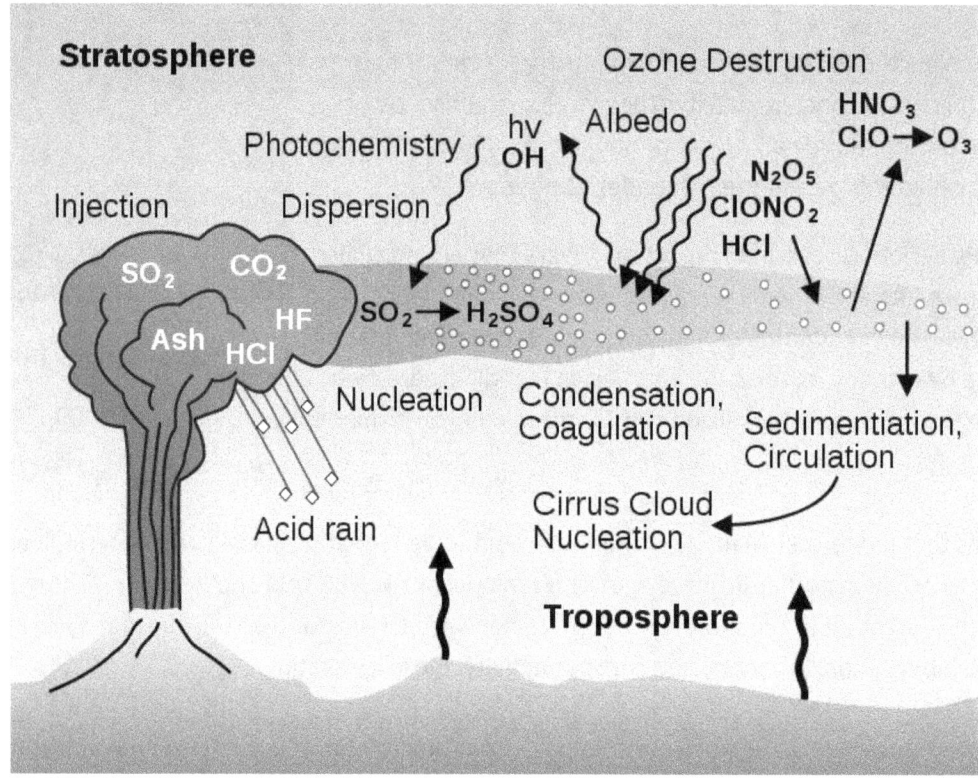

Volcanic gases

The concentrations of different volcanic gases can vary considerably from one volcano to the next. Water vapor is typically the most abundant volcanic gas, followed by carbon dioxide and sulfur dioxide. Other principal volcanic gases include hydrogen sulfide, hydrogen chloride, and hydrogen

---

[17] **References:**

- http://pubs.usgs.gov/of/of97-262/of97-262.html
- University of California – Davis (April 25, 2008). "Volcanic Eruption Of 1600 Caused Global Disruption"-*ScienceDaily*
- Supervolcano Eruption – In Sumatra – Deforested India 73,000 Years Ago". *ScienceDaily*. Nov. 24, 2009
- The new batch – 150,000 years ago". BBC – Science & Nature – The evolution of man
- "When humans faced extinction". BBC. June 9, 2003 http://news.bbc.co.uk/2/hi/science/nature/2975862.stm
- "*Volcanoes in human history: the far-reaching effects of major eruptions*". Jelle Zeilinga de Boer, Donald Theodore Sanders (2002). Princeton University Press. p.155. ISBN 0-691-05081-3
- Oppenheimer, Clive (2003). "Climatic, environmental and human consequences of the largest known historic eruption: Tambora volcano (Indonesia) 1815". *Progress in Physical Geography* **27** (2): 230–259. doi:10.1191/0309133303pp379ra.
- "Ó Gráda, C.: Famine: A Short History". Princeton University Press.
- "Yellowstone's Super Sister". Discovery Channel.
- Benton M J (2005). *When Life Nearly Died: The Greatest Mass Extinction of All Time*. Thames & Hudson. ISBN 978-0-500-28573-2.
- "Volcanic Gases and Their Effects". U.S. Geological Survey. http://volcanoes.usgs.gov/Hazards/What/VolGas/volgas.html

fluoride. A large number of minor and trace gases are also found in volcanic emissions, for example hydrogen, carbon monoxide, halocarbons, organic compounds, and volatile metal chlorides.

Large, explosive volcanic eruptions inject water vapor (H_2O), carbon dioxide (CO_2), sulfur dioxide (SO_2), hydrogen chloride (HCl), hydrogen fluoride (HF) and ash (pulverized rock and pumice) into the stratosphere to heights of 16–32 kilometres (10–20 mi) above the Earth's surface. Large injections may cause visual effects such as unusually colorful sunsets and affect global climate mainly by cooling it.

The injected ash falls rapidly from the stratosphere; most of it is removed within several days to a few weeks. Ash thrown into the air by eruptions can present a hazard to aircraft, especially jet aircraft where the particles can be melted by the high operating temperature. Dangerous encounters in 1982 after the eruption of Galunggung in Indonesia, and 1989 after the eruption of Mount Redoubt in Alaska raised awareness of this phenomenon. Nine Volcanic Ash Advisory Centers were established by the International Civil Aviation Organization to monitor ash clouds and advise pilots accordingly. The 2010 eruptions of Eyjafjallajökull caused major disruptions to air travel in Europe.

The most significant impacts from these injections come from the conversion of sulfur dioxide to sulfuric acid (H_2SO_4), which condenses rapidly in the stratosphere to form fine sulfate aerosols. The sulfate aerosols also promote complex chemical reactions on their surfaces that alter chlorine and nitrogen chemical species in the stratosphere.

This effect, together with increased stratospheric chlorine levels from chlorofluorocarbon pollution, generates chlorine monoxide (ClO), which destroys ozone (O_3). This is interesting to note, given that for the past few decades, the popular myth has been that only human-manufactured CFCs destroy the ozone layer.

As the aerosols grow and coagulate, they settle down into the upper troposphere where they serve as nuclei for cirrus clouds and further modify the Earth's radiation balance. Most of the hydrogen chloride (HCl) and hydrogen fluoride (HF) are dissolved in water droplets in the eruption cloud and quickly fall to the ground as acid rain. Gas emissions from volcanoes are a natural contributor to acid rain.

Finally, explosive volcanic eruptions release the greenhouse gas carbon dioxide and thus provide a deep source of carbon for biogeochemical cycles. Volcanic activity, on average, releases about 130 to 230 teragrams (145 million to 255 million short tons) of carbon dioxide each year. Exceptionally large explosive volcanic eruptions, of course have a larger impact. Again, this is very interesting to note, for obvious reasons.

On the positive side, volcanic eruptions also provide the benefit of adding nutrients to soil through the weathering process of volcanic rocks. These fertile soils assist the growth of plants and various crops. Volcanic eruptions can also create new islands, as the magma cools and solidifies upon contact with the water.

Effects on the Earth's Atmosphere (as per current scientific understanding)

The aerosols increase the Earth's albedo—its reflection of radiation from the Sun back into space – and thus cool the Earth's lower atmosphere or troposphere; however, they also absorb heat radiated up from the Earth, thereby warming the stratosphere.

Several eruptions during the past century have caused a decline in the average temperature at the Earth's surface of up to half a degree (Fahrenheit scale) for periods of one to three years.

Past catastrophes believed to be as a direct result of volcanic activity

Proponents of Anthropogenic Global Warming repeatedly promote the notion that current climate change is unprecedented due to human influence. However, as this book demonstrates repeatedly, catastrophic climate change due to natural factors has been common throughout human history. In fact, current changes in climate pale in comparison to past events. This section will deal with some examples of past volcanic eruptions impacting climate and species that inhabit this planet, including human beings, of course.

The biggest known volcanic eruption released energy equivalent to approximately 240 Gigatons of TNT and created the La Garita Caldera. The energy released was four thousand and eight hundred times the yield of Tsar Bomba (50 Megatons), the largest nuclear device ever detonated.

Mass extinctions

It has been suggested that volcanic activity caused or contributed to the End-Ordovician, Permian-Triassic, Late Devonian mass extinctions, and possibly others. The massive eruptive event which formed the Siberian Traps, one of the largest known volcanic events of the last 500 million years of Earth's geological history, continued for a million years and is considered to be the likely cause of the "Great Dying" about 250 million years ago, which is estimated to have killed 90% of species existing at the time.

Lake Toba

One proposed volcanic winter happened c. 70,000 years ago following the super-eruption of Lake Toba on Sumatra island in Indonesia. According to the Toba catastrophe theory to which some anthropologists and archeologists subscribe, it had global consequences, killing most humans then alive and creating a population bottleneck that affected the genetic inheritance of all humans today.

Freezing Winter of 1740-41

The freezing winter of 1740–41, which led to widespread famine in northern Europe, may also owe its origins to a volcanic eruption.

Mount Tambora

The 1815 eruption of Mount Tambora created global climate anomalies that became known as the "Year Without a Summer" because of the effect on North American and European weather. Agricultural crops failed and livestock died in much of the Northern Hemisphere, resulting in one of the worst famines of the 19th century.

Krakatoa (Indonesian: Krakatau) is a volcanic island made of lava in the Sunda Strait between the islands of Java and Sumatra in Indonesia. The name is used for the island group, the main island (also called Rakata), and the volcano as a whole. The volcano exploded in 1883, killing 40,000 people; some estimates put the death toll much higher.

The explosion is considered to be the loudest sound ever heard in modern history, with reports of it being heard nearly 3,000 miles (4,800 km) from its point of origin. The cataclysmic explosion was faintly heard as far away as Perth in Western Australia, about 1,930 miles (3,110 km) away, and the island of Rodrigues near Mauritius, about 3,000 miles (4,800 km) away.

The shock wave from the explosion was recorded on barographs around the globe.

The eruption ejected approximately 21 km3 (5.0 cu mi) of rock, ash, and pumice.

With a Volcanic Explosivity Index (VEI) of 6, the eruption was equivalent to 200 megatons of TNT (840 PJ) – about 13,000 times the nuclear yield of the Little Boy bomb (13 to 16 kt) that devastated Hiroshima, Japan, during World War II, and four times the yield of Tsar Bomba (50 Megatons), the largest nuclear device ever detonated.

This was not the first time Krakatoa had exploded in a spectacular fashion. At some point in pre-history, an earlier caldera-forming eruption occurred, leaving as remnants Verlaten, Long, Poolsche Hoed, and the base of Rakata. Later, at least two more cones (Perboewatan and Danan) formed and eventually joined with Rakata, forming the main island of Krakatoa. In the history books, explosions have been recorded in 416, 535 and 1680 AD. The Javanese Book of Kings (Pustaka Raja) records the event in 416 AD as follows:

> "A thundering sound was heard from the mountain Batuwara [now called Pulosari, an extinct volcano in Bantam, the nearest to the Sunda Strait] which was answered by a similar noise from Kapi, lying westward of the modern Bantam [Bantam is the westernmost province in Java, so this seems to indicate that Krakatoa is meant]. A great glowing fire, which reached the sky, came out of the last-named mountain; the whole world was greatly shaken and violent thundering, accompanied by heavy rain and storms took place, but not

[18] **References:**
- The Independent, May 3 2006."How Krakatoa made the biggest bang." The third explosion has been claimed as the loudest sound heard in historic times.
- "Symons, G.J. (ed) "The Eruption of Krakatoa and Subsequent Phenomena" (Report of the Krakatoa Committee of the Royal Society). London, 1888"
- "The eruption of Krakatoa, August 27, 1883". Commonwealth of Australia 2012, Bureau of Meteorology. http://www.bom.gov.au/tsunami/history/1883.shtml

only did not this heavy rain extinguish the eruption of the fire of the
mountain Kapi, but augmented the fire; the noise was fearful, at last the
mountain Kapi with a tremendous roar burst into pieces and sank into the
deepest of the earth. The water of the sea rose and inundated the land, the
country to the east of the mountain Batuwara, to the mountain Rajabasa [the
most southerly volcano in Sumatra], was inundated by the sea; the inhabitants
of the northern part of the Sunda country to the mountain Rajabasa were
drowned and swept away with all property... The water subsided but the land
on which Kapi stood became sea, and Java and Sumatra were divided into two
parts."

Did Volcanoes help cause the little Ice Age?

As mentioned earlier and will be repeated several times throughout this book, we still have scant knowledge of the multitude of factors that affect global climate, and an even more flimsy conjecture on how those factors interact to change the climate over time.

One question that still puzzles many researchers is why the medieval warm period suddenly stopped? Why did temperatures which had been rising steadily since the last Ice Age suddenly start declining significantly from around the thirteenth century, well before the Industrial Revolution? The little Ice Age that ensued is a natural climate anomaly, which has not been properly explained. After all, despite the warming of the 20th centuries, global temperature today (which are comparable to the Holocene Optimum) are still 10C below what Earth's temperature used to be 5 million years ago. There is more on past Earth temperatures in a later chapter of this book.

New research points to volcanic activity contributing to the little Ice Age, in addition to a diminishing of solar irradiance / sunspots described in the previous section. However, it should again be noted that this research is not fact, it is an educated inference. Unlike the repeated stipulations by climate organizations and, by proxy, politicians and the mainstream media to (mis)represent suppositions as incontrovertible facts; climate science is still in its infancy. Most of what is published today is subject to change, due to new research, new facts uncovered from the past and/or observed in the future.

Volcanic origin for Little Ice Age
http://www.bbc.co.uk/news/science-environment-16797075

By Richard Black
Environment correspondent, BBC News

The Little Ice Age was caused by the cooling effect of massive volcanic eruptions, and sustained by changes in Arctic ice cover, scientists conclude.

An international research team studied ancient plants from Iceland and Canada, and sediments carried by glaciers.

They say a series of eruptions just before 1300 lowered Arctic temperatures enough for ice sheets to expand.

Writing in Geophysical Research Letters, they say this would have kept the Earth cool for centuries.

The exact definition of the Little Ice Age is disputed. While many studies suggest temperatures fell globally in the 1500s, others suggest the Arctic and sub-Arctic began cooling several centuries previously.

The global dip in temperatures was less than 1C, but parts of Europe cooled more, particularly in winter, with the River Thames in London iced thickly enough to be traversable on foot.

What caused it has been uncertain. The new study, led by Gifford Miller at the University of Colorado at Boulder, US, links back to a series of four explosive volcanic eruptions between about 1250 and 1300 in the tropics, which would have blasted huge clouds of sulphate particles into the upper atmosphere.

These tiny aerosol particles are known to cool the globe by reflecting solar energy back into space.

The Little Ice Age saw an increase in cold winters in parts of Europe, but a small global change "This is the first time that anyone has clearly identified the specific onset of the cold times marking the start of the Little Ice Age," said Dr Miller.

"We have also provided an understandable climate feedback system that explains how this cold period could be sustained for a long period of time."

The scientists studied several sites in north-eastern Canada and in Iceland where small icecaps have expanded and contracted over the centuries.

When the ice spreads, plants underneath are killed and "entombed" in the ice. Carbon-dating can determine how long ago this happened.

So the plants provide a record of the icecaps' sizes at various times - and therefore, indirectly, of the local temperature.

An additional site at Hvitarvatn in Iceland yielded records of how much sediment was carried by a glacier in different decades, indicating changes in its thickness.

Putting these records together showed that cooling began fairly abruptly at some point between 1250 and 1300. Temperatures fell another notch between 1430 and 1455.

The first of these periods saw four large volcanic eruptions beginning in 1256, probably from the tropics sources, although the exact locations have not been determined.

The later period incorporated the major Kuwae eruption in Vanuatu.

Aerosols from volcanic eruptions usually cool the climate for just a few years.

When the researchers plugged in the sequence of eruptions into a computer model of climate, they found that the short but intense burst of cooling was enough to initiate growth of summer ice sheets around the Arctic Ocean, as well as glaciers.

The extra ice in turn reflected more solar radiation back into space, and weakened the Atlantic Ocean circulation commonly known as the Gulf Stream.

"It's easy to calculate how much colder you could get with volcanoes; but that has no permanence, the skies soon clear," Dr Miller told BBC News.

"And it was climate modeling that showed how sea ice exports into the North Atlantic set up this self-sustaining feedback process, and that's how a perturbation of decades can result in a climate shift of centuries."

Analysis of the later phase of the Little Ice Age also suggests that changes in the Sun's output, particularly in the ultraviolet part of the spectrum, would also have contributed cooling.

Was the Little Ice Age Triggered by Massive Volcanic Eruptions?
http://www.sciencedaily.com/releases/2012/01/120130131509.htm

ScienceDaily (Jan. 30, 2012)

A new international study may answer contentious questions about the onset and persistence of Earth's Little Ice Age, a period of widespread cooling that lasted for hundreds of years until the late 19th century.

The study, led by the University of Colorado Boulder with co-authors at the National Center for Atmospheric Research (NCAR) and other organizations, suggests that an unusual, 50-year-long episode of four massive tropical volcanic eruptions triggered the Little Ice Age between 1275 and 1300 A.D. The persistence of cold summers following the eruptions is best explained by a subsequent expansion of sea ice and a related weakening of Atlantic currents, according to computer simulations conducted for the study.

The study, which used analyses of patterns of dead vegetation, ice and sediment core data, and powerful computer climate models, provides new evidence in a longstanding scientific debate over the onset of the Little Ice Age. Scientists have theorized that the Little Ice Age was caused by decreased summer solar radiation, erupting volcanoes that cooled the planet by ejecting sulfates and other aerosol particles that reflected sunlight back into space, or a combination of the two.

"This is the first time anyone has clearly identified the specific onset of the cold times marking the start of the Little Ice Age," says lead author Gifford Miller of the University of Colorado Boulder. "We also have provided an understandable climate feedback system that explains how this cold period could be sustained for a long period of time. If the climate system is hit again and again by cold conditions over a relatively short period -- in this case, from volcanic eruptions -- there appears to be a cumulative cooling effect."

"Our simulations showed that the volcanic eruptions may have had a profound cooling effect," says NCAR scientist Bette Otto-Bliesner, a co-author of the study. "The eruptions could have triggered a chain reaction, affecting sea ice and ocean currents in a way that lowered temperatures for centuries."

The study appears this week in Geophysical Research Letters. The research team includes co-authors from the University of Iceland, the University of California Irvine, and the University of Edinburgh in Scotland. The study was funded in part by the National Science Foundation, NCAR's sponsor, and the Icelandic Science Foundation.

Scientific estimates regarding the onset of the Little Ice Age range from the 13th century to the 16th century, but there is little consensus, Miller says. Although the cooling temperatures may have affected places as far away as South America and China, they were particularly evident in northern Europe. Advancing glaciers in mountain valleys destroyed towns, and paintings from the period depict people ice-skating on the Thames River in London and canals in the Netherlands, places that were ice-free before and after the Little Ice Age.

"The dominant way scientists have defined the Little Ice Age is by the expansion of big valley glaciers in the Alps and in Norway," says Miller, a fellow at CU's Institute of Arctic and Alpine Research. "But the time in which European glaciers advanced far enough to demolish villages would have been long after the onset of the cold period."

Miller and his colleagues radiocarbon-dated roughly 150 samples of dead plant material with roots intact, collected from beneath receding margins of ice caps on Baffin Island in the Canadian Arctic. They found a large cluster of "kill dates" between 1275 and 1300 A.D., indicating the plants had been frozen and engulfed by ice during a relatively sudden event.

The team saw a second spike in plant kill dates at about 1450 A.D., indicating the quick onset of a second major cooling event.

To broaden the study, the researchers analyzed sediment cores from a glacial lake linked to the 367-square-mile Langjökullice cap in the central highlands of Iceland that reaches nearly a mile high. The annual layers in the cores -- which can be reliably dated by using tephra deposits from known historic volcanic eruptions on Iceland going back more than 1,000 years -- suddenly became thicker in the late 13th century and again in the 15th century due to increased erosion caused by the expansion of the ice cap as the climate cooled.

"That showed us the signal we got from Baffin Island was not just a local signal, it was a North Atlantic signal," Miller says. "This gave us a great deal more confidence that there was a major perturbation to the Northern Hemisphere climate near the end of the 13th century."

The team used the Community Climate System Model, which was developed by scientists at NCAR and the Department of Energy with colleagues at other organizations, to test the effects of volcanic cooling on Arctic sea ice extent and mass. The model, which simulated various sea ice conditions from about 1150 to 1700 A.D., showed several large, closely spaced eruptions could have cooled the Northern Hemisphere enough to trigger the expansion of Arctic sea ice.

The model showed that sustained cooling from volcanoes would have sent some of the expanding Arctic sea ice down along the eastern coast of Greenland until it eventually melted in the North Atlantic. Since sea ice contains almost no salt, when it melted the surface water became less dense, preventing it from mixing with deeper North Atlantic water. This weakened heat transport back to the Arctic and created a self-sustaining feedback on the sea ice long after the effects of the volcanic aerosols subsided, according to the simulations.

The researchers set solar radiation at a constant level in the climate models. The simulations indicated that the Little Ice Age likely would have occurred without decreased summer solar radiation at the time, Miller says.

Pacific Decadal Oscillation (PDO)

What is a PDO?

The Pacific Decadal Oscillation (**PDO**) is a pattern of Pacific climate variability that shifts phases on at least inter-decadal time scale, usually about 20 to 30 years. The PDO is detected as warm or cool surface waters in the Pacific Ocean, north of 20° N. During a "warm", or "positive", phase, the west Pacific becomes cool and part of the eastern ocean warms; during a "cool" or "negative" phase, the opposite pattern occurs.

The prevailing hypothesis is that the PDO is caused by a "reddening" of the El Niño-Southern Oscillation (**ENSO**) combined with stochastic atmospheric forcing.

The inter-decadal Pacific oscillation (IPO or ID) display similar sea-surface temperature (SST) and sea-level pressure (SLP) patterns, with a cycle of 15–30 years, but affects both the north and south Pacific. In the tropical Pacific, maximum SST anomalies are found away from the equator. This is quite different from the quasi-decadal oscillation (QDO) with a period of 8-to-12 years and maximum SST anomalies straddling the equator, thus resembling the ENSO.

In other words, PDO cycles impact the climate worldwide with cold and warm phases shifting every 20 to 30 years

When were PDO cycles discovered?

The Pacific (inter-)decadal oscillation was named by Steven R. Hare, who noticed it while studying salmon production pattern results in 1997.

The date of discovery of PDO is extremely important. It is not like PDO is a factor impacting global climate that has been known and studied for many decades. Quite the opposite: PDO cycles were discovered *many years after the theory of Anthropogenic Global Warming and even the first climate models were first elaborated.* In other words, the theory of AGW was initially developed without taking into account PDO cycles, because they were not known. This goes back to my point that climate science is in its infancy. There are, most likely, factors affecting global climate that we do not even know as yet.

History of PDO cycles

The PDO index has been reconstructed using tree rings and other hydrologically sensitive proxies from west North America and Asia

Several regime shifts are apparent both in the reconstructions and instrumental data, during the 20th century regime shifts associated with concurrent changes in SST, SLP, land precipitation and ocean cloud cover occurred in 1924/1925,1945/1946 and 1976/1977

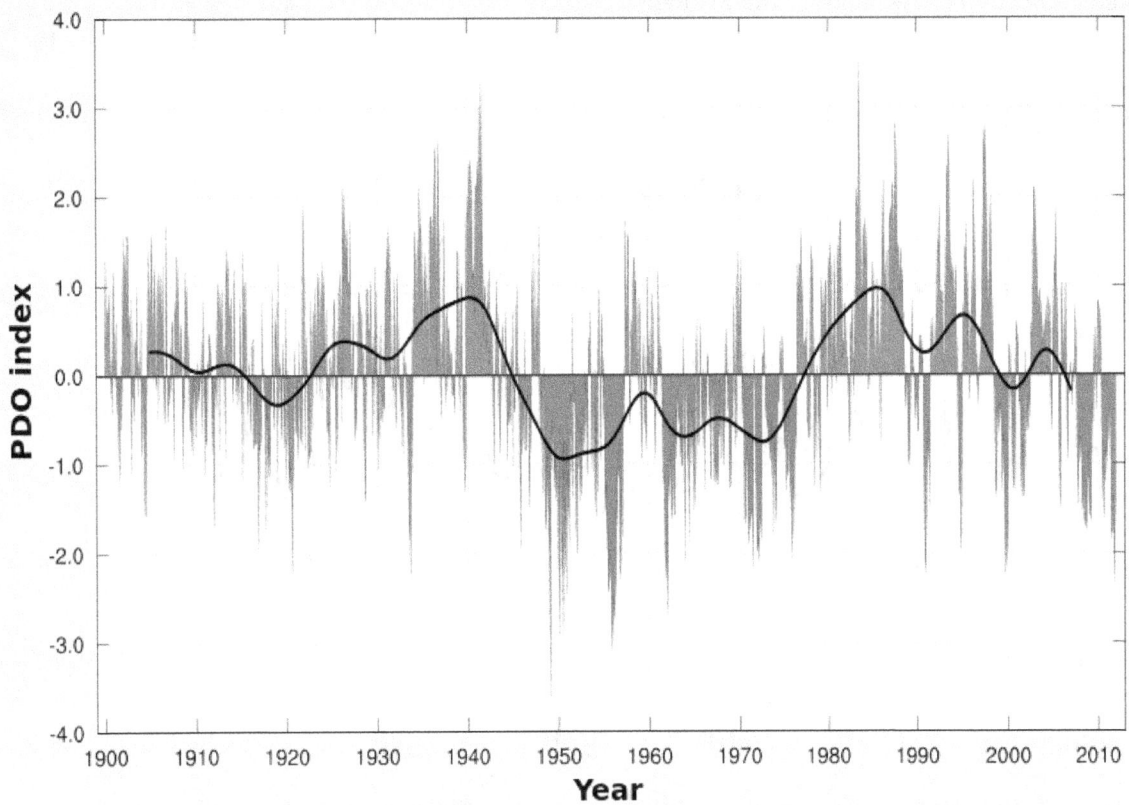

List of PDO Shifts in the 20th century

- 1924/1925: PDO changed to a "warm" phase.
- 1945/1946: The PDO changed to a "cool" phase, the pattern of this regime shift is similar to the 1970s episode with maximum amplitude in the subarctic and subtropical front but with a greater signature near the Japan while the 1970s shift was stronger near the American west coast.
- 1976/1977: PDO changed to a "warm" phase.
- 1988/1989: A weakening of the Aleutian low with associated SST changes was observed, in contrast to others regime shifts this change appears to be related to concurrent extra-tropical oscillation in the North Pacific and North Atlantic rather than tropical processes.
- 1997/1998: Several changes in Sea surface temperature and marine ecosystem occurred in the North Pacific after 1997/1998, in contrast to prevailing anomalies observed after the 1970s shift SST declined along the United States west coast and substantial changes in the populations of salmon, anchovy and sardine were observed, however the spatial pattern of the SST change was different with a meridian SST seesaw in the central and western Pacific that resemble a strong shift in the North Pacific Gyre Oscillation rather than the PDO structure, this pattern dominated much of the North Pacific SST variability after 1989.

The source of the following graphs is again from the extensive research done by Don J. Easterbrook, Professor Emeritus at Western Washington University, Bellingham, WA. You will find a complete list of Easterbrook's publications at the following link:

http://myweb.wwu.edu/dbunny/pdfs/dje_publications.pdf

Graph showing correlation of PDO Index with Global Climate

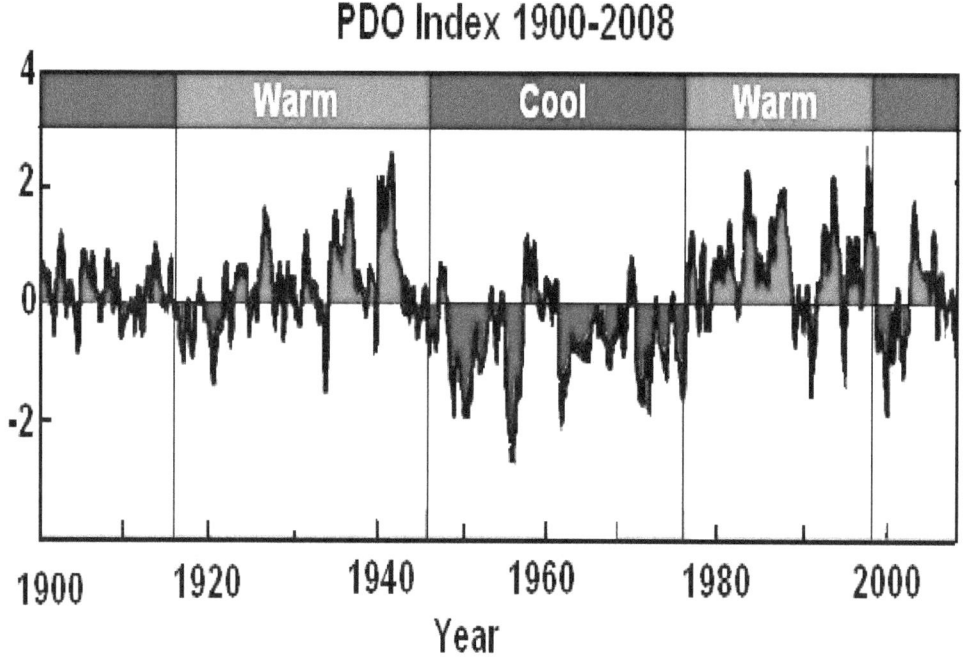

PDO COLD MODE (1945-77)

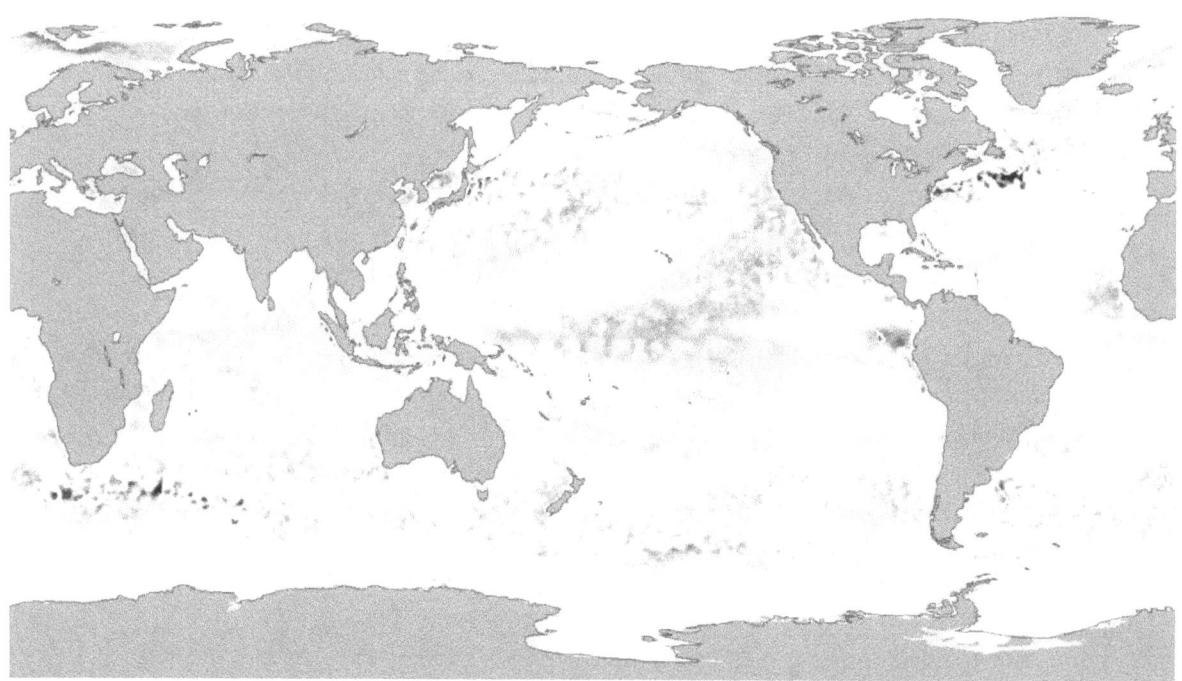

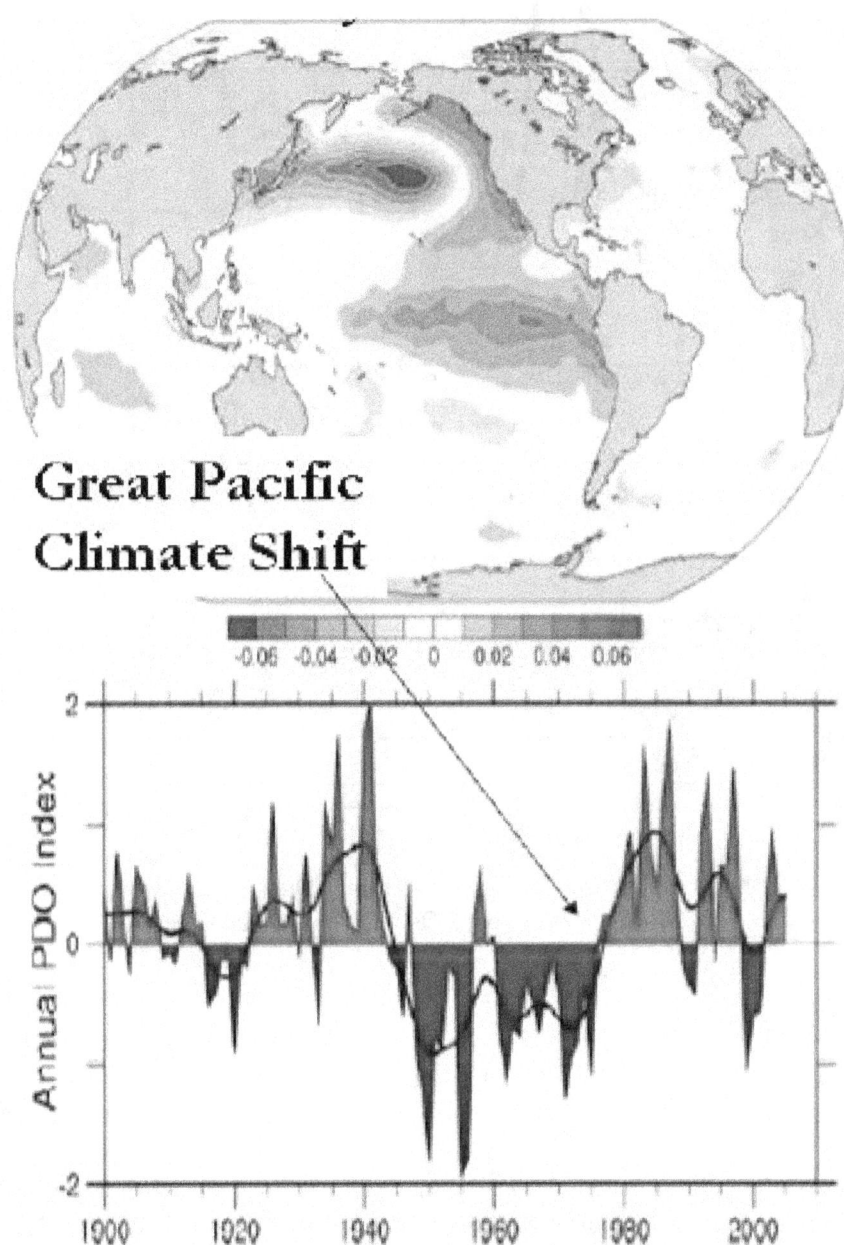

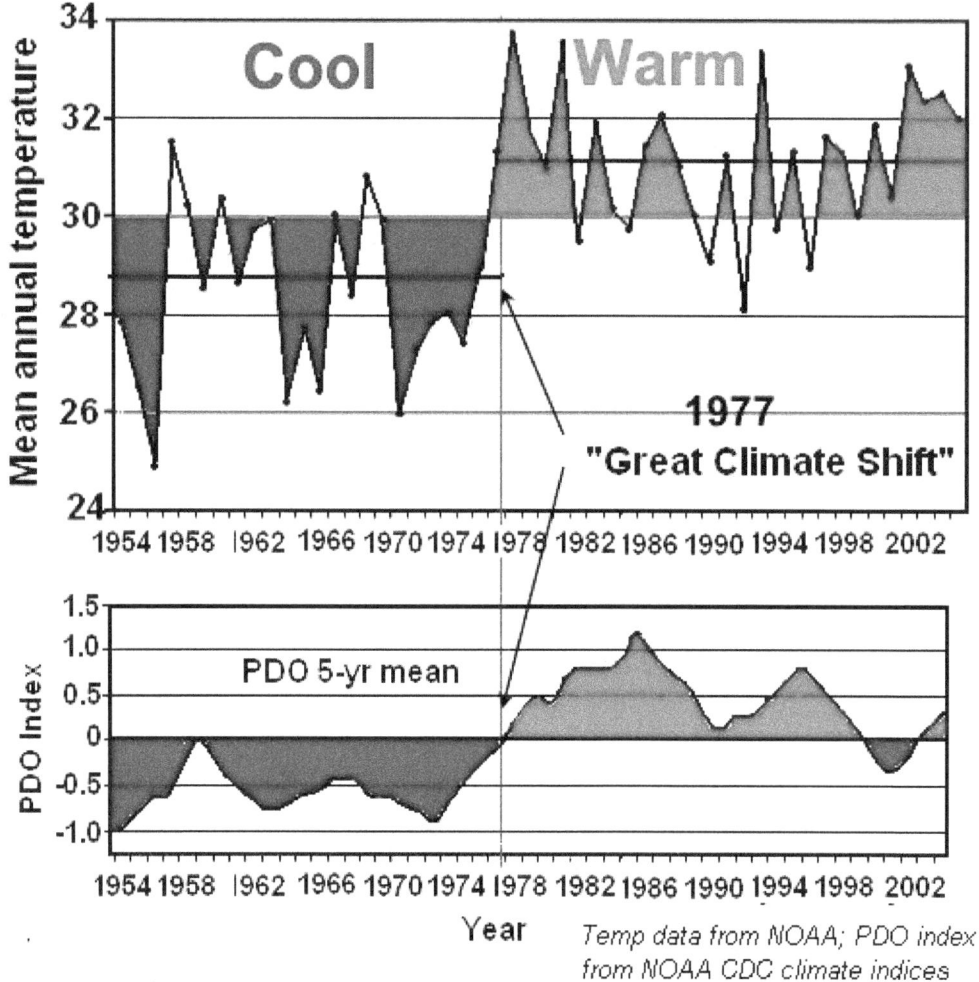

The last Millennium

A PDO signal has been reconstructed to 1661 through tree-ring chronologies in the Baja California area

MacDonald and Case reconstructed the PDO back to 993 using tree rings from California and Alberta. The index shows a 50-70 year periodicity but this is a strong mode of variability only after 1800, a persistent negative phase occurred during medieval times (993-1300) which is consistent with la nina conditions reconstructed in the tropical Pacific and multi-century droughts in the South-West United States.

In 1750, PDO displayed an unusually strong oscillation

The following graph shows the reconstruction of PDO for the entire second Millennium:

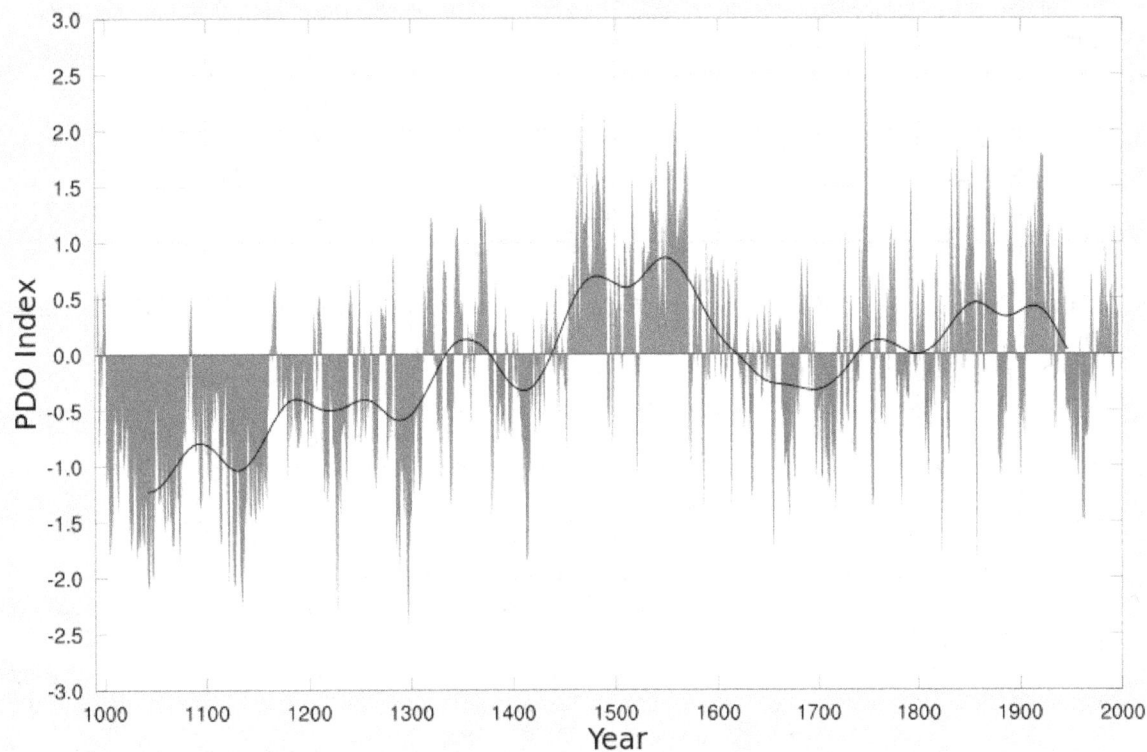

Predictability

NOAA's forecast use a linear inverse modeling (LIM) method to predict the PDO, LIM assumes that the PDO can be separated into a linear deterministic component and a non-linear component represented by random fluctuations.

Much of the LIM PDO predictability arises from ENSO and the global trend rather than extra-tropical processes and is thus limited to ~4 season, the prediction is consistent with the seasonal foot-printing mechanism in which an optimal SST structure evolve into the ENSO mature phase 6–10 months later that subsequently impact the North Pacific Ocean SST via the atmospheric bridge.

Skills in predicting decadal PDO variability could arise from taking into account the impact of the externally forced and internally generated Pacific variability.

References

- Mantua, Nathan J. et al (1997). "A Pacific interdecadal climate oscillation with impacts on salmon production". *Bulletin of the American Meteorological Society* **78** (6): 1069–1079. Bibcode 1997BAMS...78.1069M. doi:10.1175/1520-0477(1997)078<1069:APICOW>2.0.CO;2. http://www.atmos.washington.edu/~mantua/abst.PDO.html.
- Newman, M.; Compo, G.P.; Alexander, Michael A. (2003). "ENSO-Forced Variability of the Pacific Decadal Oscillation". *Journal of Climate* **16** (23): 3853–3857. doi:10.1175/1520-0442(2003)016<3853:EVOTPD>2.0.CO;2.
- Biondi, Franco; Gershunov, Alexander; Cayan, Daniel R. (2001). "North Pacific Decadal Climate Variability since 1661". *Journal of Climate* **14** (1): 5–10. Bibcode 2001JCli...14....5B. doi:10.1175/1520-0442(2001)014<0005:NPDCVS>2.0.CO;2. http://www.ngdc.noaa.gov/paleo/pubs/biondi2001/biondi2001.html.

- Vimont, Daniel J. (2005). "The Contribution of the Interannual ENSO Cycle to the Spatial Pattern of Decadal ENSO-Like Variability". *Journal of Climate* **18** (12): 2080–2092. doi:10.1175/JCLI3365.1.
- Schneider, Niklas; Bruce D. Cornuelle (2005). "The Forcing of the Pacific Decadal Oscillation". *Journal of Climate* **18** (8): 4355–4372. doi:10.1175/JCLI3527.1.
- Qiu, Bo; Niklas Schneider, Shuiming Chen (2007). "Coupled Decadal Variability in the North Pacific: An Observationally Constrained Idealized Model". *Journal of Climate* **20** (14): 3602–3620. doi:10.1175/JCLI4190.1. http://journals.ametsoc.org/doi/full/10.1175/JCLI4190.1
- Shen, Caiming; Wei-Chyung Wang; Wei Gong; Zhixin Hao (2006). "A Pacific Decadal Oscillation record since 1470 AD reconstructed from proxy data of summer rainfall over eastern China". *Geophys. Res. Lett.* **33** (3). Bibcode 2006GeoRL..3303702S. doi:10.1029/2005GL024804. http://www.agu.org/journals/ABS/2006/2005GL024804.shtml.
- D'arrigo, R.; Wilson R. (2006). "On the Asian Expression of the PDO". *International Journal of Climatology* **26** (12): 1607–1617. doi:10.1002/joc.1326.
- MacDonald, G.M.; Case R.A. (2005). "Variations in the Pacific Decadal Oscillation over the past millennium". *Geophys. Res. Lett.* **32** (8). Bibcode 2005GeoRL..3208703M. doi:10.1029/2005GL022478. http://www.agu.org/journals/ABS/2005/2005GL022478.shtml. Retrieved 2010-10-26.
- Rein, Bert; Andreas Lückge; Frank Sirocko (2004). "AA major Holocene ENSO anomaly during the Medieval period". *Geophys. Res. Lett.* **31** (17). Bibcode 2004GeoRL..3117211R. doi:10.1029/2004GL020161. http://www.agu.org/journals/ABS/2004/2004GL020161.shtml. Retrieved 2010-10-26.
- Seager, Richard; Graham, Nicholas; Herweijer, Celine; Gordon, Arnold L.; Kushnir, Yochanan; Cook, Ed (2007). "Blueprints for Medieval hydroclimate". *Quaternary Science Reviews* **26** (19–21): 2322–2336. doi:10.1016/j.quascirev.2007.04.020.
- Deser, Clara; Phillips, Adam S.; Hurrell, James W. (2004). "Pacific Interdecadal Climate Variability: Linkages between the Tropics and the North Pacific during Boreal Winter since 1900". *Journal of Climate* **17** (15): 3109–3124. doi:10.1175/1520-0442(2004)017<3109:PICVLB>2.0.CO;2.
- Minobe, Shoshiro; Atsushi Maeda (2005). "A 1° monthly gridded sea-surface temperature dataset compiled from ICOADS from 1850 to 2002 and Northern Hemisphere frontal variability". *International Journal of Climatology* **25** (7): 881–894. doi:10.1002/joc.1170.
- Hare, Steven R.; Mantua, Nathan J. (2000). "Empirical evidence for North Pacific regime shifts in 1977 and 1989". *Progress in Oceanography* **47** (2–4): 103–145. Bibcode 2000PrOce..47..103H. doi:10.1016/S0079-6611(00)00033-1.
- Trenberth, Kevin; Hurrell, James W. (1994). "Decadal atmosphere-ocean variations in the Pacific". *Climate Dynamics* **9** (6): 303–319. Bibcode 1994ClDy....9..303T. doi:10.1007/BF00204745. http://www.springerlink.com/content/m5711482u6554132/.
- Yasunaka, Sayaka; Kimio Hanawa (2003). "Regime Shifts in the Northern Hemisphere SST Field: Revisited in Relation to Tropical Variations". *Journal of the Meteorological Society of Japan* **81** (2): 415–424. doi:10.2151/jmsj.81.415. http://www.jstage.jst.go.jp/article/jmsj/81/2/81_415/_article/-char/en
- Chavez, Francisco P; John Ryan, Salvador E. Lluch-Cota, Miguel Ñiquen C. (2003). "From Anchovies to Sardines and Back: Multidecadal Change in the Pacific Ocean". *Science* **299** (5604): 217–221. doi:10.1126/science.1075880. http://www.sciencemag.org/cgi/content/short/299/5604/217.
- Bond, N.A.; J. E. Overland; M. Spillane; P. Stabeno (2003). "Recent shifts in the state of the North Pacific". *Geophys. Res. Lett* **30** (23). Bibcode 2003GeoRL..30wCLM1B. doi:10.1029/2003GL018597. http://europa.agu.org/?uri=/journals/gl/gl0323/2003GL018597/2003GL018597.xml&view=article
- http://jisao.washington.edu/pdo/PDO.latest
- PDO reconstruction 993-1996, black line 101 years smooth. Data source:ftp://ftp.ncdc.noaa.gov/pub/data/paleo/treering/reconstructions/pdo-macdonald2005.txt
- Don J. Easterbrook: http://myweb.wwu.edu/dbunny/pdfs/dje_publications.pdf

Clouds[19]

In meteorology, a cloud is a visible mass of liquid droplets or frozen crystals made of water and/or various chemicals suspended in the atmosphere above the surface of a planetary body. They are also known as aerosols. Clouds in earth's atmosphere are studied in the cloud physics branch of meteorology. Two processes, possibly acting together, can lead to air becoming saturated: cooling the air or adding water vapor to the air. In general, precipitation will fall to the surface; an exception is virga, which evaporates before reaching the surface.

On Earth the condensing substance is typically water vapor, which forms small droplets or ice crystals, typically 0.01 mm in diameter. When surrounded by billions of other droplets or crystals they become visible as clouds.

Why are clouds important to weather and climate?

Clouds can cast shadows.

Dense deep clouds exhibit a high reflectance (70% to 95%) throughout the visible range of wavelengths: they thus appear white, at least from the top. Cloud droplets tend to scatter light efficiently, so that the intensity of the solar radiation decreases with depth into the gases, hence the gray or even sometimes dark appearance of the clouds at their base.

Thin clouds may appear to have acquired the color of their environment or background, and clouds illuminated by non-white light, such as during sunrise or sunset, may be colored accordingly. In the near-infrared range, clouds would appear darker because the water that constitutes the cloud droplets strongly absorbs solar radiation at those wavelengths.

Cloud Albedo[20] [21]

Cloud albedo is a measure of the how much solar radiation is reflected by a cloud. Higher values mean that the cloud reflects more solar radiation.

Cloud albedo varies from less than 10% to more than 90% and depends on drop sizes, liquid water or ice content, thickness of the cloud, and the sun's zenith angle. Usually, clouds with smaller drops and greater liquid water content have higher cloud albedo.

Low, thick clouds (such as stratocumulus) primarily reflect incoming solar radiation and, therefore, have a high albedo.

On the other hand, high, thin clouds (such as Cirrus) tend to transmit solar radiation to the surface and trap outgoing infrared radiation, thus resulting in low albedo.

Low Cloud Albedo = Greenhouse Effect

Popular myth, often reinforced by the mainstream press and science journals to the layperson, is that Carbon Dioxide and Methane are the main greenhouse gases. This is not quite true. Water vapor

---

[19] Glossary of Meteorology (June 2000). "Adiabatic Process". American Meteorological Society. http://amsglossary.allenpress.com/glossary/search?id=adiabatic-process1

[20] "EO Library: Clouds & Radiation Fact Sheet". NASA http://earthobservatory.nasa.gov/Library/Clouds/

[21] http://www.cgd.ucar.edu/cms/cchen/Latham_et_al_2008.pdf

constitutes over 1% (10,000 parts per million) of the Earth's atmosphere, as compared to a concentration of 390 parts per million for Carbon Dioxide. As anyone looking at, or, even better these days, flying through the sky will notice, a significant portion of this water vapor manifests itself as high, thin clouds. By transmitting solar radiation to the surface and trapping outgoing infrared radiation, high, thin clouds are a very effective contributor to the greenhouse effect. In other words, water vapor is a key greenhouse gas. And that is not a bad thing. If less heat were trapped by greenhouse gases is the Earth's atmosphere, the planet would probably have been too cold to sustain life.

Nephology[22] [23]

You have probably never heard of the term nephology before. The term does not even exist in the U.S. English dictionary provided with Microsoft Word.

A long time ago, nephology was the study of clouds and cloud formation. In the early 19th century, British meteorologist Luke Howard was a major researcher within this field, establishing a cloud classification system. Luke Howard has been called "the father of meteorology" because of his comprehensive recordings of weather in the London area from 1801 to 1841 and his writings, which transformed the science of meteorology.

Nephology fell out of common use by the middle of the twentieth century. Meteorologists became more interested in what clouds did *after* they were formed, rather than *how* they were formed and what their composition was, except to know whether or not they would cause rain.

Well, guess what? Thanks to the controversy over Anthropogenic Global Warming, since the late 1990s, the study of nephology has suddenly been abruptly reincarnated

In modern times, different nephologists have differing views about clouds and their impact on global climate. Again, this goes back to my point that climate science is in its infancy. This is definitely true when it concerns clouds, with educated guesses, none of which should be taken as fact.

- Some nephologists have suggested that high solar activity lowers levels of cosmic rays, thus reducing high albedo cloud cover and warming the planet. [24]
- Others nephologists say that there is no statistical evidence for such an effect
- Some nephologists believe that an increase in global temperature could decrease the thickness and brightness (ability to reflect light energy), which would further increase global temperature.[25]

---

[22] "Nephology". *Oxford English Dictionary*. Oxford University Press. DRAFT REVISION September 2003. http://dictionary.oed.com/cgi/entry/00323129
[23] Fluffy Thinking *Financial Times*
[24] http://www.scientificamerican.com/article.cfm?id=cloud-formation-may-be-linked-to-cosmic-rays
[25] Clouds' role in global warming studied *CNN website*

Are cosmic rays causing climate change?

Recently, research has been going on at CERN's CLOUD facility to study the effects of the solar cycle and cosmic rays on cloud formation[26]

Below is a very interesting article from Nature Magazine:
http://www.nature.com/news/2011/110824/full/news.2011.504.html

Cloud formation may be linked to cosmic rays
Experiment probes connection between climate change and radiation bombarding the atmosphere.
Geoff Brumfiel

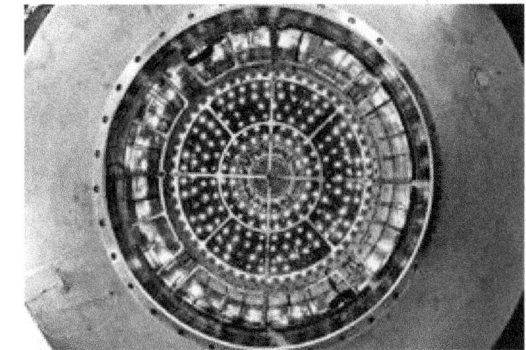

The CLOUD experiment is studying whether cosmic rays play a role in cloud formation. It sounds like a conspiracy theory: 'cosmic rays' from deep space might be creating clouds in Earth's atmosphere and changing the climate. Yet an experiment at CERN, Europe's high-energy physics laboratory near Geneva, Switzerland, is finding tentative evidence for just that.

The findings, published today in Nature1, are preliminary, but they are stoking a long-running argument over the role of radiation from distant stars in altering the climate.

For a century, scientists have known that charged particles from space constantly bombard Earth. Known as cosmic rays, the particles are mostly protons blasted out of supernovae. As the protons crash through the planet's atmosphere, they can ionize volatile compounds, causing them to condense into airborne droplets, or aerosols. Clouds might then build up around the droplets.

The number of cosmic rays that reach Earth depends on the Sun. When the Sun is emitting lots of radiation, its magnetic field shields the planet from cosmic rays. During periods of low solar activity, more cosmic rays reach Earth.

Scientists agree on these basic facts, but there is far less agreement on whether cosmic rays can have a large role in cloud formation and climate change. Since the late 1990s, some have suggested that when high solar activity lowers levels of cosmic rays, that in turn reduces cloud cover and warms the planet. Others say that there is no statistical evidence for such an effect.

Polarizing lens

"People are far too polarized, and in my opinion there are huge, important areas where our understanding is poor at the moment," says Jasper Kirkby, a physicist at CERN. In particular, he says, little controlled research has been done on exactly what effect cosmic rays can have on atmospheric chemistry.

---

[26] "Cloud formation may be linked to cosmic rays" (Press release). Nature News. August 24, 2011.
http://www.nature.com/news/2011/110824/full/news.2011.504.html

To find out, Kirkby and his team are bringing the atmosphere down to Earth in an experiment called Cosmics Leaving Outdoor Droplets (CLOUD). The team fills a custom-built chamber with ultrapure air and chemicals believed to seed clouds: water vapor, sulfur dioxide, ozone and ammonia. They then bombard the chamber with protons from the same accelerator that feeds the Large Hadron Collider, the world's most powerful particle smasher. As the synthetic cosmic rays stream in, the group carefully samples the artificial atmosphere to see what effect the rays are having.

Early results seem to indicate that cosmic rays do cause a change. The high-energy protons seemed to enhance the production of nanometer-sized particles from the gaseous atmosphere by more than a factor of ten. But, Kirkby adds, those particles are far too small to serve as seeds for clouds. "At the moment, it actually says nothing about a possible cosmic-ray effect on clouds and climate, but it's a very important first step," he says.

Scientists on both sides of the debate welcome the findings, although they draw differing conclusions. "Of course there are many things to explore, but I think the cosmic-ray/cloud-seeding hypothesis is converging with reality," says Henrik Svensmark, a physicist at the Technical University of Denmark in Copenhagen, who claims a link between climate change and cosmic rays.

Others disagree. The CLOUD experiment is "not firming up the connection", counters Mike Lockwood, a space and environmental physicist at the University of Reading, UK, who is skeptical. Lockwood says that the small particles may not grow fast enough or large enough to be important in comparison with other cloud-forming processes in the atmosphere.

"I think it's an incredibly worthwhile and overdue experiment," says Piers Forster, a climatologist at the University of Leeds, UK, who studied the link between cosmic rays and climate for the latest scientific assessment by the International Panel on Climate Change. But for now at least, he says that the experiment "probably raises more questions than it answers".

Kirkby hopes that the experiment will eventually answer the cosmic-ray question. In the coming years, he says, his group is planning experiments with larger particles in the chamber, and they hope eventually to generate artificial clouds for study. "There is a series of measurements that we will have to do that will take at least five years," he says. "But at the end of it, we want to settle it one way or the other."

Now we are getting spooky, some would say in the realm of science fiction. But research from CERN is current science, is it not? Below is an excerpt:

Report on the session "PS and non accelerator physics" at the New Opportunities on the Physics Landscape at CERN. Convenor: T. Sloan 1. Comments from the Convenor
indico.cern.ch/materialDisplay.py?materialId=12&confId=51128

```
1.2 Accelerator experiments.

The second abstract was the CLOUD experiment which described its plans for
the coming years. While few people believe that there is a strong link
between global warming and cosmic rays, nevertheless it is interesting from
the point of view of both meteorology and climatology to establish whether or
not ionization from cosmic rays plays a part in cloud formation. A similar
experiment is being performed by a Danish group. However, to obtain credible
results it is necessary to have data from both groups.

All abstracts are summarized below and more details are given in the
speakers' slides which are available on the workshop web site...

2.2 Searches for Dark Matter (DM) (abstracts 13,23,72,73) Laura Baudis
reviewed these abstracts of which 4 concerned DM detection and one included
also a search for neutrinoless double beta decay. Such searches were
highlighted in Ed Witten's first overview talk and represent frontier
research topics in our field. The proposed experiments all aimed to make
larger and larger detectors (ton scale) which are needed to achieve the
improved sensitivity over current detectors. This will lead to more stringent
limits and hopefully in the long term to positive detection of DM."

2.3 Axion Searches (abstract 73) Axions were highlighted in the opening talk
by Ed Witten. They have been postulated to prevent the neutron having too
large an electric dipole moment in the Standard Model. They are also
predicted in string theories and they might be a window to them. They are a
candidate for the dark matter of the Universe, being connected also to the
dark energy. Thomas Papaevangelou described the plans of the CAST
collaboration, which is the leading experiment in the field of solar axions.
They request support at the same level as in previous years for their
anticipated data taking with upgraded detectors also in 2011-12with improved
sensitivity. They then look to increasing the magnetic field strength with
further detector and X-ray optics upgrades with plans starting in 2013 and
continuing beyond 2016. This should lead to a large part of the QCD favoured
model region being investigated including the otherwise non-accessible sub-
keV range."
```

Searching for answers in the dark?

In the past few sections, we learned that solar irradiance does have an effect on the global climate but we still have little knowledge about what the Sun will get up to next. We also learned that volcanoes have had past impact on the climate, but we are currently studying if the anomalism of the Little Ice Age was linked to volcanic activity. We also learned that PDO Cycles were discovered well after the theory of Anthropogenic Global Warming was first hypothesized. We also learned that we are still relatively clueless on the factors that influence cloud formation.

For sure, Carbon Dioxide does have a greenhouse effect. That is proven science. But so does water vapor which is at least 25 times more plentiful in our atmosphere. And the sun, volcanoes, PDO cycles, cloud formation, and possibly other factors not known to us at the current time all play a major role in shaping our past, current and future climate. Therefore, what level of human arrogance and deception is required for some to conclude and promote definitively that the increase in 100 parts per million of CO2 over the past century (which is only partially human-contributed) has caused the warming noticeable from 1980 to the early 2000s. In addition, the warming is by no means unprecedented or unusual, especially given that the world had previously cooled considerably during the little Ice age from the thirteenth to the beginning of the twentieth century; more about global temperatures in various periods of the Earth's long history in a later Chapter.

The global warming of 1980 to the early 2000s can be compared to some people deciding to play a prank on a sleeping friend by transporting the person into an unlit dark sauna. Soon afterwards, the person wakes up with a sweat and starts making educated guesses about why it's so hot? Did he/she just have a fever while sleeping which just abated? Is it only the person who is sweating, but hey the floor and walls seem wet too? One of the doors opens to a colder place ("Little Ice Age") while the other side of another door adjoining to the next sauna booth is as hot as here? Etc.

In other words, climate scientists today are searching for answers in the dark, while at the same time learning more about the Earth's long past from the fossil, tree-ring and other proxy data.

The Sun's Magnetic Field

The sun's magnetic field flips every 11 years, adding further credence to the stipulation that natural variations in solar activity have a major effect on global climate. The Sun has not remained unchanged for a long time as some proponents of Anthropogenic Global Warming have attempted to disseminate. Quite to the contrary, the Sun has a very short cycle of change.

The below article from CNN.com provides more details on the impact of solar magnetic fluctuations on our home planet.

The sun's magnetic poles flip like clockwork about every 11 years
February 16, 2001
Web posted at: 3:51 p.m. EST (2051 GMT)
http://archives.cnn.com/2001/TECH/space/02/16/sun.flips/index.html

The poles of the sun's powerful magnetic field have reversed, signaling a time of peak solar activity that could spell trouble for planet Earth, astronomers said this week.

The sun's magnetic north and south poles were in their respective northern and southern hemisphere several months ago. Yet since then they have migrated to the opposite sides of the star.

The bipolar flip did not take astronomers by surprise. It takes place like clockwork on the sun, once about every 11 years.

"This always happens around the time of solar maximum," NASA physicist David Hathaway said in a statement. "The magnetic poles exchange takes place at the peak of the sunspot cycle. In fact, it's a good indication that Solar Max is really here."

During Solar Max or solar maximum, the sun tends to exhibit more sunspots and eruptions, and jettison more solar flares, which send powerful streams of charged particles into the solar system.

Solar blasts heading toward Earth can stoke up beautiful auroras in the nighttime sky, and wreak havoc on communications satellites and electrical power grids.

Astronomers consider the current solar maximum more powerful than average, but less intense than the last two in 1989 and 1979.

Earth's magnetic fields also change places but with much less predictability and frequency. The reversals take place between intervals lasting from 5,000 to 5 million years. The last one happened 740,000 years ago, scientists estimate. Some suggest the planet is overdue for another one.

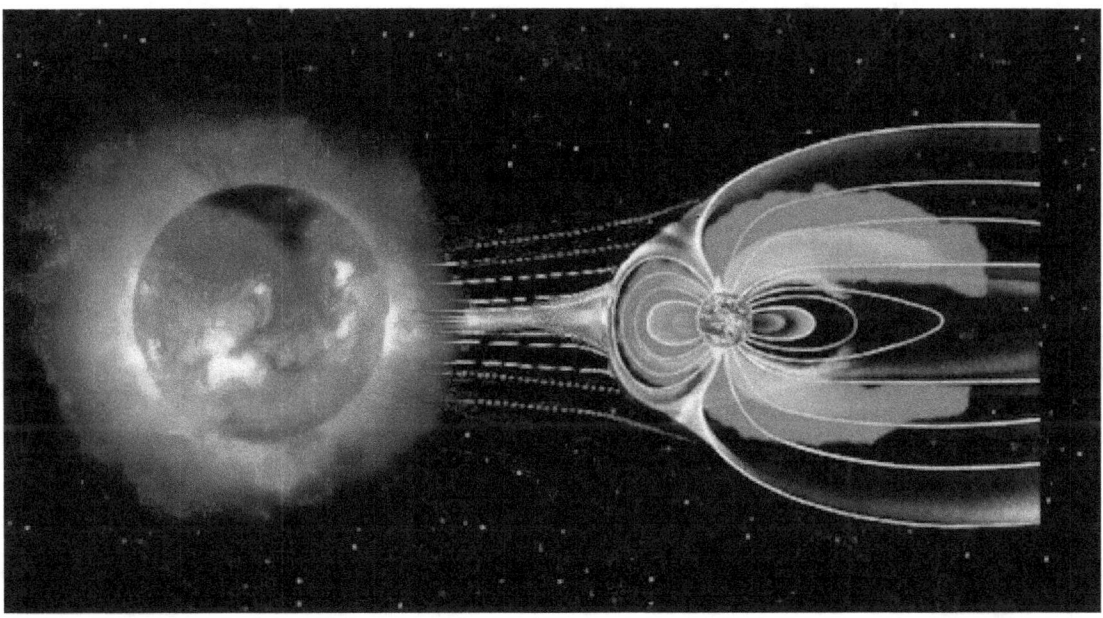

The Earth's Magnetic Field

Global Warming and Cooling

Below is a summary by Eugene D. and David R. Richard

http://www.earths-magnetic-field-and-how-it-reverses-and-more.com/PageSixteen.htm

Some of the heat on the surface of the earth is a result of the energy contained within the earth. The magnetic field experienced on the surface of the earth is a result of the internal energy within the earth. The internal heat of the earth causes ionization within the earth and separation of the electrons from the donor atoms. If the separation is great enough and the number of atoms ionized great enough, the result will be sufficient to cause a magnetic field within, on, and above the surface of the earth. This field is known as a dipole field.

The magnetic field of the earth has been decreasing and we are approaching the end of a warming cycle and will be entering a cooling cycle with the onset of a magnetic reversal. How are these three things related, namely, warming, cooling, and magnetic field of the earth? The controlling mechanism is the magnetic field of the earth. When global warming is great enough, then the electrons will tend to migrate closer to the surface of the earth where the observer is located. When the electrons are at the surface of the earth then there is no relative motion between the observer and the electrons and no magnetic field will be produced by these electrons. In order for charges to produce a magnetic field there has to be relative motion between the observer and the charges. If the charges and the observer are at the same distance from the center of the earth, then there is no relative motion, in spite of the fact that they are both rotating at a high speed. Now electrons are not expected to migrate to the surface but rather will move towards the surface with a resulting decrease in magnetic field. By the same token, the positive charges which are locked into position at or near the center of the earth will experience a large relative motion which will produce a large magnetic field due to their high relative velocity. If the magnetic field is decreasing with an increasing of internal heat of the earth, then one can expect a reversal of the earth's magnetic field. This could occur within a decade, a century, or a millennium[27].

With a reversal, one can expect global cooling, which will bring on a glacial period that will last possibly for a million years. This global cooling will cause the electrons to move inward due to electrostatic forces between positive charges in the center of the earth and the electrons, which were moved away from the center due to the heat of the earth.

As the electrons move nearer to the center of the earth, their relative velocity will increase and result in a greater magnetic field due to electrons. When the electrons have moved near enough to the center of the earth, the resulting magnetic field will be great enough to overcome the magnetic field produced by the positive charges. The result will be a reversal from a reversed field to a normal field.

The two paragraphs above explain how a normal magnetic field can change to a reverse magnetic field and then can be made to reverse again to a normal field. This process has been going on for at least the last 65 million years. Prior to that time the intervals between normal and reverse was much greater; however the theory is still applicable. The evidence is scant; however there is evidence of glacial periods as far back as 2.2 billion years ago[28].

From the above, one can theorize that the normal magnetic field such as we are in today is a global warming period and the reverse field would be a global cooling period (glacial period). What is the cause of this switching process from warm to cool? From the above paragraphs the switching process is activated by the earth's magnetic field. If the earth gets too hot the magnetic field causes a switch. If the earth gets too cold then the earth's magnetic field reverses and the heating of earth begins again. How the earth's magnetic field controls the internal heat of the earth so that switching can occur is unknown.

The above material is covered by copyrights issued to Eugene D. Richard.
The computer model referenced above is covered by a copyright issued to David R. Richard

---

[27] E-book and computer models located on website, www.earthsgeomotor.com
[28] Physical Geology by Sheldon Judson and Marvin E. Kauffman eighth edition p 348

Let us now get the results of a Danish study led by Henrik Svensmark.. Svensmark is a physicist at the Danish National Space Center in Copenhagen who studies the effects of cosmic rays on cloud formation. His work presents hypotheses about solar activity as an indirect cause of global warming; his research has suggested a possible link through the interaction of the solar wind and cosmic rays. His conclusions have been controversial as the prevailing scientific opinion on climate change considers solar activity unlikely to be a major contributor to recent warming. As a consequence, proponents of Anthropogenic Global Warming have launched a visceral campaign to discredit the Danish scientist.[29] Unfortunately, this is now the fate that any scientist with some integrity to challenge the "party line" of pre-conceptual AGW can expect by default. Such tactics are an effective mechanism to silence dissent in the science community, but a travesty for the veracity and credibility of science itself.

The earth's magnetic field impacts climate: Danish study
illustration only
by Staff Writers
Copenhagen (AFP) Jan 12, 2009
http://www.terradaily.com/reports/The_earths_magnetic_field_impacts_climate_Danish_study_999.html

The earth's climate has been significantly affected by the planet's magnetic field, according to a Danish study published Monday that could challenge the notion that human emissions are responsible for global warming.

"Our results show a strong correlation between the strength of the earth's magnetic field and the amount of precipitation in the tropics," one of the two Danish geophysicists behind the study, Mads Faurschou Knudsen of the geology department at Aarhus University in western Denmark, told the Videnskab journal.

He and his colleague Peter Riisager, of the Geological Survey of Denmark and Greenland (GEUS), compared a reconstruction of the prehistoric magnetic field 5,000 years ago based on data drawn from stalagmites and stalactites found in China and Oman.

The results of the study, which has also been published in US scientific journal Geology, lend support to a controversial theory published a decade ago by Danish astrophysicist Henrik Svensmark, who claimed the climate was highly influenced by galactic cosmic ray (GCR) particles penetrating the earth's atmosphere.

Svensmark's theory, which pitted him against today's mainstream theorists who claim carbon dioxide (CO2) is responsible for global warming, involved a link between the earth's magnetic field and climate, since that field helps regulate the number of GCR particles that reach the earth's atmosphere.

"The only way we can explain the (geomagnetic-climate) connection is through the exact same physical mechanisms that were present in Henrik Svensmark's theory," Knudsen said.

---

[29] For example, please refer to http://ossfoundation.us/projects/environment/global-warming/myths/henrik-svensmark

"If changes in the magnetic field, which occur independently of the earth's climate, can be linked to changes in precipitation, then it can only be explained through the magnetic field's blocking of the cosmetic rays," he said.

The two scientists acknowledged that CO2 plays an important role in the changing climate, "but the climate is an incredibly complex system, and it is unlikely we have a full overview over which factors play a part and how important each is in a given circumstance," Riisager told Videnskab.

National Geographic

Finally, let us review the summary by National Geographic, one of the most respected journals on the environment for the past century:

Earth's magnetic field is fading.Today it is about 10 percent weaker than it was when German mathematician Carl Friedrich Gauss started keeping tabs on it in 1845, scientists say
"Mini Ice Age" May Be Coming Soon, Sea Study Warns.
http://news.nationalgeographic.com/news/2004/09/0909_040909_earthmagfield.html

The ice age theory contradicts the global warming theory. But, I think connecting some dots may show that the warming precedes the icing trend. So both happen and both are correct.

The Earth's frequency is known as Schumann Resonance[30]. This is a more accurate explanation than what you find on metaphysical sites. It does include 7.8 or (7.83hz).

Supposedly as the field gets weaker, the frequency is increasing so that the 7.83hz is climbing up towards 13hz and is supposedly going to be there in 2012 (Dec 21, 2012) time frame. It sure wouldn't surprise me if that is true.

So look at this overall picture so far.
Strong magnetic field, livable condition on earth.

Field starts to weaken to approach flipping and the frequency increases, more solar radiation penetrates, Earth warms up, people start to debate if it is really warming up and what caused it, ice caps melt, water levels rise, wind and under water currents change. It gets hotter, populations die, some survive, field flips and the wind hits the fan, the field is flipped and starts to strengthen and the field starts to slow down, it gets colder and ice age for a while, gets to maximum strength and then starts to weaken more solar gets in, melts ice, make it habitable for life for a long time, gets weaker until nice medium, field strengthens and slows down, stays like this long time, people and animals and plants are happy, then field weakens, starts to flip, increases frequency, etc... repeat. That is my personal idea of what is happening mixed with some facts and of course my imagination filling in the gaps and connecting the dots. Could be way off, but this model seems to make enough sense to me to my satisfaction.

Meteorites - Asteroids, Comets, etc.

---

[30] http://image.gsfc.nasa.gov/poetry/ask/q768.html

Asteroid and comet impacts to planets do not only happen in movies. Between approximately 3.8 billion and 4.1 billion years ago, numerous asteroid impacts during the Late Heavy Bombardment caused significant changes to the greater surface environment of the Earth.

In the past half-billion years, there have been five major mass extinctions. The most recent such event was 65 million years ago, when an asteroid impact triggered the extinction of the (non-avian) dinosaurs and other large reptiles, but spared some small animals such as mammals, which then resembled shrews. Since this cataclysmic event, mammalian life has diversified, and several million years ago an African ape-like animal such as *Orrorin tugenensis* gained the ability to stand upright. This enabled tool use and encouraged communication that provided the nutrition and stimulation needed for a larger brain, which allowed the evolution of the human race. The development of agriculture, and then civilization, allowed humans to influence the Earth in a short time span as no

other life form had, affecting both the nature and quantity of other life forms.[31]

Chicxulub crater[32] [33]

The Chicxulub crater is a prehistoric impact crater buried underneath the Yucatán Peninsula in Mexico. The center of the crater is located near the town of Chicxulub, after which the crater is named. The crater is more than 110 miles in diameter, making the feature one of the largest confirmed impact structures on Earth. The extraterrestrial body that impacted the Earth and formed the crater was at least 6 miles in diameter.

The age of the rocks shows that this impact structure dates from

Chicxulub Crater

Yucatán Peninsula

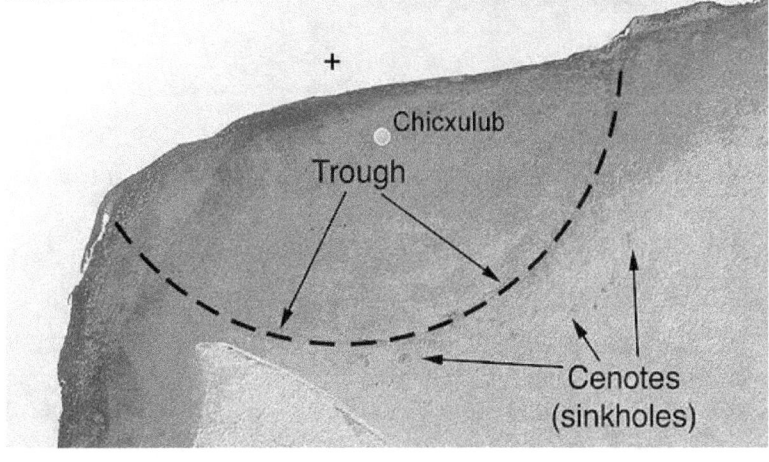

Chicxulub

Trough

Cenotes (sinkholes)

the end of the Cretaceous Period, roughly 65 million years ago. Recent evidence suggests that the impacting body may have been a piece of a much larger asteroid that broke up in a collision in distant space more than 160 million years ago.

[31] Raup, D. M.; Sepkoski, J. J. (1982). "Mass Extinctions in the Marine Fossil Record". *Science* **215** (4539): 1501–1503. Bibcode 1982Sci...215.1501R. doi:10.1126/science.215.4539.1501. PMID 17788674

[32] ^ "PIA03379: Shaded Relief with Height as Color, Yucatan Peninsula, Mexico". *Shuttle Radar Topography Mission*. NASA. http://photojournal.jpl.nasa.gov/catalog/PIA03379

[33] ^ "Chicxulub". *Earth Impact Database*. University of New Brunswick. http://www.passc.net/EarthImpactDatabase/chicxulub.html

The impact associated with the crater is implicated in causing the extinction of the dinosaurs, although some critics argue that the impact was not the sole reason, and others debate whether there was a single impact or whether the Chicxulub asteroid fragment was one of several that may have struck the Earth at around the same time.

Effects of the Chicxulub Impact[34]

The Chicxulub asteroid fragment delivered an estimated energy equivalent of 96 teratons of TNT (4×10^{23} J), a total explosive power many times stronger than the totality of all nuclear devices ever constructed by human beings. By contrast, the most powerful man-made explosive device ever detonated, the Tsar Bomba, had a yield of only 50 megatons of TNT, making the Chicxulub impact two million times more powerful.

The impact would have caused some of the largest mega-tsunamis in Earth's history, reaching thousands of meters high. A cloud of super-heated dust, ash and steam would have spread from the crater, as the asteroid fragment burrowed underground in less than a second. Excavated material along with pieces of the asteroid fragment, ejected out of the atmosphere by the blast, would have been heated to incandescence upon re-entry, broiling the Earth's surface and possibly igniting global wildfires; meanwhile, colossal shock waves would have triggered global earthquakes and volcanic eruptions. The emission of dust and particles could have covered the entire surface of the Earth for several years, possibly a decade, creating a harsh environment for living things. The shock production of carbon dioxide caused by the destruction of carbonate rocks would have led to a sudden *greenhouse effect*. Over a longer period, sunlight would have been blocked from reaching the surface of the earth by the dust particles in the atmosphere, cooling the surface dramatically. Photosynthesis by plants would also have been interrupted, affecting the entire food chain. A model

[34] **References:**

- Covey *et al*
- Adamsky and Smirnov, 19
- Mason, *et al*
- Melosh, interview
- Melosh. "On the ground, you would feel an effect similar to an oven on broil, lasting for about an hour [...] causing global forest fires."
- Hildebrand, Penfield, *et al.*; 5
- Perlman
- Pope, Ocampo, *et al*
- Lomax, B.; Beerling D., Upchurch Jr G. & Otto-Bliesner B. (2001). "Rapid (10-yr) recovery of terrestrial productivity in a simulation study of the terminal Cretaceous impact event". *Earth and Planetary Science Letters* **192** (2): 137–144. Bibcode 2001E&PSL.192..137L. doi:10.1016/S0012-821X(01)00447-2. http://www.sciencedirect.com/science?_ob=ArticleURL&_udi=B6V61-4441439-3&_user=10&_coverDate=10%2F15%2F2001&_rdoc=1&_fmt=high&_orig=search&_sort=d&_docanchor=&view=c&_searchStrId=1327486115&_rerunOrigin=google&_acct=C000050221&_version=1&_urlVersion=0&_userid=10&md5=1f071c79677e3cb01503970a31509a1d
- Marc Airhart (January 1, 2008). "Seismic Images Show Dinosaur-Killing Meteor Made Bigger Splash". http://www.jsg.utexas.edu/news/2008/01/seismic-images-show-dinosaur-killing-meteor-made-bigger-splash/.
- Hildebrand, Penfield, *et al.*; 1
- Hildebrand, Penfield, *et al.*; 3
- Grieve.
- Hildebrand, Penfield, *et al.*; 4
- Kring, "Discovering the Crater"
- Sigurdsson

of the event developed by Lomax et al. (2001) suggests that net primary productivity (NPP) rates may have increased to higher than pre-impact levels over the long term because of the high carbon dioxide concentrations.

In February 2008, a team of researchers led by Sean Gulick at the University of Texas at Austin's Jackson School of Geosciences used seismic images of the crater to determine that the asteroid fragment landed in deeper water than was previously assumed. They argued that this would have resulted in increased sulfate aerosols in the atmosphere. According to the press release, that "could have made the impact deadlier in two ways: by altering climate (sulfate aerosols in the upper atmosphere can have a cooling effect) and by generating acid rain (water vapor can help to flush the lower atmosphere of sulfate aerosols, causing acid rain).

Anyone who has been to Cancun on holiday and observed the very unique and fine sand on the beaches will now perhaps understand why the beautiful sand is "out of this world".

Chances of another existential catastrophe[35]

Existential risks pose unique challenges to prediction, even more than other long-term events, because of observation selection effects. Unlike with most events, the failure of catastrophic events to occur in the past is not evidence against their likelihood in the future, because every world that has experienced one has no observers, so regardless of their frequency, no civilization observes existential risks in its history.

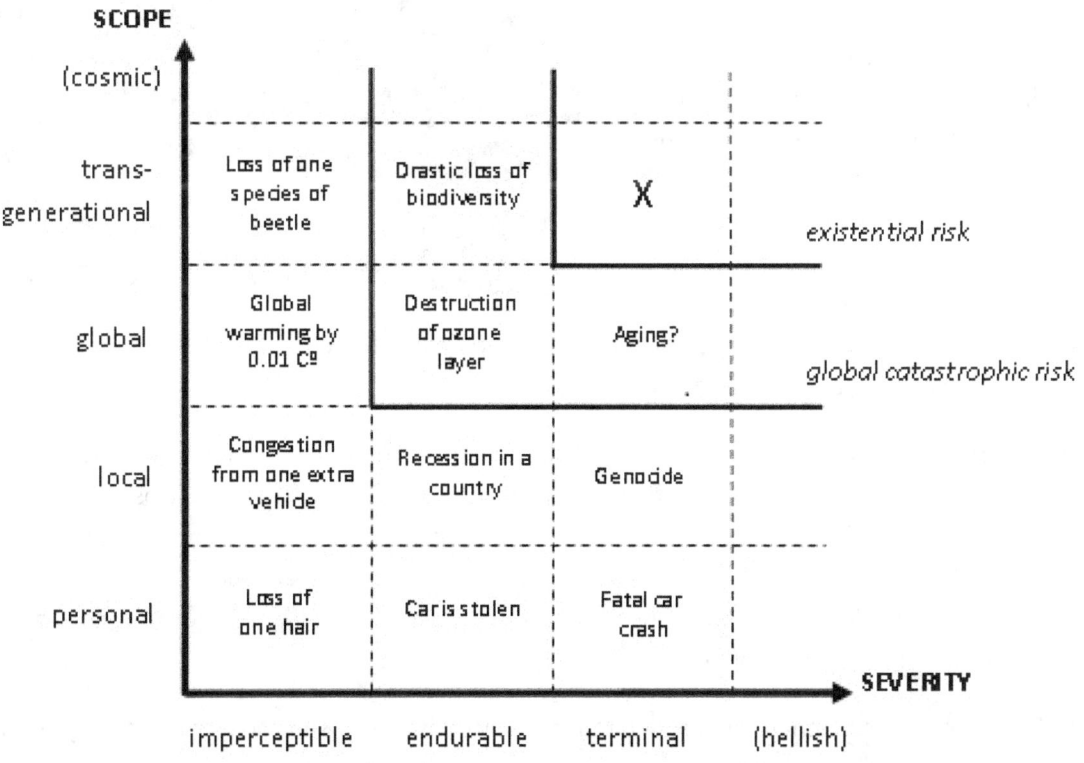

---

[35] Bostrom, Nick (March 2002). "Existential Risks: Analyzing Human Extinction Scenarios and Related Hazards". *Journal of Evolution and Technology*

Fortunately, the estimates generally conclude that existential risk, such as that from asteroid impact, is a one-in-a-million chance of causing our extinction in the next century. However, later research suggested the actual rate of large impacts could be much higher than predicted. By comparison, the frequency of volcanic eruptions of sufficient magnitude to cause catastrophic climate change, similar to the Toba Eruption mentioned previously, which almost caused the extinction of the human race, has been estimated at about 1 in every 50,000 years. The Toba Eruption occurred 70,000 years ago, so humanity is well past the "sell by" date.

Asteroid and comet impacts are more frequent than one thinks.

Comet Shoemaker–Levy 9[36]

Comet Shoemaker–Levy 9 (formally designated D/1993 F2) was a comet that broke apart and

collided with Jupiter in July 1994. The first impact occurred at 20:13 UTC on July 16, 1994, when fragment A of the nucleus slammed into Jupiter's southern hemisphere at a speed of about 40 miles per second. Instruments on the Galileo spacecraft detected a fireball which reached a peak temperature of about 24,000 (K)elvins, compared to the typical Jovian cloud top temperature of about 130 K, before expanding and cooling rapidly to about 1500 K after forty seconds. The plume from the fireball quickly reached a height of almost 2,000 miles. A few minutes after the impact fireball was detected, Galileo measured renewed heating, probably due to ejected material falling back onto the planet. Earth-based observers detected the fireball rising over the limb of the planet shortly after the initial impact.

Astronomers had expected to see the fireballs from the impacts, but did not have any idea in advance how visible the atmospheric effects of the impacts would be from Earth. Observers soon saw a huge dark spot after the first impact. The spot was visible even in very small telescopes, and was about 3,700 miles (one Earth radius) across. This and subsequent dark spots were thought to have been caused by debris from the impacts, and were markedly asymmetric, forming crescent shapes in front of the direction of impact.

Over the next six days, twenty-one distinct impacts were observed, with the largest coming on July 18 at 07:33 UTC when fragment G struck Jupiter. This impact created a giant dark spot almost 8,000 miles across, and was estimated to have released an energy equivalent to 6 Gigatons of TNT (six hundred times the world's nuclear arsenal). Two impacts twelve hours apart on July 19th created

---

[36] "Comet Shoemaker–Levy 9 Collision with Jupiter". National Space Science Data Center. February 2005
http://nssdc.gsfc.nasa.gov/planetary/comet.html.

impact marks of similar size to that caused by fragment G, and impacts continued until July 22nd, when fragment W struck the planet.

2009 Jupiter impact event[37]

A 2003 paper estimated comets with a diameter larger than 1.5 kilometers impact Jupiter about

every 90 to 500 years. However, this did not stop a comet from hitting Jupiter, barely 15 years after Comet Shoemaker- Levy 9

The 2009 Jupiter impact event, occasionally referred to as the Wesley impact, was a July 2009 impact on Jupiter that caused a black spot in the planet's atmosphere. The spot was similar in area to the planet's Little Red Spot, approximately the size of the Pacific Ocean. The impacting object, a comet is estimated to have been only about 200 to 500 meters in diameter

2010 Jupiter impact event[38]

And again in 2010, Jupiter had an impact event. It is fortunate that no life form exists on the planet. The impacting object, estimated to be about 8–13 meters in diameter, may have been an asteroid, comet, centaur, extinct comet, or temporary satellite capture. The object had a mass between 500-2000 metric tons. Jupiter probably gets hit by several objects of this size each year.

[37] Jupiter pummeled, leaving bruise the size of the Pacific Ocean University of California, Berkeley press release, July 21, 2009
[38] Hueso, R.; Wesley; Perez-Hoyos; Wong; et al. (2010). "First Earth-based Detection of a Superbolide on Jupiter". *The Astrophysical Journal* **721** (2). 1009.1824. Bibcode 2010ApJ...721L.129H. doi:10.1088/2041-8205/721/2/L129.

Life on Earth has been fortunate not to have to endure an extinction-level extraterrestrial impact since the last event some 65 million years ago. But 65 million years is but a tiny speck in Earth's long history, and there is no guarantee that our good luck will continue forever.

Also, as most people probably already know, celestial objects (commonly referred to as *meteors*) are frequently bombarding the Earth. A meteor is the visible path of a meteoroid that has entered the Earth's atmosphere. Meteors typically occur in the mesosphere, and most range in altitude from 50 to 60 miles. *Millions of meteors occur in the Earth's atmosphere every day*. Most *meteoroids* that cause meteors are about the size of a pebble.

Thanks to Earth's oxygen-rich atmosphere, most meteoroids burn up before they can impact the surface. But the left-over debris called meteoric dust can persist in the atmosphere for up to several months. These particles *occasionally affect climate*, both by scattering electromagnetic radiation and by catalyzing chemical reactions in the upper atmosphere

Most, but not all meteoroids burn up. A meteorite or shooting/falling star is a portion of a meteoroid or asteroid that survives its passage through the atmosphere and impact with the ground without being destroyed. Meteorite craters pockmark the landscape all over the world to remind us that meteor impacts have been a frequent occurrence, even during the relatively short period when human beings have existed on this planet.

As an example, in the American Southwest alone, beginning in the mid-1990s, amateur meteorite hunters began scouring the arid areas of the southwestern United States. To date, meteorites numbering possibly into the thousands have been recovered from the Mojave, Sonoran, Great Basin, and Chihuahuan Deserts, with many being recovered on dry lake beds. Perhaps the most notable find in recent years has been the Los Angeles meteorite, a martian meteorite that was reportedly found by Robert Verish. Several of the meteorites found recently are currently on display in the Griffith Observatory in Los Angeles.[40]

A couple of examples of the more significant, larger meteorite impacts are now described:

---

[39] Beech, M.; Steel, D. I. (September 1995). "On the Definition of the Term Meteoroid". *Quarterly Journal of the Royal Astronomical Society* **36** (3): 281–284. Bibcode 1995QJRAS..36..281B.)

[40] **For a more detailed list of meteorite finds all over the world, please refer to:**

- Bland, P.A.; Artemieva, Natalya A. (2006). "The rate of small impacts on Earth". *Meteoritics and Planetary Science* **41** (4): 607–631. Bibcode 2006M&PS...41..607B. doi:10.1111/j.1945-5100.2006.tb00485.x.
- Sears, D. W. (1978). *The Nature and Origin of Meteorites*. New York: Oxford Univ. Press. ISBN 978-0-85274-374-4.
- Ward, Henry L. (1917). "A new meteorite". *Science* **46** (1185): 262–263. Bibcode 1917Sci....46..262W. doi:10.1126/science.46.1185.262-a. PMID 17844300. – a report of the Colby (Wisconsin) fall
- Krot, A.N.; Keil, K.; Scott, E.R.D.; Goodrich, C.A.; Weisberg, M.K. (2007). "1.05 Classification of Meteorites". In Holland; Turekian, Karl K.. *Treatise on Geochemistry*. **1**. Elsevier Ltd. pp. 83–128. doi:10.1016/B0-08-043751-6/01062-8. ISBN 978-0-08-043751-4.
- The NHM Catalogue of Meteorites. Internt.nhm.ac.uk
- MetBase. Metbase.de. Retrieved on 2011-12-17

THE CLOVIS COMET

Part I:
Evidence for a Cosmic Collision 12,900 Years Ago

OR REASONS still not entirely understood, most of the large animals in the New World became extinct at the end of the Pleistocene epoch. For decades, their passing has been a source of wonder and contention among the researchers who study them. Natural vegetation shifts, climate changes, over-hunting by humans, plagues, and various combinations thereof have been put forward as proximate causes of the extinctions, though no definitive consensus has been reached.

As it turns out, all those theories might be further

Allen West sampling a backhoe trench profile at the Big Pine Tree site for Clovis-age sediments.

off the mark than previously realized. If the authors of a study published in the 7 October 2007 issue of the *Proceedings of the National Academy of Science* (*PNAS*) are correct—and over a dozen converging lines of evidence argue that they are—then a comet hit North America 12,900 years ago, dooming the Pleistocene megafauna and decimating the local human population.

Allen West, whose research was the impetus for the *PNAS* study, realizes that some observers won't like the theory. But he's convinced it's the explanation that best fits the facts. The signs are well documented and copious; when taken to-

gether, they form a composite "smoking gun" that strongly suggests that something came out of the stars and hit us at the beginning of the Holocene. Following standard scientific protocol, West and his research team have entertained many other theories in search of a viable alternative. But in the end, he says, "We sure can't think of one."

Back to the deep freeze

Just as things were warming up after the long Pleistocene Ice Age, an abrupt temperature reversal plunged the Northern Hemisphere into a thousand-year cold spell known as the Younger Dryas (YD) interval. The beginning of the YD in North America, Greenland, and Western Europe is well established at 12,900 CALYBP. At about the same time, the last of the Pleistocene megafauna were disappearing in North America and the Clovis culture was breathing its last.

A recent reexamination of the Clovis time range by geochronologist Tom Stafford and geoarchaeologist Mike Waters (MT 22-3, -4, "Clovis Dethroned: A New Perspective on the First Americans") makes it clear that, among other things, Clovis came to an end in the extraordinarily brief period 12,800–12,925 CALYBP. Their conclusion is consistent with data compiled by C. Vance Haynes, who has demonstrated the presence of dark organic deposits, known to scientists as "black mats," that mark the end of the Clovis era at more than 50 different archaeological sites across North America. They form a boundary that's easily identified in about one-third of the known Clovis sites, and the best explanation for them is that they represent significant organic enrichment of the local sediments via algal blooms or a sudden infusion of charcoal or soot. The formation of black mats dates conclusively to the beginning of the YD interval.

The sudden onset of the Younger Dryas is implicated, therefore, in the decline of Clovis. But what triggered the YD in the first place? Traditional explanations center on a sudden influx of glacial meltwater into the North Atlantic, which would have disrupted the saline density and interfered with established patterns of ocean circulation that contributed to the warming of the Northern Hemisphere. This explanation seems reasonable, since it's well known that modern England, for example, would be significantly colder without the Gulf Stream. But oddly enough, it didn't happen during any previous interglacial, so some random event must have triggered the abrupt climate change.

That random event might have been the impact of a relatively small highly fragmented comet or asteroid, particularly one that exploded in the upper atmosphere, igniting fires over a large area. Such an event would fill the atmosphere with soot and dust that would block out significant amounts of solar radiation for weeks or months, resulting in a "nuclear winter" effect. Even after the skies

http://conservationreport.com/2009/06/28/nephology-mysterious-noctilucent-cloud-phenomenon-increases-noctilucent-cloud-formation-linked-to-space-shuttle-activity-and-possibly-tunguska-explosion-of-1908/

A recent study suggests "the Tunguska explosion of 1908 was caused by a comet hitting Earth[41] . . . based on the behavior of water vapor from the space shuttle's exhaust" forming noctilucent clouds

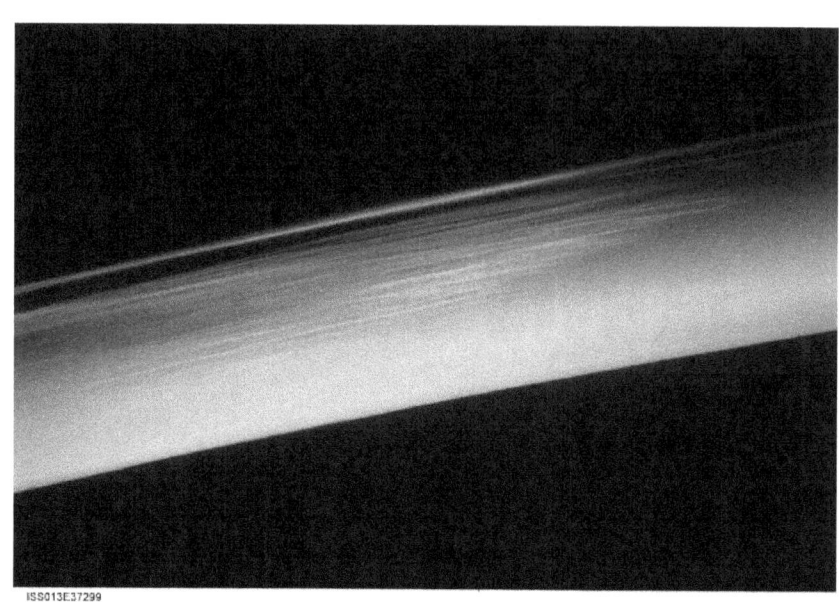

The research, accepted for publication (June 24, 2009) by the journal Geophysical Research Letters, published by the American Geophysical Union, connects the two events by what followed each about a day later: brilliant, night-visible clouds, or noctilucent clouds, that are made up of ice particles and only form at very high altitudes and in extremely cold temperatures.

"It's almost like putting together a 100-year-old murder mystery," said Michael Kelley, the James A. Friend Family Distinguished Professor of Engineering at Cornell who led the research team. "The evidence is pretty strong that the Earth was hit by a comet in 1908." Previous speculation had ranged from comets to meteors.

The researchers contend that the massive amount of water vapor spewed into the atmosphere by the comet's icy nucleus was caught up in swirling eddies with tremendous energy by a process called two-dimensional turbulence, which explains why the noctilucent clouds formed a day later many thousands of miles away.

In summary, a mysterious and completely new noctilucent cloud phenomenon in the 20th century is probably linked to a comet hitting Earth.

---

[41] There is still debate on whether a comet or a large meteoroid caused the Tunguska event, see:

- Shoemaker, Eugene (1983). "Asteroid and Comet Bombardment of the Earth". Annual Review of Earth and Planetary Sciences (US Geological Survey, Flagstaff, Arizona: Annual Review of Earth and Planetary Sciences) 11 (1): 461. Bibcode 1983AREPS..11..461S. doi:10.1146/annurev.ea.11.050183.002333. http://www.annualreviews.org/doi/abs/10.1146/annurev.ea.11.050183.002333?prevSearch=Tunguska.
- Longo, G.; Serra R., Cecchini S. and Galli M., (1994). "Search for microremnants of the Tunguska Cosmic Body". Planetary and Space Science (UK: Elsevier Science Ltd) 42 (2): 163–177. Bibcode 1994P&SS...42..163L. doi:10.1016/0032-0633(94)90028-0. http://www-th.bo.infn.it/tunguska/papers/planetspace.html.

The rapid advancement of human beings in a relatively short period of time (from a geological perspective) has resulted in a high sense of arrogance that we can control our destiny. We cannot. By design, the universe is a dangerous place. During our short existence on this planet, humans have almost been rendered extinct at least once and in severe peril as a result of major natural disasters many times. A near-extinction event occurred 70,000 years ago when the super-volcanic eruption in Lake Toba in Sumatra, Indonesia had global consequences wiping out most humans across the Indian Ocean. Another example of a major natural disaster: a massive, unprecedented drought in ancient times pushed human beings out of Africa.

The more advanced humanity has become, the more our arrogance has grown. When the Maunder solar Minimum. which started at the beginning of the fourteenth century, helped trigger bitterly cold and longer winters, cool and rainy summers, violent storms, recurrent flooding, massive crop failures and famine in mid-to high latitudes, the population did not concoct a theory that humans somehow caused the drastic change in climate. About a third of the population of Europe perished, but they *accepted* the severe weather as acts of nature (in those days referred to as an Acts of God), *adapted*, *learned* from their adversity and *prospered*. The start of pre-eminence of European civilization and the start of the Maunder Minimum may not be coincidental after all.

Sound familiar in today's context? Only this time around, almost every severe or unusual weather event is hyped by the media and others to be a consequence of human activity, using as justification, a theory we invented. At the same time, we delude ourselves that nature is tranquil – after all beautiful Mother Nature could never sanction such terrible and severe "climate weirding". Really?

About the only natural disasters that we cannot yet blame on human beings are major earthquakes and volcanoes, but that is not for lack of trying. The colossal Indian Ocean Earthquake and tsunami at the end of 2004, zooming across nearly a quarter of the planet at 500 miles per hour, and bringing immense death and destruction to several countries, should have been a wake-up call that humanity controls neither this planet nor this universe. But it was not. It only took a year or so for this fast-moving world to mainly forget about this massive natural disaster.

As Darwin's Theory of Evolution teaches us, species that survive are those which are fittest at adapting to change. For many millions of years, far longer than humans have inhabited this planet, dinosaurs arrogantly trotted the world believing that they could remain in charge for eternity. Then in the blink of an asteroid, they vanished because they could not adapt to the new Earth reality. Small and insignificant mammals, whom the dinosaurs had nothing but disdain for – the "insects" of the dinosaur era, took over the planet. I have no doubt that this can happen again, this time to human beings, especially if our arrogance keeps us in the delusion that the planet and the universe are constant, and that any negative change that occurs is our doing. That mode of human operation is fine only up to the point when nature strikes with a global catastrophe once again, whether it is from an asteroid, comet, volcano, earthquake, solar event or other. And our advanced technological state means that we are much more vulnerable than previous generations of human beings. How many people do you know who have the skills to survive without electricity, running water, supermarkets for food, public transport or cars for mobility, hospitals for healthcare, etc? What if a global natural catastrophe such as Lake Toba wiped all of that out? A catastrophic scenario similar to that described in the movie *The Day after Tomorrow* is possible. But it is far more likely to be triggered by a natural catastrophe than by human-induced global warming.

Since 1990, more than $100 billion has been pumped into supporting research by thousands of scientists to try to prove the preconception of human impact on global impact.[42] After all of that expenditure, we are still unable to isolate and measure human influence on global temperature. That influence remains buried deeply in the noise and natural variation of the multitude of factors that affect the Earth's climate system, as described in the previous sections.

By contrast, a small fraction of research dollars and scientists has been assigned to large meteorite (e.g. asteroid or comet) impact avoidance. Even if the probability of such an impact is relatively low, there is no doubt that even one such impact will have immediate and catastrophic consequences to all life on this planet.

One wonders if we really have the priorities straight.

Anthropogenic explanation for global temperatures plateauing in the first decade of the 21ˢᵗ century

We are coming to the end of this chapter. Therefore, it is time for some more dark humor. Proponents of anthropogenic global warming have been baffled by the plateauing of global temperatures and the return of cold winters in recent years. Despite the preponderance of evidence to the contrary, and consistent with their strong preconception that Mother Nature by default is tranquil and stable, the proponents have now concocted an anthropogenic explanation as to why the rate of warming has slowed lately when compared to the 1980s and the 1990s. Details are provided below, pasted verbatim without modification.

http://www.guardian.co.uk/environment/2011/jul/04/sulphur-pollution-china-coal-climate

Sulphur from Chinese power stations 'masking' climate change

Research reveals decade of global warming from China's coal power stations has partly been offset by 'cooling' effect of sulphur pollution

The huge increase in coal-fired power stations in China has masked the impact of global warming in the last decade because of the cooling effect of their sulphur emissions, new research has revealed. But scientists warn that rapid warming is likely to resume when the short-lived sulphur pollution – which also causes acid rain – is cleaned up and the full heating effect of long-lived carbon dioxide is felt.

The last decade was the hottest on record and the 10 warmest years have all occurred since 1998. But within that period, global surface temperatures did not show a rising trend, leading some to question whether climate change had stopped. The new study shows that while greenhouse gas emissions continued to rise, their warming effect on the climate was offset by the cooling produced by the rise in sulphur pollution. This combined with the sun entering a less intense part of its 11-year cycle and the peaking of the El Niño climate warming phenomenon.

---

[42] http://www.cfact.org/a/1845/Cool-it-with-all-the-research-dollars

The number of coal-fired power stations in China multiplied enormously in that period: the electricity-generating capacity rose from just over 10 gigawatts (GW) in 2002 to over 80GW in 2006 (a large plant has about 1GW capacity).

But rather than suggesting that cutting carbon emissions is less urgent due to the masking effect of the sulphur, Prof Robert Kaufman, at Boston University and who led the study, said: "If anything the paper suggests that reductions in carbon emissions will be more important as China installs scrubbers [on its coal-fired power stations], which reduce sulphur emissions. This, and solar insolation increasing as part of the normal solar cycle, [will mean] temperature is likely to increase faster."

Prof Joanna Haigh, at Imperial College London, commented: "The researchers are making the important point that the warming due to the CO_2 released by Chinese industrialisation has been partially masked by cooling due to reflection of solar radiation by sulphur emissions. On longer timescales, with cleaner emissions, the warming effect will be more marked."

The cooling effect of sulphur pollution on climate has long been recognised by scientists studying volcanic eruptions, which have, for example, caused failed crops and famines in the past. Sulphur dioxide forms droplets of sulphuric acid in the stratosphere, which increases the reflection of the Sun's heat back to space, cooling the Earth's surface.

The effect also explains the lack of global temperature rise seen between 1940 and 1970: the effect of the sulphur emissions from increased coal burning outpaced that of carbon emissions, until acid rain controls were introduced, after which temperature rose quickly. Some have even proposed sulphur dioxide could used to geoengineer the planet by deliberately injecting millions of tonnes into the atmosphere to combat warming.

The new study, published in the Proceedings of the National Academy of Sciences on Monday, analysed possible reasons for the flat 1998-2008 temperature trend using climate models and concluded that it was unlikely to be due simply to the random variation inherent in the planet's climate system. Instead it found the effect of sulphur, the sun and El Niño dominated, with the El Niño climate phase peaking in 1998 – the hottest year ever recorded – then moving into a phase dominated by its cooler mirror image, La Niña. The scientists ruled out changes in water vapour or carbon soot in the atmosphere as significant factors.

They emphasised the rapid increase in coal burning in Asia, and in China in particular, noting that Chinese coal consumption doubled between 2002 and 2007: the previous doubling had taken 22 years.

Michael E Mann, at Pennsylvania State University and not part of the research team, said the study was "a very solid, careful statistical analysis" which reinforces research showing "there is a clear impact of human activity on ongoing warming of our climate". It demonstrated, Mann said, that "the claim that 'global warming has stopped' is simply false."

****

My satirical comment: Therefore," to be logical", wouldn't a simple and inexpensive solution to stop the proclaimed "warming" permanently be to request China to burn a lot more coal? A little acid rain may be a small price to pay for the "catastrophic warming" predicted.

IV.   History of Global Temperature

Historical Summary

Over the past few decades, proponents of Anthropogenic Global Warming have frequently been directly or indirectly implying that the Earth is undergoing an *unprecedented* period of warming never before experienced. Whatever the manner and tonality of communication, this myth has been propagated widely and very effectively through the mainstream media, to the extent that many millions of people, including many educated people, politicians, etc., believe the myth as incontrovertible gospel.

A second myth that has been widely disseminated and now widely believed is that the world's climate has been relatively stable over billions of years, and it is only human activity that has triggered warming, and, thus, a dramatic increase of freak weather events, which are unnatural and not witnessed in Earth's past.

Nothing can be further from the truth.

As the graph on the right clearly illustrates, today's Earth is a very cold, frigid "Ice House" compared to the much higher temperatures in most of Earth's history, including the world of the dinosaurs when this planet was teeming with life.

Source: http://www.scotese.com/climate.htm

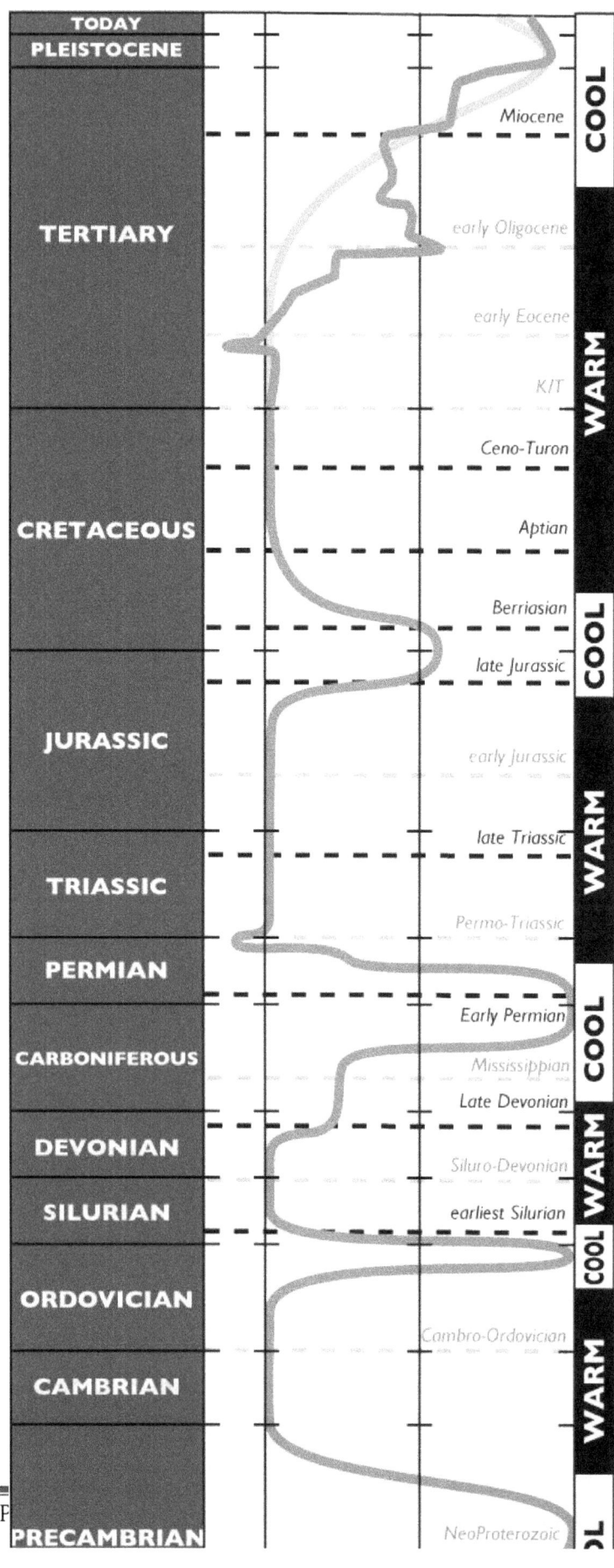

The Earth has cooled up to 6 °C in the past 5 million years

Estimates vary but it is now generally accepted in the scientific community that the world has cooled significantly in the past 5 million years.

Please refer to the below graph:

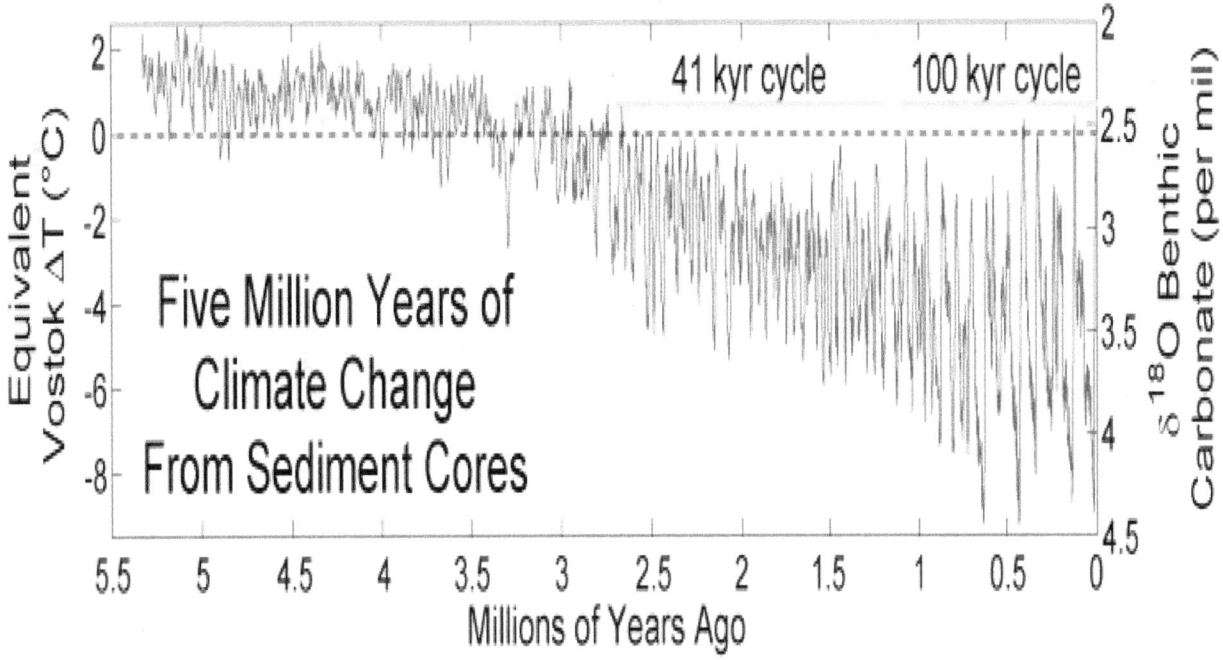

Sources:

- Climate record of climate record of Lisiecki and Raymo (2005): http://lorraine-lisiecki.com/LisieckiRaymo2005.pdf / http://dx.doi.org/10.1029/2004PA001071
- Petit, J. R.; Jouzel, J.; Raynaud, D.; Barkov, N. I.; Barnola, J. M.; Basile, I.; Bender, M.; Chappellaz, J.; Davis, J.; Delaygue, G.; Delmotte, M.; Kotlyakov, V. M.; Legrand, M.; Lipenkov, V.; Lorius, C.; Pépin, L.; Ritz, C.; Saltzman, E.; Stievenard, M. (1999). "Climate and Atmospheric History of the Past 420,000 years from the Vostok Ice Core, Antarctica". *Nature* **399**: 429-436. DOI:10.1038/20859

Notes:

- Lisiecki and Raymo concluded their research in 2005, well after the hype of anthropogenic global warming was at it is zenith. It is unfortunate that the IPCC and other climate-related international organizations have not sufficiently taken later research into consideration and adapted the content and tone of their reports accordingly.
- A horizontal line at 0 °C indicates modern temperatures (circa 1950"). You will see a similar variation repeating itself many times over the past 3 million years, spiking up to 0 °C or higher and spiking down increasingly from about -4 °C about 2.5 million years ago to -8 °C just a few hundred years ago at the nadir of the Little Ice Age, displaying a consistent cooling trend over millions of years.

Let us now cross-check the above information with the research done by Columbia University:

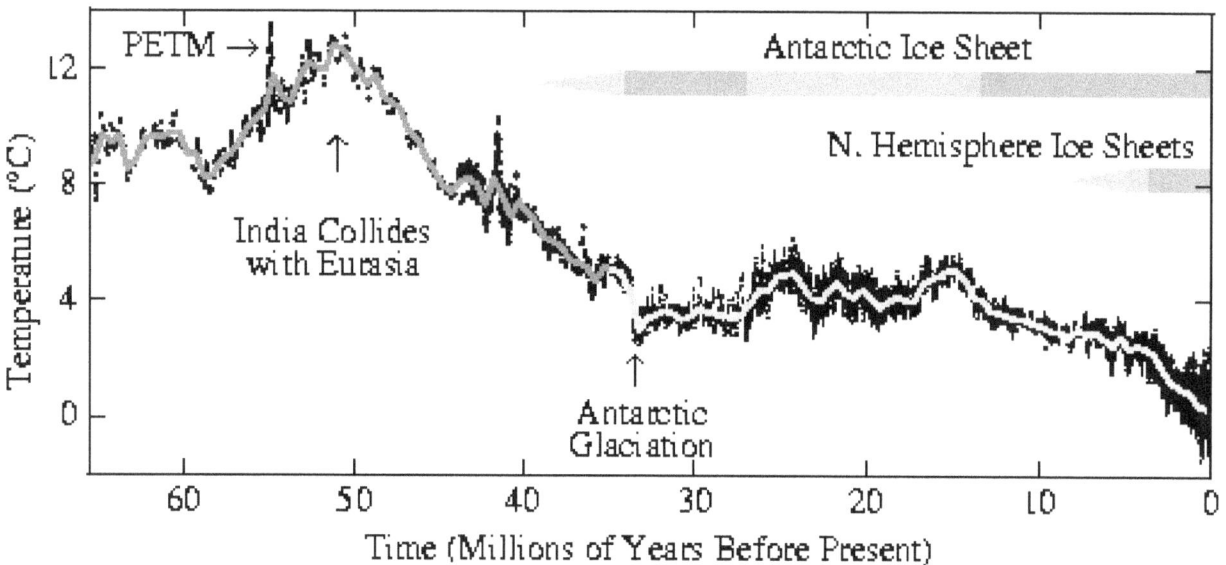

Source:

- http://www.columbia.edu/~mhs119/Storms/Storms_Fig.18.gif

Observations :

- Global temperatures today are the lowest they have been in over 60 million years
- Antarctic Ice Sheets were formed 40 million years ago when global temperatures were more than 10 °C warmer than today
- Antarctic Glaciation occurred about 35 million years ago when temperatures were still nearly 4°C warmer than today
- Northern Hemisphere Ice Sheets were formed about 10 million years ago when temperatures were nearly 4°C warmer than today.

The following sections will describe the Earth's climate during various periods in its very long history.

One will easily observe that:

- Climate change is natural and has occurred almost continuously during Earth's history
- The so-called "severe" weather incidents hyped by the mainstream media pale in comparison to the really serious climate change that the Earth has experienced in the past. Species that have survived are the ones which have anticipated changes in climate and adapted quickly.

Cambrian Climate (540 to 520 million years ago)

The climate of the Cambrian is not well known. It was probably not very hot, nor very cold. There is no evidence of ice at the poles.

Early Cambrian

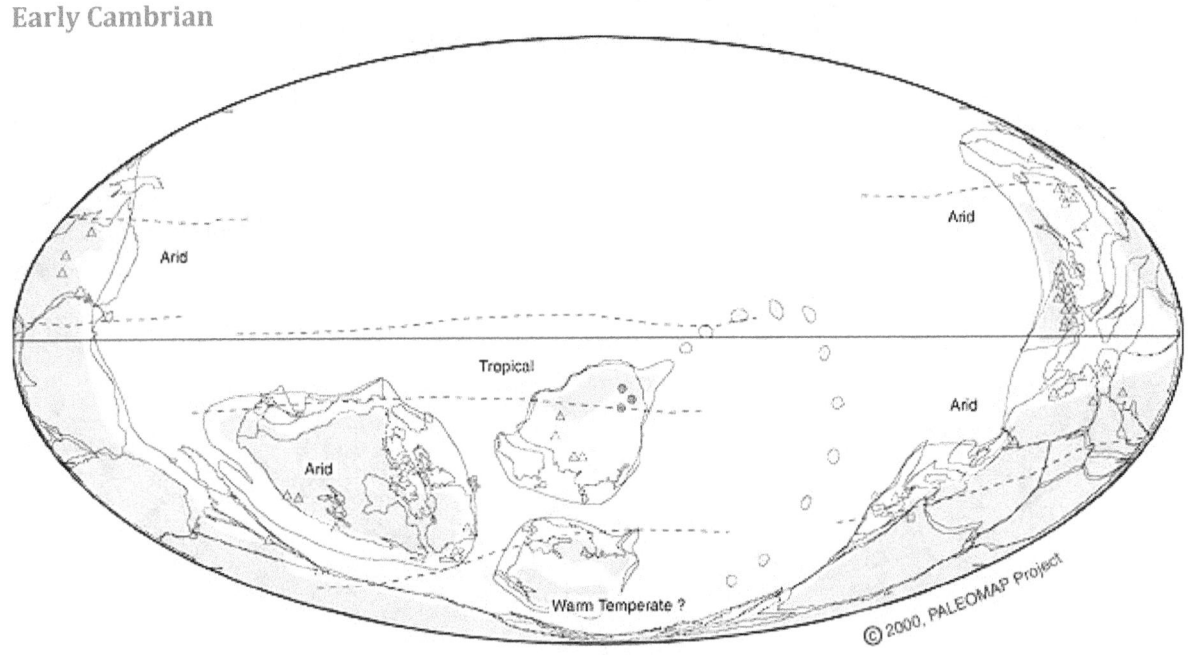

Lower Cambrian

Source: http://www.scotese.com/ecambcli.htm

Middle and Upper Cambrian

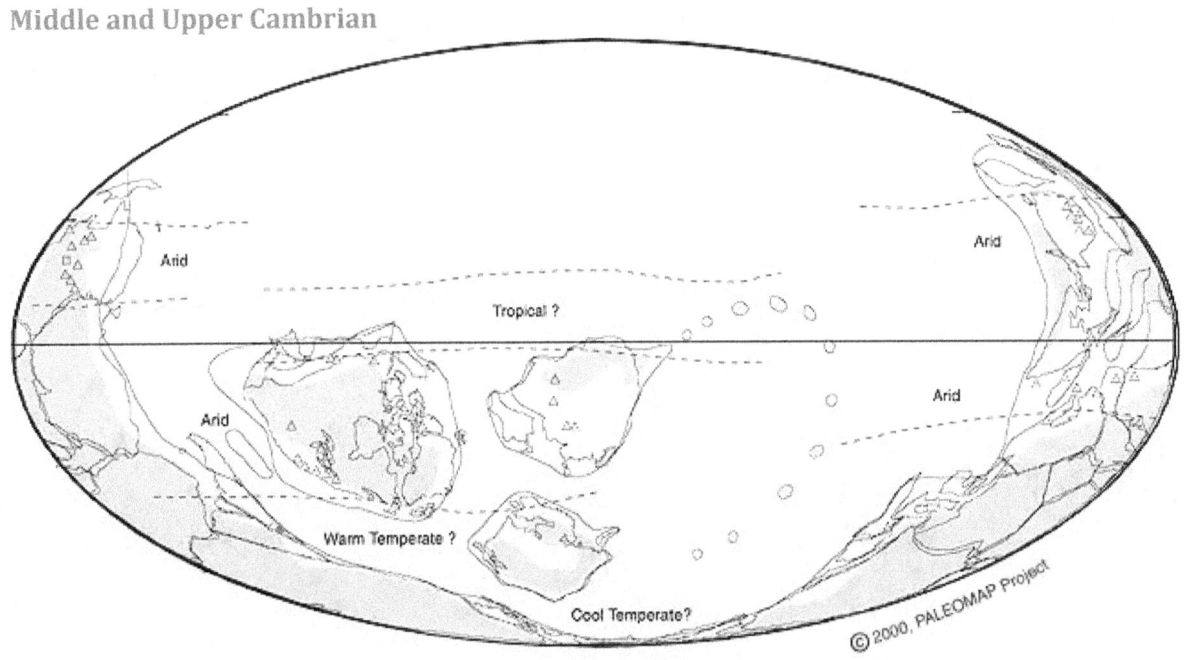

Middle & Upper Cambrian

Source: http://www.scotese.com/mlcambcl.htm

Early Ordovician Climate (480 million years ago)

Mild climates probably covered most of the globe. The continents were flooded by the oceans creating warm, broad tropical
seaways.

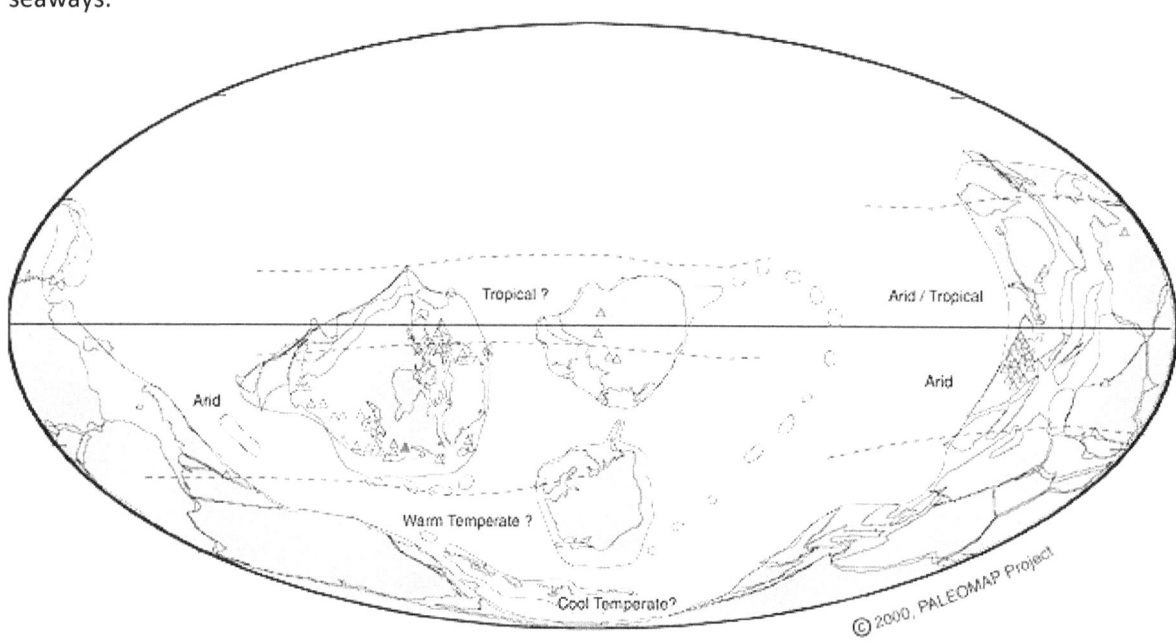

Lower Ordovician

Source: http://www.scotese.com/eordclim.htm

Middle & Late Ordovician (440 million years ago)

The Late Ordovican was an **Ice House World.** The South Polar Ice Cap covered much of Africa and South America. The climate in North America, Europe, Siberia and the eastern part of Gondwana was warm and
sunny

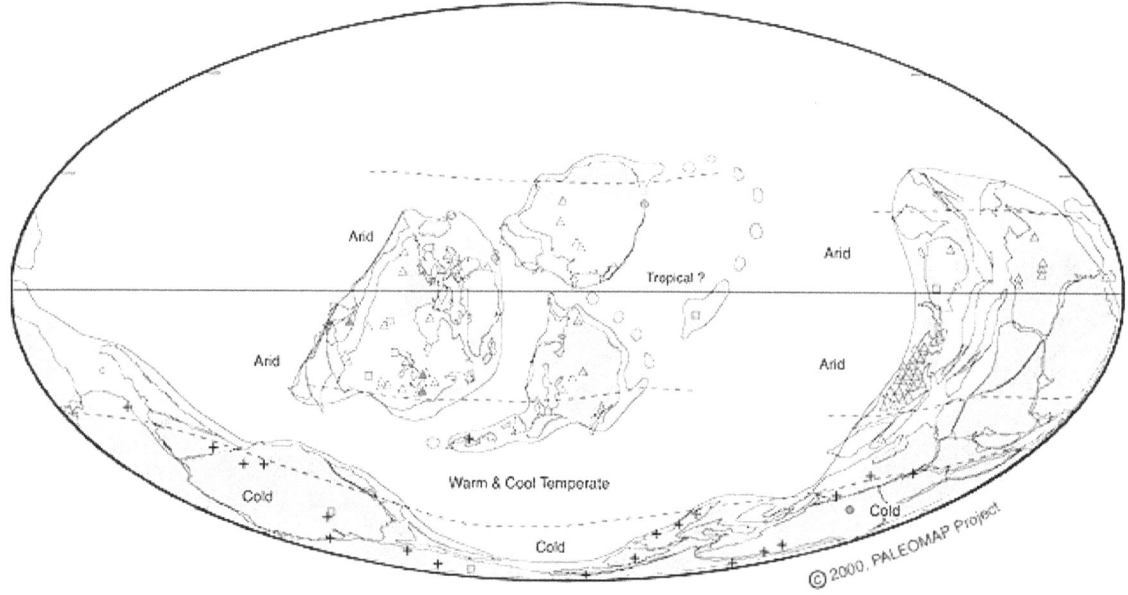

Middle & Upper Ordovician

Silurian Climate (420 million years ago)

Coral reefs thrived in the clear sunny skies of the southern Arid belt which stretched across North America and northern Europe.   Lingering glacial conditions prevailed near the South Pole.

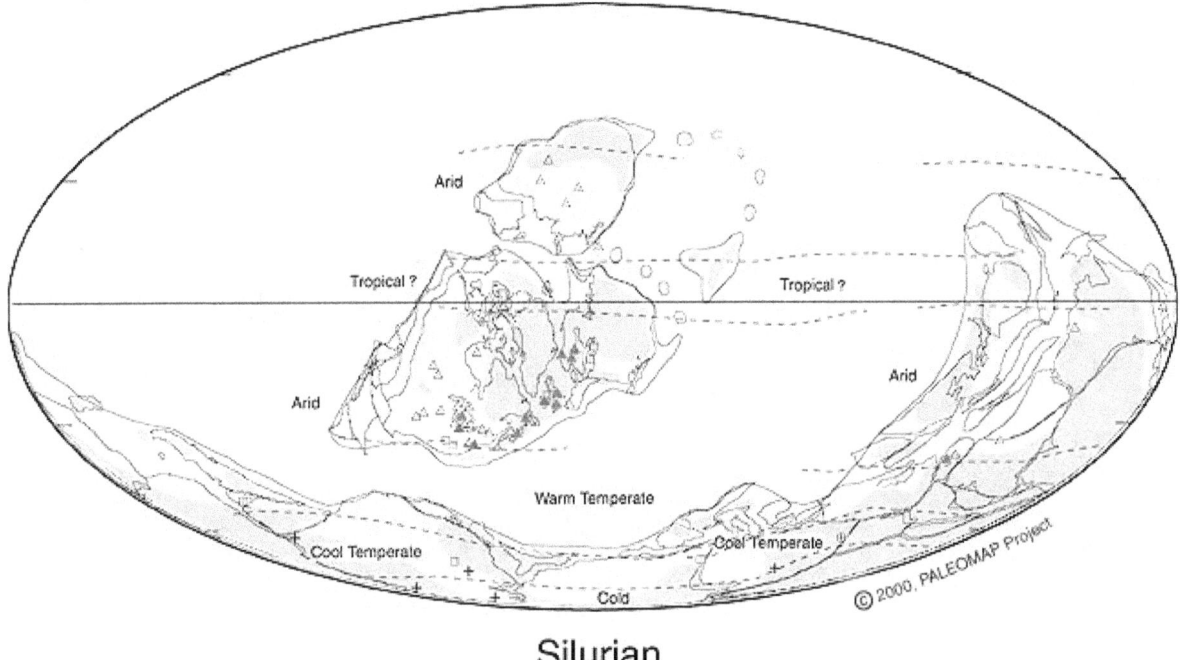

Silurian

Early Devonian Climate (400 million years ago)

Generally dry conditions prevailed across much of North America, Siberia, China and Australia during the early Devonian.   South America and Africa were covered by cool, temperate seas.

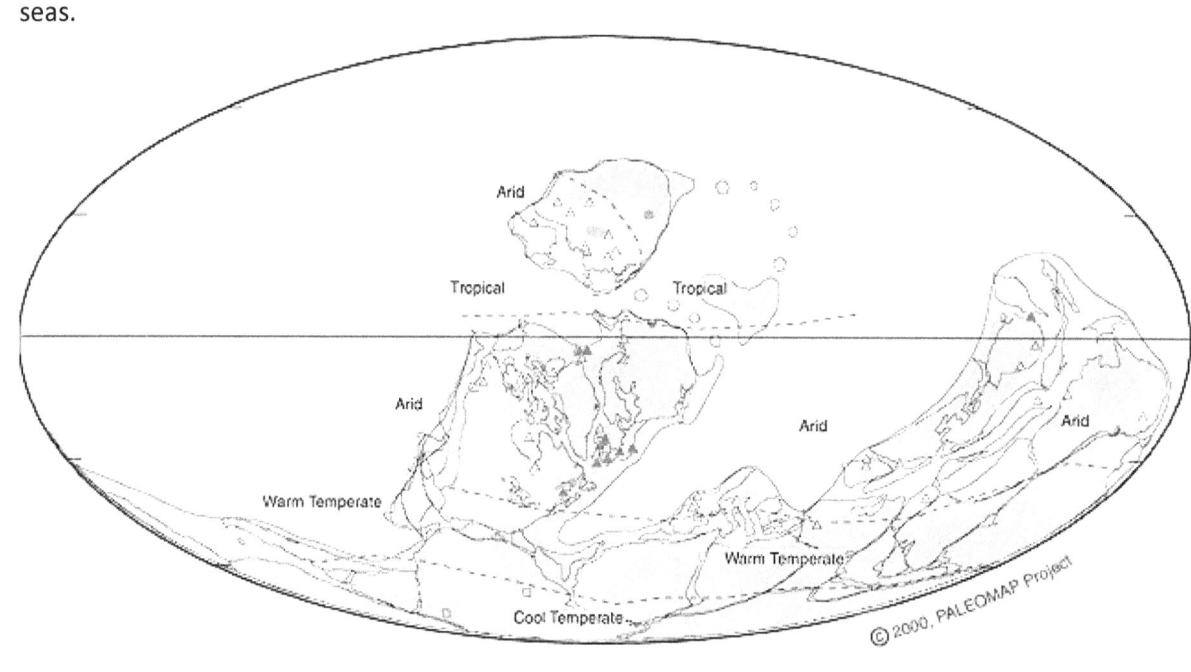

Lower Devonian

Source: http://www.scotese.com/edevclim.htm

Middle Devonian Climate (380 million years ago)

During the Middle Devonian the Equator ran through Arctic Canada. Coals began to accumulate as land plants flourished in the equatorial rainy belt. Warm shallow seas, under cloudless skies, covered much of North America, Siberia and
Australia.

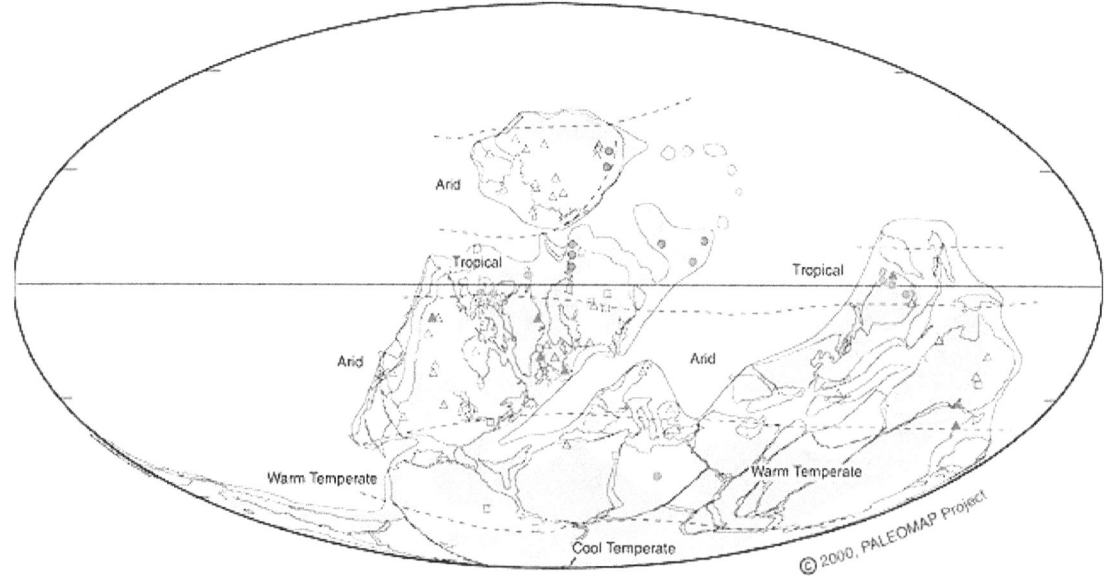

Middle Devonian

Late Devonian Climate (360 million years ago)

During the Late Devonian Pangea began to assemble.   Thick coals formed for the first time in the tropical rainforests in the Canadian Arctic and in southern China.   Glaciers covered parts of the Amazon Basin, which was located close to the South

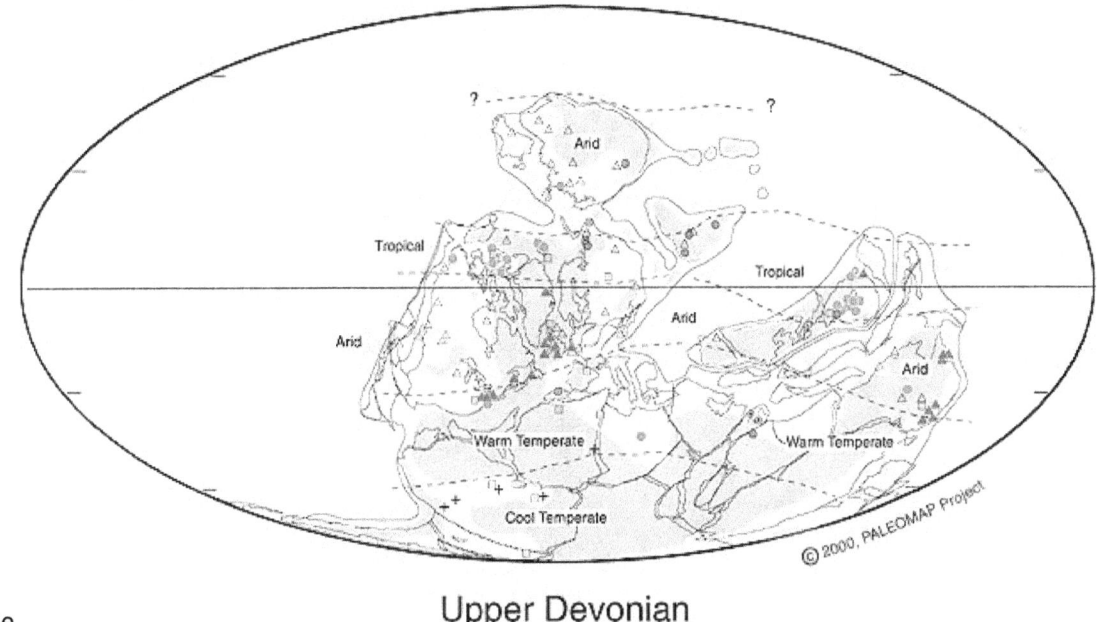

Pole.

Upper Devonian

Early Carboniferous Climate (360 to 330 million years ago)

As Pangea moves northward, the climatic belts move southward. Tropical rainforests cross from Arctic Canada to Newfoundland and Western Europe. The desert regions in mid-North America begin to contract. The Southern Hemisphere begins to cool off.

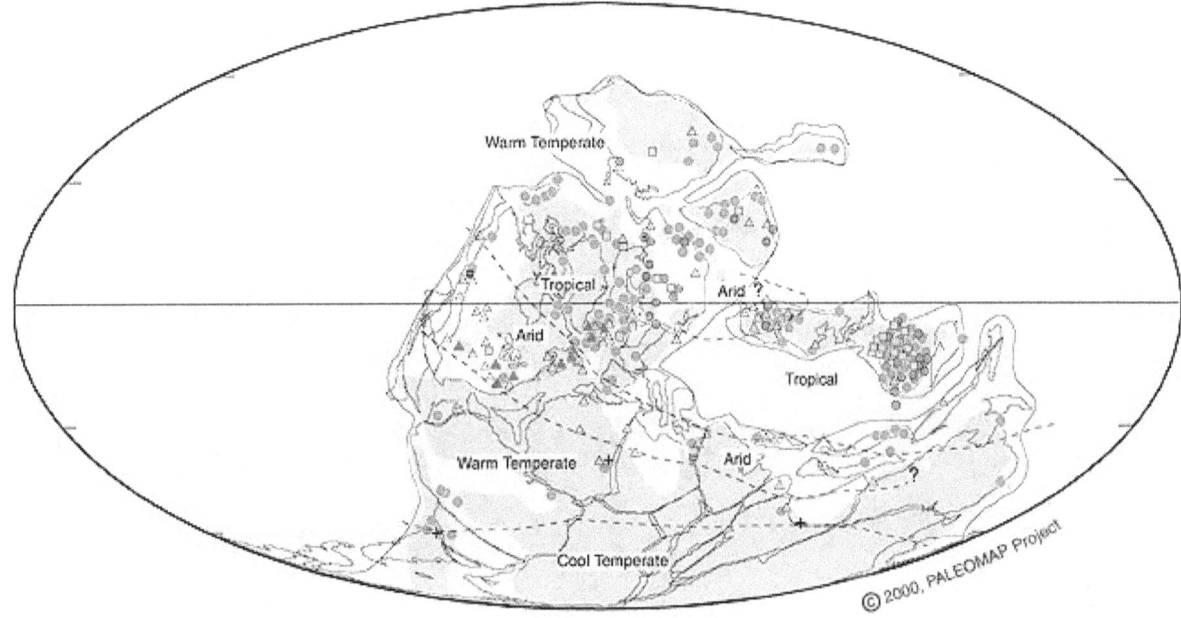

Lower Carboniferous (Tournaisian - Visean)

Early Late Carboniferous Climate (330 to 320 million years ago)

Rainforests covered the tropical regions of Pangea which was bounded to the north and south by deserts. An ice cap began to expand northward from the South Pole.

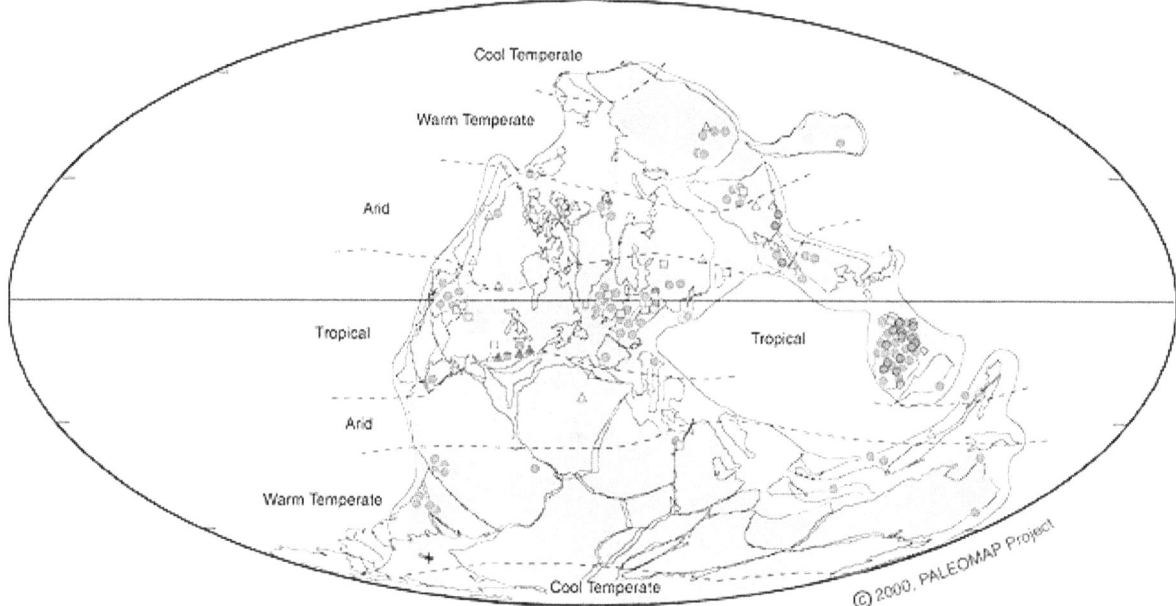

Lower Carboniferous (Serpukhovian)

Late Carboniferous Climate (320 to 305 million years ago)

Extensive rainforests covered the tropical regions of Pangea which was bounded to the north and south by deserts. An ice cap covered the South

Pole.

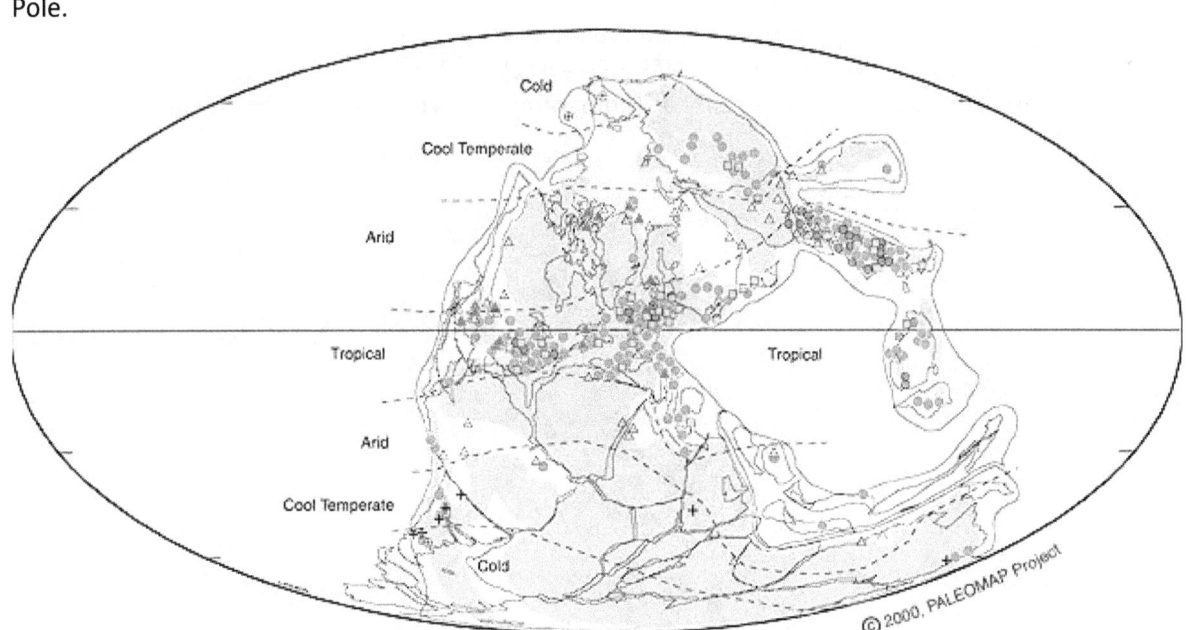

Upper Carboniferous (Bashkirian - Moscovian)

Source: http://www.scotese.com/bashclim.htm

Latest Carboniferous (Gzelian) Climate (305 to 300 million years ago)

Extensive rainforests covered the tropical regions of Pangea which was bounded to the north and south by deserts. An ice cap covered the South
Pole.

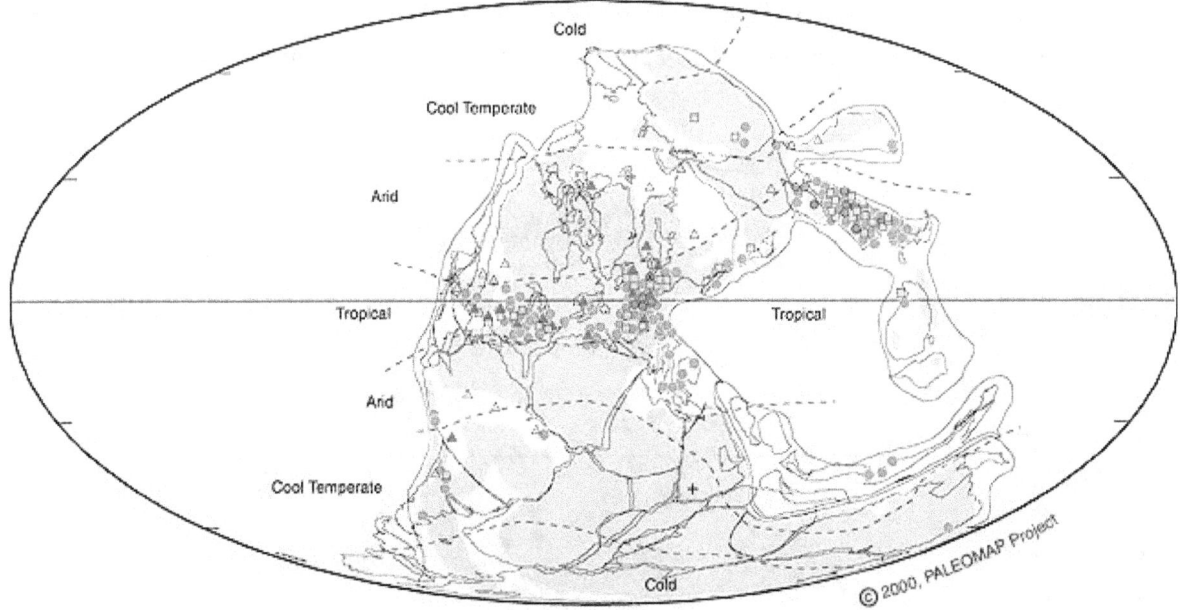

Upper Carboniferous (Gzelian)

Source: http://www.scotese.com/gzelclim.htm

Early Permian Climate (280 million years ago)

Much of the Southern Hemisphere was covered by ice as glaciers pushed northward.   Coal was produced in both Equatorial rainforests and in Temperate forests during the warmer "Interglacial" periods.

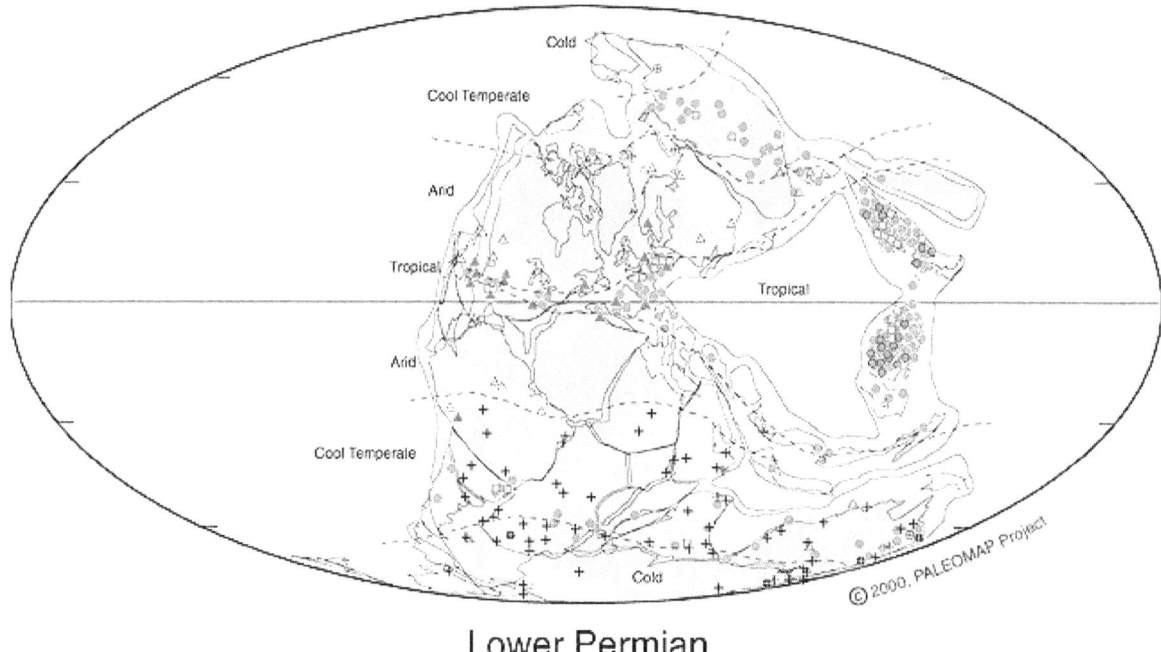

Lower Permian

Source: http://www.scotese.com/epermcli.htm

Middle & Late Permian Climate (270 to 250 million years ago)

Equatorial rainforest disappeared as deserts spread across central Pangea. Though the southern ice sheets were gone, an ice cap covered the North Pole. Rainforests covered South China as it crossed the
Equator.

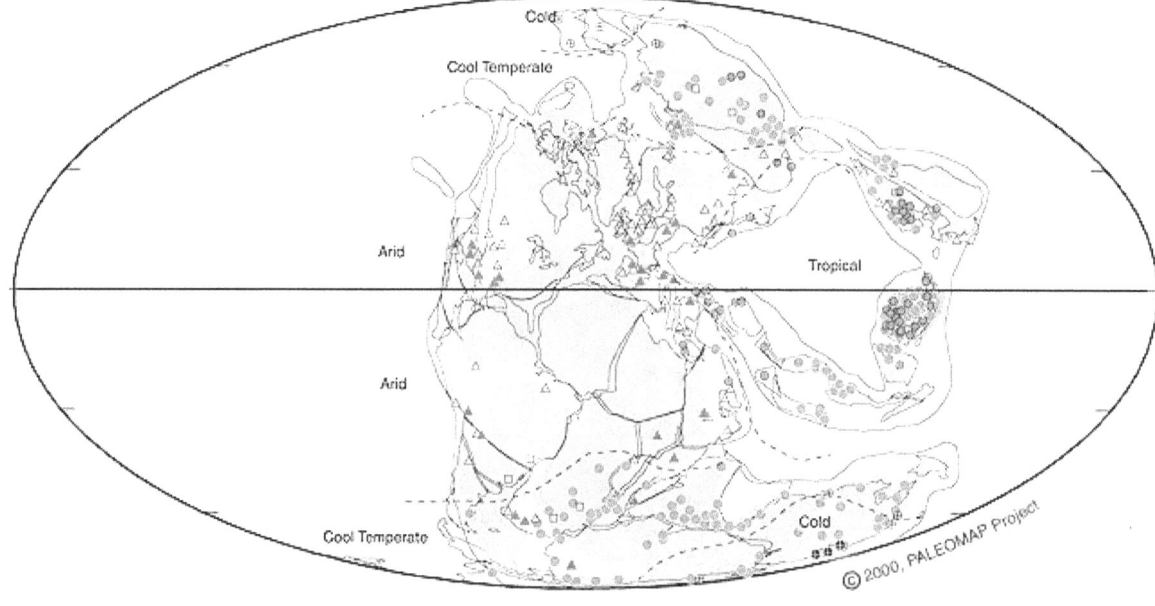

Middle & Upper Permian

Early Triassic Climate (250 million years ago)

The interior of Pangea was hot and dry during the Triassic. Warm Temperate climates extended to the Poles. This may have been one of the hottest times in Earth history. Rapid Global Warming at the very end of the Permian may have created a super - "Hot House" world that caused the great Permo-Triassic extinction. 99% of all life on Earth perished during the Permo-Triassic extinction.

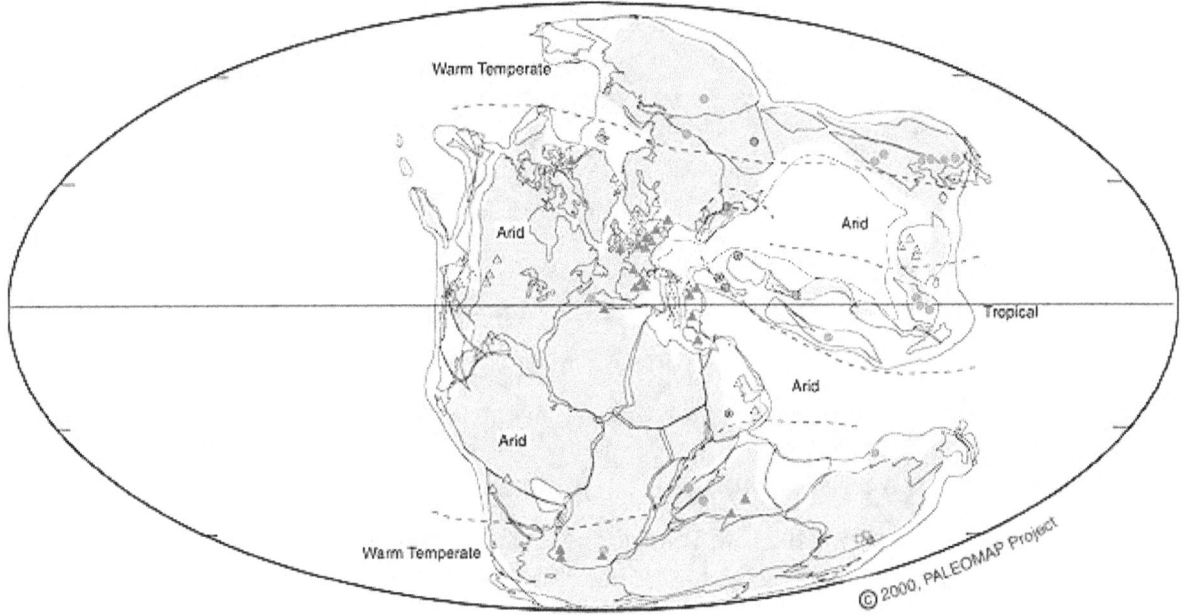

Lower Triassic

Source: http://www.scotese.com/etriascl.htm

Middle Triassic Climate (245 to 230 million years ago)

The interior of Pangea was dry during the Triassic. **Polar regions were warm, even during the**

winter.

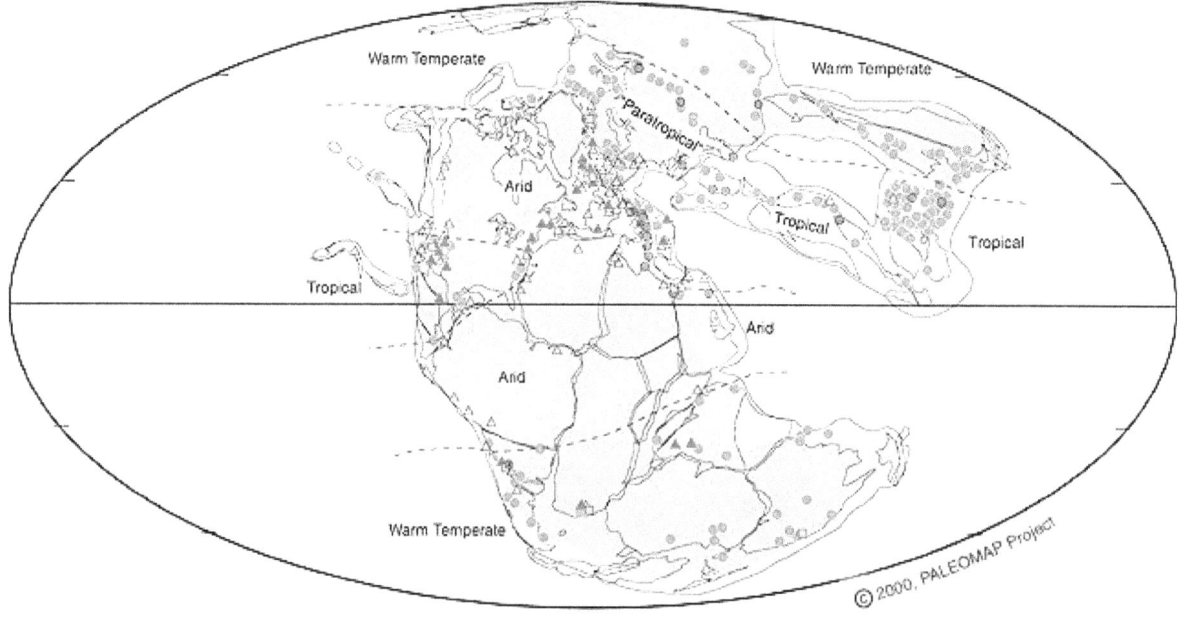

Middle Triassic

Source: http://www.scotese.com/mtriascl.htm

Late Triassic Climate (230 to 200 million years ago)

Global climate was warm during the Late Triassic. There was no ice at either North or South Poles. Warm Temperate conditions extended towards the poles.

Upper Triassic

Source: http://www.scotese.com/ltriascl.htm

Early & Middle Jurassic Climate (200 to160 million years ago)

The Pangean Mega-monsoon was in full swing during the Early and Middle Jurassic. The interior of Pangea was very arid and hot. Deserts covered what is now the Amazon and Congo rainforests. China, surrounded by moisture bearing winds was lush and verdant.

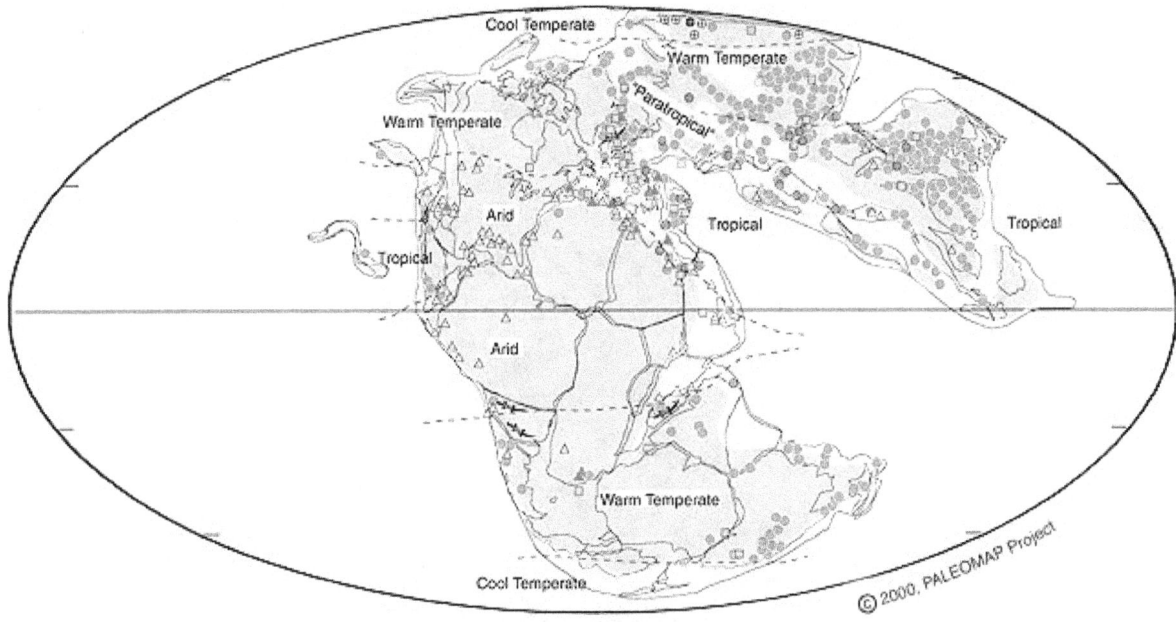

Lower Jurassic

Source: http://www.scotese.com/ejurclim.htm

Late Jurassic Climate (160 to 145 million years ago)

During the Late Jurassic the global climate began to change due to breakup of Pangea. The interior of Pangea became less dry, and seasonal snow and ice frosted the polar regions.

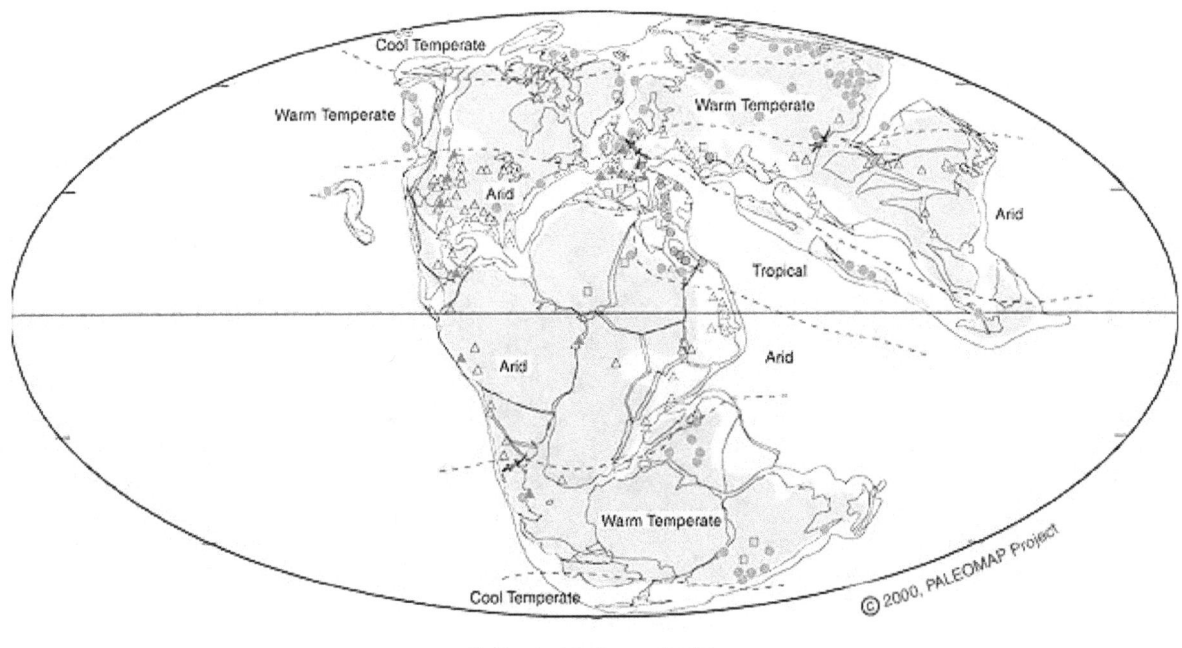

Upper Jurassic

Early Cretaceous Climate (145 million to 140 million years ago)

The Early Cretaceous was a mild "Ice House" world. There was snow and ice during the winter seasons, and Cool Temperate forests covered the polar regions.

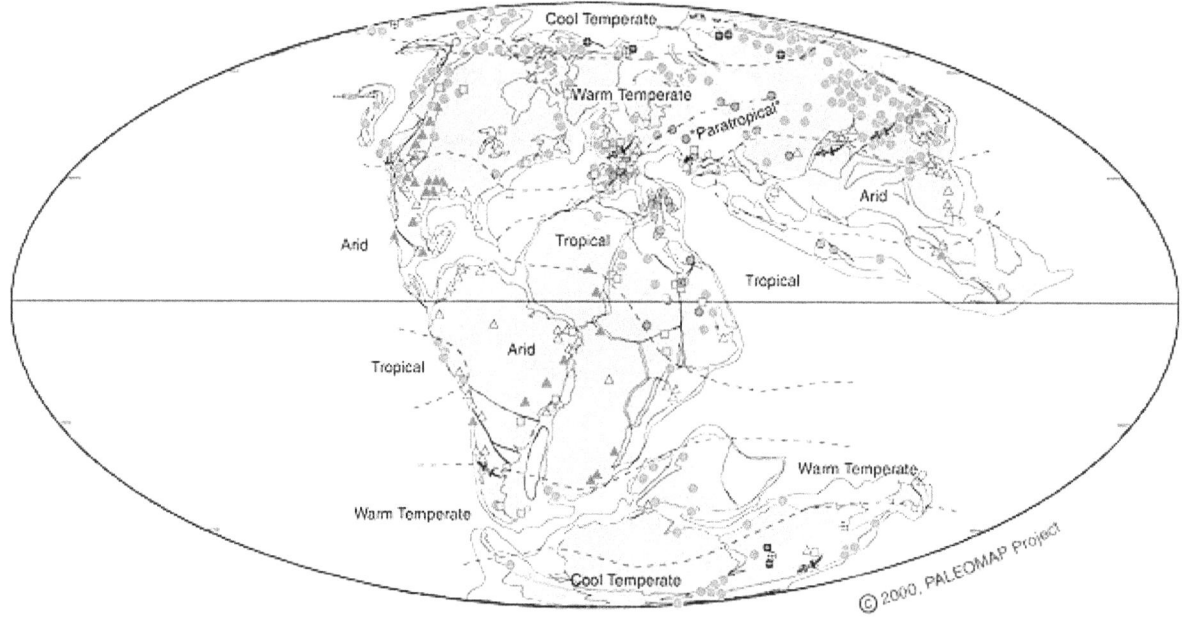

Lower Cretaceous

Late Cretaceous Climate (140 to 65 million years ago)

During the Late Cretaceous the global climate was warmer than today's climate. No ice existed at the Poles. Dinosaurs migrated between the Warm Temperate and Cool Temperate Zones as the seasons changed.

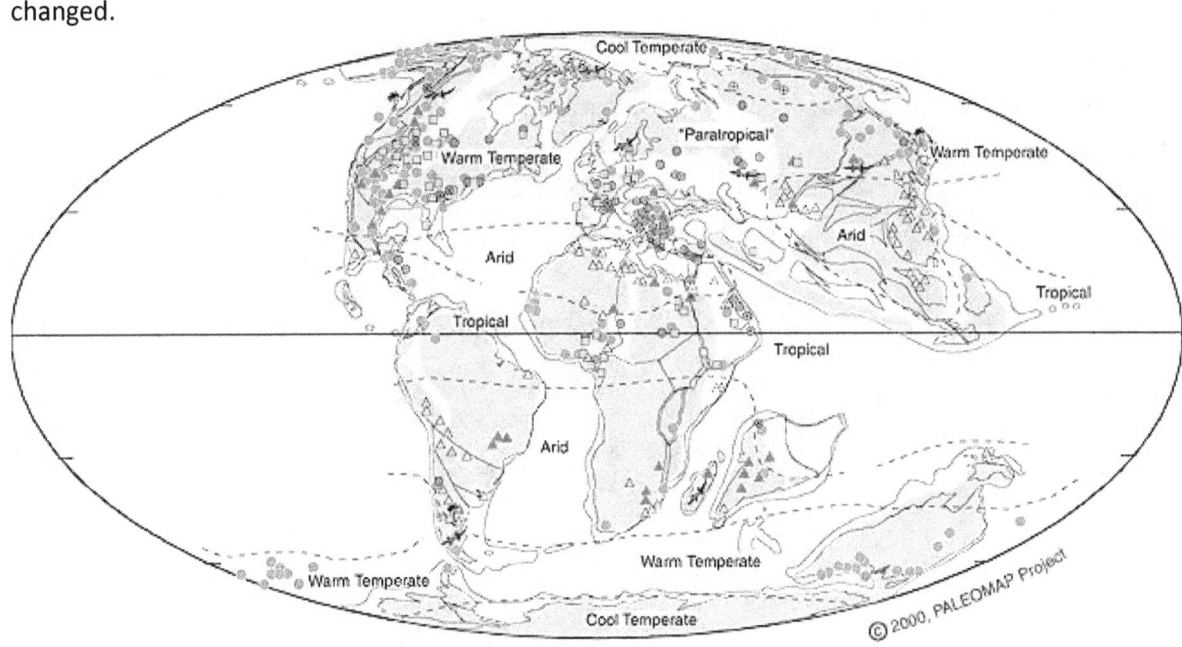

Upper Cretaceous

Source: http://www.scotese.com/lcretcli.htm

Paleocene Climate (65 to 56 million years ago)

The climate during the Paleocene was much warmer than today. Palm trees grew in Greenland and Patagonia. The Mangrove swamps of southern Australia were located at 65 degrees south latitude.

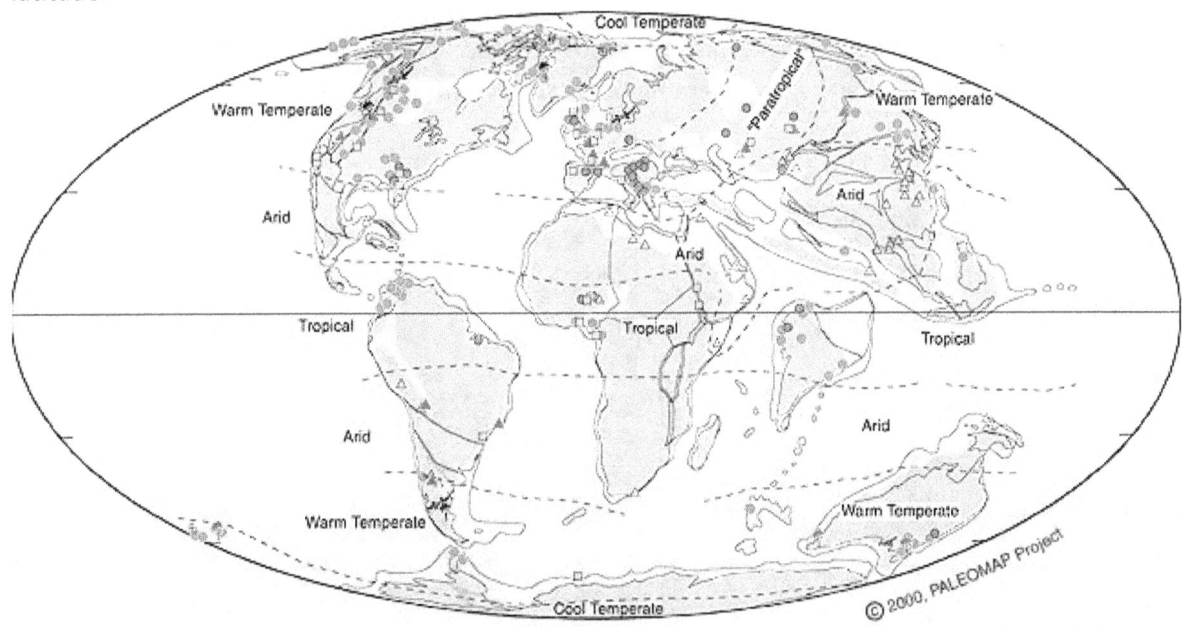

Paleocene

Source: http://www.scotese.com/paleocen.htm

Early Eocene Climate (56 to 45 million years ago)

During the Early Eocene alligators swam in swamps near the North Pole, and palm trees grew in southern Alaska. Much of central Eurasia was warm and humid.

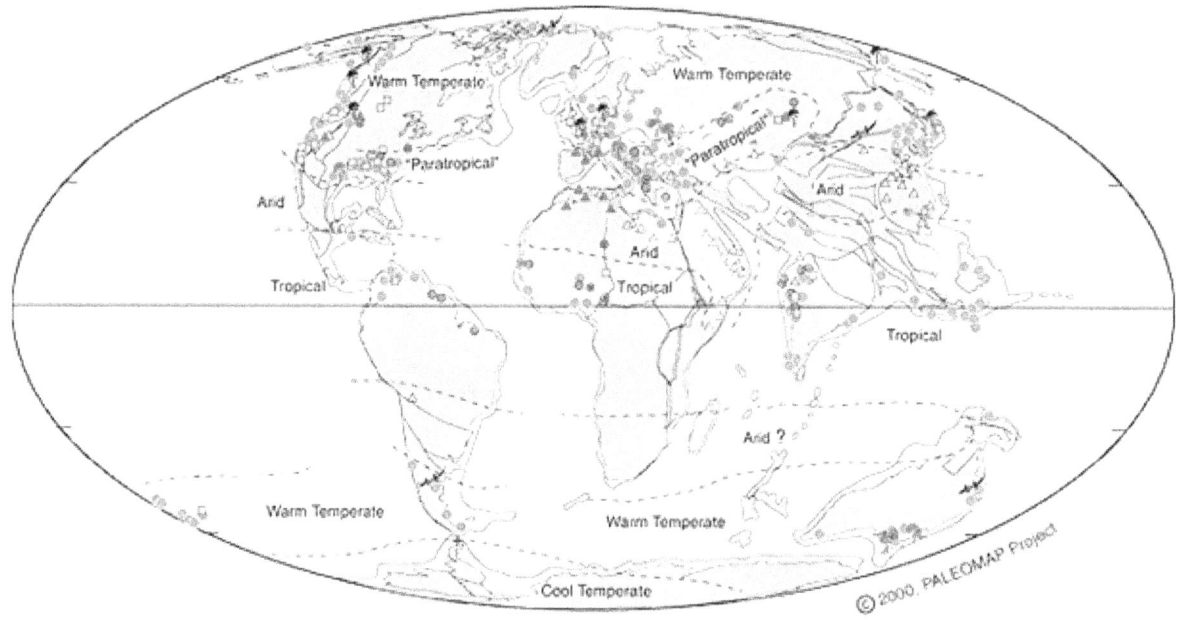

Lower Eocene

Source: http://www.scotese.com/earlyeoc1.htm

Middle & Late Eocene Climate (45 to 34 million years ago)

Global climate during the Late Eocene was warmer than today. Ice had just begun to form at the South Pole. India was covered by tropical rainforests, and Warm Temperate forests covered much of Australia.

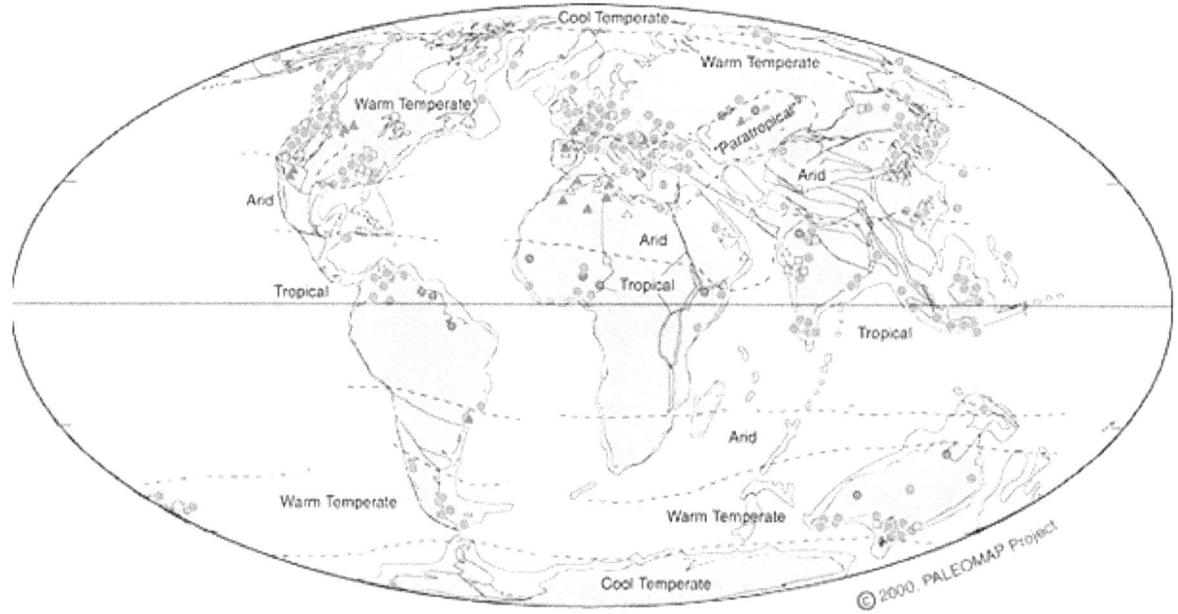

Middle & Upper Eocene

During the Oligocene, ice covered the South Pole but not the North Pole.

Warm Temperate forests covered northern Eurasia and North America.

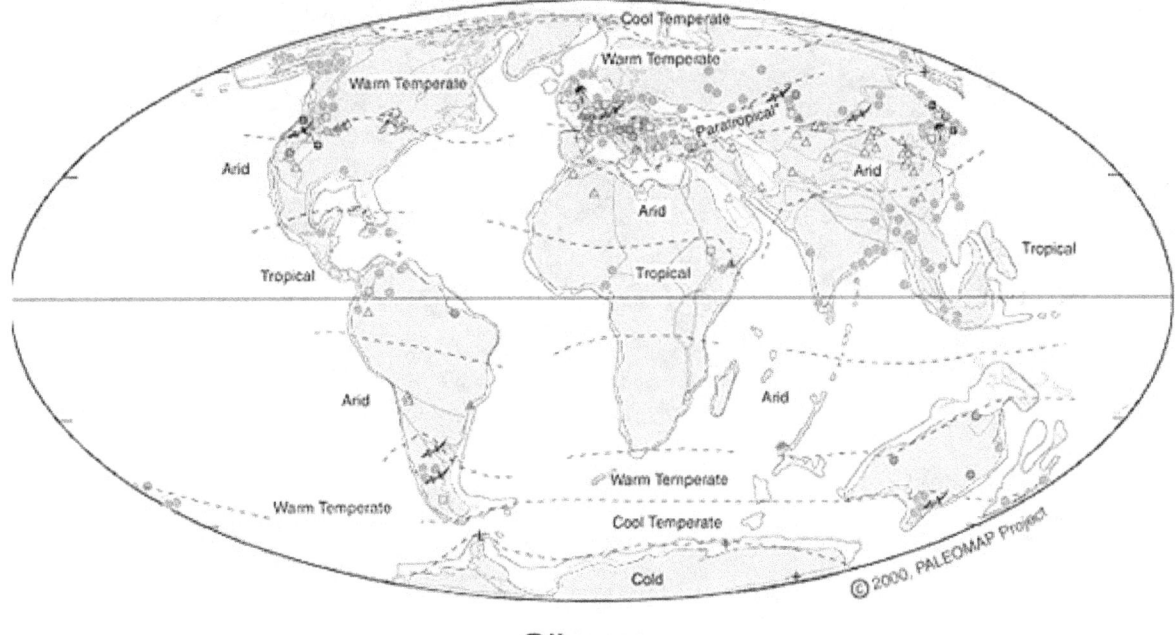

Oligocene

Miocene Climate (23 to 5 million years ago)

The climate during the Miocene was similar to today's climate, but warmer.

Well-defined climatic belts stretched from Pole to Equator, however, there were palm trees and alligators in England and Northern Europe. Australia was less arid than it is now.

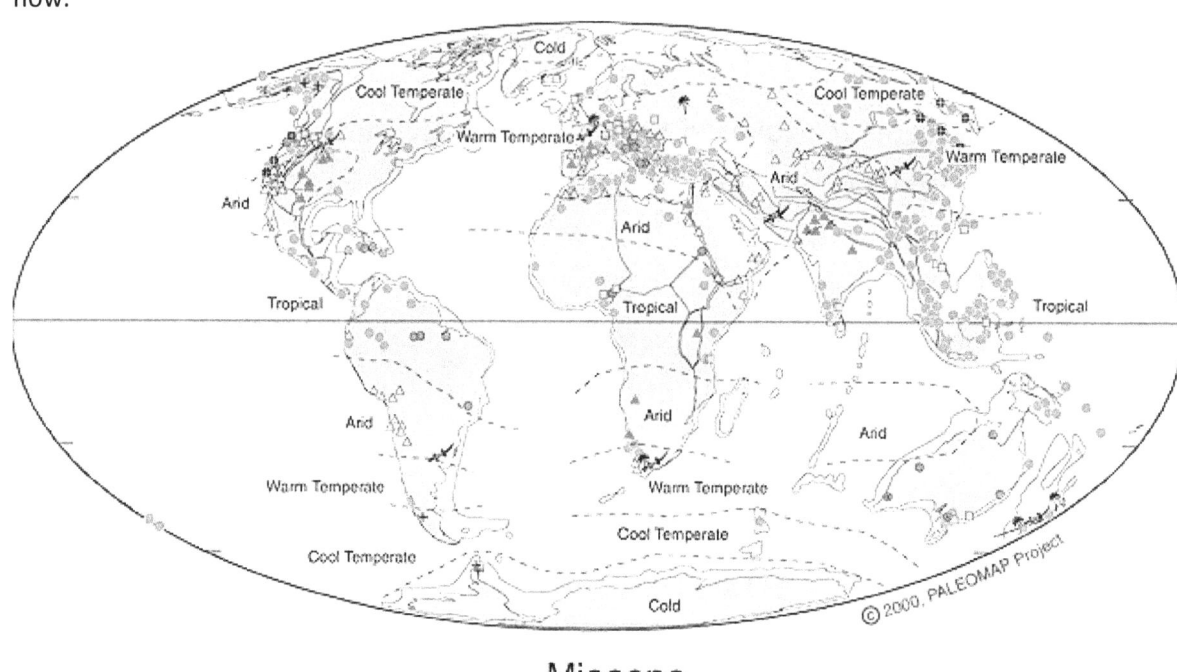

Miocene

Source: http://www.scotese.com/miocene1.htm

Previous Earth Periods: Climate versus CO2 Concentration Matrix

Now let us take some time to compare the climate of some previous periods of the Earth's history versus the CO2 concentration where readily available.

The table below shows is that the greenhouse effect CO2 concentration is only one of the factors affecting global climate, and quite clearly not the dominant factor, as is often portrayed by proponents of Anthropogenic Global Warming.

Not only does the mean CO2 concentration display a correlation with the mean surface temperature, but also the significant climatic variations within each period demonstrate clearly that climate can vary widely from warm to cool with the same and often much higher level of CO2 concentration than we have today.

This graph shows estimates of the changes in carbon dioxide concentrations during the Phanerozoic. Three estimates are based on geochemical modeling: GEOCARB III (Berner and Kothavala 2001), COPSE (Bergmann et al. 2004) and Rothman (2001). These are compared to the carbon dioxide measurement database of Royer et al. (2004) and a 30 Myr filtered average of those data. Error envelopes are shown when they were available. The right hand scale shows the ratio of these measurements to the estimated average for the last several million years (the Quaternary).

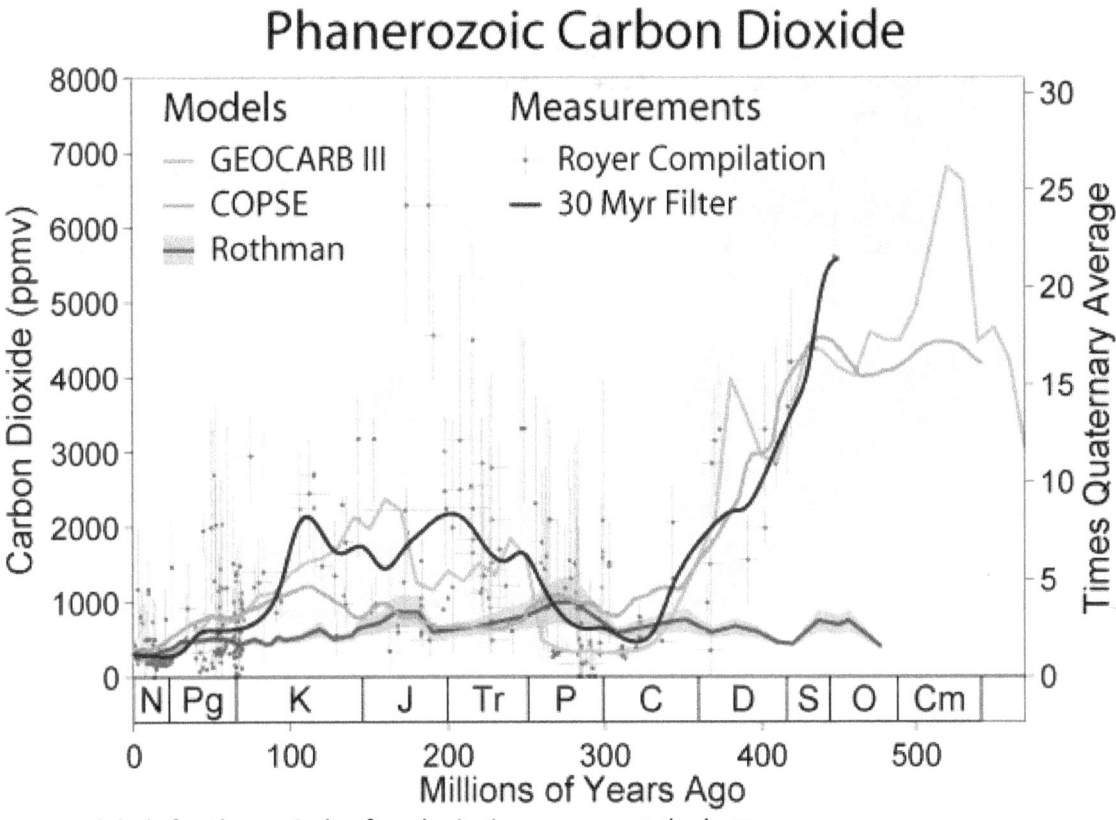

Customary labels for the periods of geologic time appear at the bottom.

Direct determination of past carbon dioxide levels relies primarily on the interpretation of carbon isotopic ratios in fossilized soils (paleosols) or the shells of phytoplankton and through interpretation of stomatal density in fossil plants. Each of these is subject to substantial systematic uncertainty.

Estimates of carbon dioxide changes through geochemical modeling instead rely on quantifying the geological sources and sinks for carbon dioxide over long time scales particularly: volcanic inputs, erosion and carbonate deposition. As such, these models are largely independent of direct measurements of carbon dioxide.

Both measurements and models show considerable uncertainty and variation; however, all point to carbon dioxide levels in the past that have been signifcantly higher than they are at present. While the GEOCARB Carbon dioxide levels in the most part of the Phanerzoic Eon shows a fit and resultíng climate sensitivity similar to todays values, the early Phanerozoic includes a global ice age during the

Ordovician age combined with high atmospheric carbon contents based on the same project. There have been different speculations about the reasons but no acknowledged mechanism so far.

Sources

- Bergman, Noam M., Timothy M. Lenton, and Andrew J. Watson (2004). "COPSE: A new model of biogeochemical cycling over Phanerozoic time". *American Journal of Science* **301**: 182-204.
- Berner, RA and Z. Kothavala (2001). "GEOCARB III: A revised model of atmospheric CO_2 over Phanerozoic time". *American Journal of Science* **304**: 397–437.
- Gradstein, FM and JG Ogg (1996). "A Phanerozoic time scale". *Episodes* **19**: 3-5.
- Gradstein, FM, JG Ogg and AG Smith (2005) *A geologic time scale 2004* Cambridge University Press ISBN 0521786738
- Rothman, Daniel H. (2001). "Atmospheric carbon dioxide levels for the last 500 million years". *Proceedings of the National Academy of Sciences* **99** (7): 4167-4171.
- Royer, Dana L., Robert A. Berner, Isabel P. Montañez, Neil J. Tabor, and David J. Beerling (2004). "CO_2 as a primary driver of Phanerozoic climate". *GSA Today* **14** (3): 4-10. doi:10.1130/1052-5173(2004)014<4:CAAPDO>2.0.CO;2
- Veizer, J., Godderis, Y. & Francois. L.M., Evidence for decoupling of atmospheric CO2 and global climate during the Phanerozoic eon. Nature 408, 698-701 (2000) Link
- Nir J. Shaviv, Ján Veizer: Celestial driver of Phanerozoic climate?, Geological Society of America Vol. 13, Issue 7 (Juliy 2003), S. 4–10, Online (pdf 454 KByte) Link

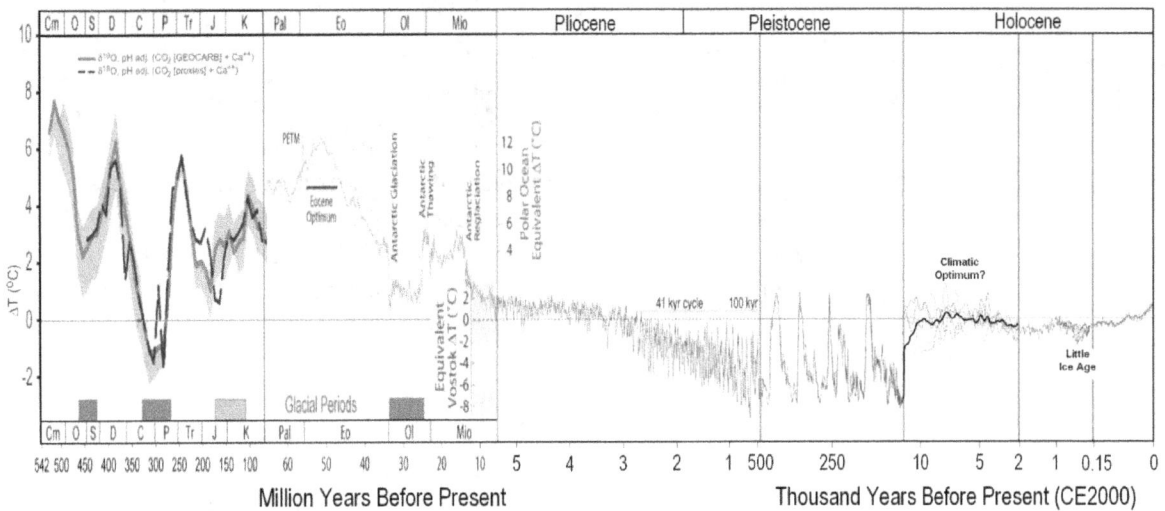

Sources

- 540 - 65 Myr BP : Royer, Dana L. and Robert A. Berner, Isabel P. Montañez, Neil J. Tabor, David J. Beerling (2004) CO_2 as a primary driver of Phanerozoic climate *GSA Today* July 2004, volume 14, number 3, pages 4-10, doi:10.1130/1052-5173(2004)014<4:CAAPDO>2.0.CO;2

- 65 - 5.5 Myr BP : Zachos, James, Mark Pagani, Lisa Sloan, Ellen Thomas, and Katharina Billups (2001). "Trends, Rhythms, and Aberrations in Global Climate 65 Ma to Present". *Science* **292** (5517): 686–693. doi:10.1126/science.1059412

- 5.5 Myr - 420 kyr BP : Lisiecki, L. E., and M. E. Raymo (2005), A Pliocene-Pleistocene stack of 57 globally distributed benthic δ18O records, *Paleoceanography*, 20, PA1003, doi:10.1029/2004PA001071. Link

- 420 kyr - 12 kyr BP : Petit J.R., Jouzel J., Raynaud D., Barkov N.I., Barnola J.M., Basile I., Bender M., Chappellaz J., Davis J., Delaygue G., Delmotte M., Kotlyakov V.M., Legrand M., Lipenkov V., Lorius C., Pépin L., Ritz C., Saltzman E., Stievenard M. (1999) Climate and Atmospheric History of the Past 420,000 years from the Vostok Ice Core, Antarctica, *Nature*, 399, pp.429-436, doi:10.1038/20859

- 12 kyr – Today: see later sections of this chapter

Period	Mean CO2 Concentration	Mean Surface Temperature	How long ago?	Climate	Comments
Ordovician	4200 parts per million (10.5 times that of today)	16 °C (2 degrees above current)	480 to 420 million years ago	The Ordovician saw the highest sea levels of the Paleozoic, and the low relief of the continents led to many shelf deposits being formed under hundreds of meters of water. Sea level rose more or less continuously throughout the Early Ordovician, leveling off somewhat during the middle of the period. Locally, some regressions occurred, but sea level rise continued in the beginning of the Late Ordovician. A change was soon on the cards, however, and sea levels fell steadily in accord with the cooling temperatures for the ~30 million years leading up to the Hirnantian glaciation. Within this icy stage, sea level seems to have risen and dropped somewhat, but despite much study the details remain unresolved.[43][44] At the beginning of the period, around 480 million years ago, the climate was very hot due to high levels of CO2, which gave a strong greenhouse effect. The marine waters are assumed to have been around 45°C, which restricted the diversification of complex multi-cellular organisms. But over time, the climate become cooler, and around 460 million years ago, the ocean temperatures became comparable to those of present day equatorial waters.	The climate cooled into an Ice Age despite CO2 concentration averaging 4,200 parts per million for the entire period.

[43] Munnecke, A.; Calner, M.; Harper, D. A. T.; Servais, T. (2010). "Ordovician and Silurian sea-water chemistry, sea level, and climate: A synopsis". *Palaeogeography, Palaeoclimatology, Palaeoecology* **296** (3–4): 389–413 doi:10.1016/j.palaeo.2010.08.001.

[44] Explosion in marine biodiversity explained by climate change

Period	Mean CO2 Concent-ration	Mean Surface Tempera-ture	How long ago?	Climate	Comments
				As with North America and Europe, Gondwana was largely covered with shallow seas during the Ordovician. Shallow clear waters over continental shelves encouraged the growth of organisms that deposit calcium carbonates in their shells and hard parts. The Panthalassic Ocean covered much of the northern hemisphere, and other minor oceans included Proto-Tethys, Paleo-Tethys, Khanty Ocean, which was closed off by the Late Ordovician, Iapetus Ocean, and the new Rheic Ocean. As the Ordovician progressed, we see evidence of glaciers on the land we now know as Africa and South America. At the time these land masses were sitting at the South Pole, and covered by ice caps.	
Carboniferous	800 parts per million (twice that of today)	14 °C (same as modern level)	360 to 330 million years ago	The early part of the Carboniferous was mostly warm; in the later part of the Carboniferous, the climate cooled. Glaciations in Gondwana, triggered by Gondwana's southward movement, continued into the Permian and because of the lack of clear markers and breaks, the deposits of this glacial period are often referred to as Permo-Carboniferous in age. The cooling and drying of the climate led to the Carboniferous Rainforest Collapse (CRC). Tropical rainforests fragmented and then were eventually devastated by climate change. [45]	

[45] Sahney, S., Benton, M.J. & Falcon-Lang, H.J. (2010). "Rainforest collapse triggered Pennsylvanian tetrapod diversification in Euramerica" (PDF). *Geology* **38** (12): 1079–1082. doi:10.1130/G31182.1. http://geology.geoscienceworld.org/cgi/content/abstract/38/12/1079.

Period	Mean CO2 Concentration	Mean Surface Temperature	How long ago?	Climate	Comments
Permian	900 parts per million	16 °C	280 to 250 million years ago	The climate in the Permian was quite varied. At the start of the Permian, the Earth was still at the grip of an Ice Age from the Carboniferous. Glaciers receded around the mid-Permian period as the climate gradually warmed, drying the continent's interiors. In the late Permian period, the drying continued although the temperature cycled between warm and cool cycles[46]	
Triassic	1750 parts per million (4.5 times that of today)	17 °C	250 to 200 million years ago	The Triassic climate was generally hot and dry, forming typical red bed sandstones and evaporites. There is no evidence of glaciation at or near either pole; in fact, the polar regions were apparently moist and temperate, a climate suitable for reptile-like creatures. Pangaea's large size limited the moderating effect of the global ocean; its continental climate was highly seasonal, with very hot summers and cold winters. It probably had strong, cross-equatorial monsoons.[47]	
Jurassic	1950 parts per million (5 times that of today)	16.5°C (lower than Triassic period)	200 to 145 million years ago	*Early & Middle:* "The Pangean Mega-monsoon was in full swing during the Early and Middle Jurassic. The interior of Pangea was very arid and hot. Deserts covered what is now the Amazon and Congo rainforests. China, surrounded by moisture bearing winds was lush and verdant. Late: "During the Late Jurassic the global climate began to change due to breakup of Pangea. The interior of Pangea became less dry, and seasonal snow and ice frosted the polar regions."	Climate summary is from www.scotese.com as already described earlier

---

[46] http://www.palaeos.com/Paleozoic/Permian/Permian.htm
[47] Stanley, 452-3

Period	Mean CO2 Concentration	Mean Surface Temperature	How long ago?	Climate	Comments
Cretaceous	1700 parts per million (4.5 times that of today)	18°C (higher than Jurassic period even if CO2 conc. was 250 ppm lower)	145 to 65 million years ago	The Berriasian epoch showed a cooling trend that had been seen in the last epoch of the Jurassic. There is evidence that snowfalls were common in the higher latitudes and the tropics became wetter than during the Triassic and Jurassic. Glaciation was however restricted to alpine glaciers on some high-latitude mountains, though seasonal snow may have existed farther south. Rafting by ice of stones into marine environments occurred during much of the Cretaceous but evidence of deposition directly from glaciers is limited to the Early Cretaceous of the Eromanga Basin in southern Australia.[48] After the end of the Berriasian, however, temperatures increased again, and these conditions were almost constant until the end of the period. This trend was due to intense volcanic activity which produced large quantities of carbon dioxide.[49] The production of large quantities of magma, variously attributed to mantle plumes or to extensional tectonics, further pushed sea levels up, so that large areas of the continental crust were covered with shallow seas. The Tethys Sea connecting the tropical oceans east to west also helped in warming the global climate. Warm-adapted plant fossils are known from localities as far north as Alaska and Greenland, while dinosaur fossils	

---

[48] The Berriasian Age

[49] Foulger, G.R. (2010). *Plates vs. Plumes: A Geological Controversy*. Wiley-Blackwell. ISBN 978-1-4051-6148-0. http://www.wiley.com/WileyCDA/WileyTitle/productCd-1405161485.html

Period	Mean CO2 Concentration	Mean Surface Temperature	How long ago?	Climate	Comments
				have been found within 15 degrees of the Cretaceous south pole.[50] A very gentle temperature gradient from the equator to the poles meant weaker global winds, contributing to less upwelling and more stagnant oceans than today. This is evidenced by widespread black shale deposition and frequent anoxic events.[51] Sediment cores show that tropical sea surface temperatures may have briefly been as warm as 42 °C (107 °F), 17 °C (31 °F) warmer than at present, and that they averaged around 37 °C (99 °F). Meanwhile deep ocean temperatures were as much as 15 to 20 °C (27 to 36 °F) higher than today's.[52] [53]	
Paleocene	Not readily available	Not readily available	65 to 56 million years ago	The early Paleocene was cooler and dryer than the preceding Cretaceous, though temperatures rose sharply during the Paleocene–Eocene Thermal Maximum. The climate became warm and humid worldwide towards the Eocene boundary, with subtropical vegetation growing in Greenland and Patagonia, crocodiles swimming off the coast of Greenland, and early primates evolving in tropical palm forests of northern Wyoming. The Earth's poles were cool and temperate; North America, Europe, Australia and southern South America were warm and temperate; equatorial	

[50] Stanley, pp. 480–2
[51] Stanley, pp. 481–2
[52] "Warmer than a Hot Tub: Atlantic Ocean Temperatures Much Higher in the Past" PhysOrg.com.
[53] Skinner, Brian J., and Stephen C. Porter. *The Dynamic Earth: An Introduction to Physical Geology.* 3rd ed. New York: John Wiley & Sons, Inc., 1995. ISBN 0-471-59549-7. p. 557

Period	Mean CO2 Concentration	Mean Surface Temperature	How long ago?	Climate	Comments
				areas had tropical climates; and north and south of the equatorial areas, climates were hot and arid. [54] [55]	
Eocene	Not readily available See note at the end of the table	Not readily available	56 to 34 million years go	One of the unique features of the Eocene's climate as mentioned before was the equable and homogeneous climate that existed in the early parts of the Eocene. The Eocene Epoch contained a wide variety of different climate conditions that includes the warmest climate in the Cenozoic Era and ends in an icehouse climate. The evolution of the Eocene climate began with warming after the end of the Palaeocene-Eocene Thermal Maximum (PETM) at 56 million years ago to a maximum during the Eocene Optimum at around 49 million years ago. During this period of time, little to no ice was present on Earth with a smaller difference in temperature from the equator to the poles. Following the maximum, was a descent into an icehouse climate from the Eocene Optimum to the Eocene-Oligocene transition at 34 million years ago. During this decrease ice began to reappear at the poles, and the Eocene-Oligocene transition is the period of time where the Antarctic ice sheet began to rapidly expand. A multitude of proxies support the presence of a warmer equable climate being present during this period of time. A few of these proxies include the	

---

[54] Science Notes 2003:
[55] PaleoMap Project: Paleocene Climate

Period	Mean CO2 Concentration	Mean Surface Temperature	How long ago?	Climate	Comments
				presence of fossils native to warm climates, such as crocodiles, located in the higher latitudes, the presence in the high-latitudes of frost-intolerant flora such as palm trees which cannot survive during sustained freezes, and fossils of snakes found in the tropics that would require much higher average temperatures to sustain them. Using isotope proxies to determine ocean temperatures indicate sea surface temperatures in the tropics as high as 35 °C (95 °F) and bottom water temperatures that are 10 °C (18 °F) higher than present day values. With these bottom water temperatures, temperatures in areas where deep-water forms near the poles are unable to be much cooler than the bottom water temperatures.[56] [57] [58] **An issue arises, however, when trying to model the Eocene and reproduce the results that are found with the proxy data. Using all different ranges of greenhouse gases that occurred during the early Eocene, models were unable to produce the warming that was found at the poles and the reduced seasonality that occurs with winters at the poles being substantially warmer.** [59]	

[56] Sloan, L. C., and D. K. Rea, 1995: Atmospheric carbon dioxide and early Eocene climate: a general circulation modeling sensitivity study. *Paleogeo. Paleoclim. Paleoeco.* 119, 275-292

[57] Huber, M., 2009: Snakes tell a torrid tale. *Nature*, 457, 669-671

[58] Huber, M., and R. Caballero, 2011: The early Eocene equable climate problem revisited. *Clim. Past Discuss*. 6, 241-304

[59] Sloan, L. C., and E. J. Barron, 1990: "Equable" climates during Earth history? *Geology*, 18, 489-492

Period	Mean CO2 Concentration	Mean Surface Temperature	How long ago?	Climate	Comments
				The models, while accurately predicting the tropics, tend to produce significantly cooler temperatures of up to 20 °C (36 °F) underneath the actual determined temperature at the poles. This error has been classified as the "equable climate problem". To solve this problem, the solution would involve finding a process to warm the poles without warming the tropics.	
Oligocene	Not readily available See note at the end of the table	Not readily available	34 to 23 million years ago	The Paleogene Period general temperature decline is interrupted by an Oligocene 7 million year stepwise climate change. A deeper 8.2 °C, 400,000 year temperature depression leads the 2 °C, 7 million year stepwise climate change 33.5 Ma (Million years ago). The stepwise climate change began 25.5Ma and lasted through 32.5 Ma, as depicted in the PaleoTemps chart. The Oligocene climate change was a global increase in ice volume and a 55 M (181 feet) decrease in sea level (35.7-33.5 Ma) with a closely related (25.5-32.5 Ma) temperature depression. The 7 million year depression abruptly terminated within 1-2 million years of the La Garita Caldera eruption at 28-26 Ma. A deep 400,000 year glaciated Oligocene Miocene boundary event is recorded at McMurdo Sound and King George Island.[60] [61] [6	

---

[60] A.Zanazzi (et al.) 2007 'Large Temperature Drop across the Eocene Oligocene in central North America' Nature, Vol. 445, 8 February 2007

Period	Mean CO2 Concentration	Mean Surface Temperature	How long ago?	Climate	Comments
Miocene	Not readily available	Not readily available	23 to 5 million years ago	The earth went from the Oligocene Epoch through the Miocene and into the Pliocene as it cooled into a series of Ice Ages. The Miocene boundaries are not marked by a single distinct global event but consist rather of regional boundaries between the warmer Oligocene and the cooler Pliocene. [64] [65]	
Today	390 parts per million	14 °C			

[61] C.R.Riesselman (et al.) 2007 'High Resolution stable isotope and carbonate variability during the early Oligocene climate transition: Walvis Ridge (ODP Site 1263) USGS OF-2007-1047

[62] Lorraine E. Lisiecki Nov 2004; *A Pliocene-Pleistocene stack of 57 globally distributed benthic δ¹⁸O records* Brown University, PALEOCEANOGRAPHY, VOL. 20

[63] Kenneth G. Miller Jan-Feb 2006; *Eocene–Oligocene global climate and sea-level changes St. Stephens Quarry, Alabama* GSA Bulletin, Rutgers University, NJ, Link

[64] Robert A. Rohde (2005). "GeoWhen Database". http://www.stratigraphy.org/bak/geowhen/stages/Miocene.html

[65] Susanne S. Renner (2011). "Living fossil younger than thought". *Science* **334** (6057): 766–767. doi:10.1126/science.1214649. PMID 22076366

Tim Patterson, Pubs paleoclimatologist and Professor of Geology at Carleton University in Canada:

"There is no meaningful correlation between CO2 levels and Earth's temperature over this [geologic] time frame. In fact, when CO2 levels were over ten times higher than they are now, about 450 million years ago, the planet was in the depths of the absolute coldest period in the last half billion years. **On the basis of this evidence, how could anyone still believe that the recent relatively small increase in CO2 levels would be the major cause of the past century's modest warming?"**

R. Timothy Patterson is a professor of geology at Carleton University, where he is Director of the Ottawa-Carleton Geoscience Centre in Ottawa, Ontario, Canada. He is also a Senior Visiting Fellow in the School of Geography, Queen's University of Belfast, Northern Ireland. He holds a B.Sc. in Biology, B.A. in Geology, both from Dalhousie University, Halifax, Nova Scotia and a Ph.D. in geology from the University of California at Los Angeles (UCLA).

Patterson serves as Canadian leader of UNESCO International Geological Correlation Program (IGCP) Project 495 "Quaternary Land-Ocean interactions", which is mandated to study the record of sea level change past and future and has been Principal Investigator of large Natural Sciences and Engineering Research Council of Canada (NSERC) and Canadian Foundation for Climate and Atmospheric Sciences (CFCAS) projects, examining high-resolution climate records from marine basins off the west coast of Canada.

He was a founding editor of the journal Palaeontologia Electronica (Executive Editor, 1998–2000), is presently Associate Editor for the Journal of Foraminiferal Research and is past associate editor of the journal Micropaleontology.

Sources
- http://http-server.carleton.ca/~tpatters/
- http://timpaterson-brown.timpaterson-brown.com/
- Patterson, R.T., Prokoph, A., Reinhardt, E., and Roe, H., 2007. "Climate cyclicity in anoxic marine sediments from the Seymour-Belize Inlet Complex, British Columbia". Marine Geology.
- Patterson, R.T., Dalby, A.P., Roe, H.M., Guilbault, J.-P., Hutchinson, I., and Clague, J.J. 2005. "Relative utility of foraminifera, diatoms and macrophytes as high resolution indicators of paleo-sea level". Quaternary Science Reviews, v. 24, p. 2002-2014.
- Chang, A.S., and Patterson, R.T. 2005. "Climate shift at 4400 years BP: Evidence from high-resolution diatom stratigraphy, Effingham Inlet, British Columbia, Canada". Palaeogeography, Palaeclimatology, Palaeoecology. v. 226, 72-92.
- Patterson, R.T., Prokoph, A., and Chang, A.S. 2004. "Late Holocene sedimentary response to solar and cosmic ray activity influenced climate variability in the NE Pacific". Sedimentary Geology. 172, p. 67-84.
- Prokoph, A., and Patterson, R.T. 2004. "Application of wavelet and discontinuity analysis to trace temperature changes: Eastern Ontario as a case study. Atmosphere Ocean". v. 42, p. 201-212.
- Patterson, R.T., Fowler, A.D., and Huber, B., 2004. "Evidence of Hierarchical Organization in the Planktic Foraminiferal Evolutionary Record". Journal of Foraminiferal Research, v. 34 (2), p. 85-95.
- Patterson, R.T., Prokoph A.,Wright, C., Chang, A.S., Thomson, R.E., and Ware, D.M., 2004. "Holocene Solar Variability and Pelagic Fish Productivity in the NE Pacific". Palaeontologia Electronica, v. 6 (1). 17 pp.
- Gehrels, W.R., Milne, G.A., Jason R. Kirby, J.R., Patterson, R.T., and Belknap, D.F., 2004. "Late Holocene sea-level changes and isostatic crustal movements in Atlantic Canada. Quaternary International". v. 120, p. 79-89.
- http://www.canadafreepress.com/2006/harris061206.htm

Note on CO2 concentration during the Eocene and Oligocene periods

Science 22 July 2005, volume 309 pp. 600-603.

The relation between the partial pressure of atmospheric carbon dioxide (pCO2) and Paleogene climate is poorly resolved. We used stable carbon isotopic values of di-unsaturated alkenones extracted from deep sea cores to reconstruct pCO2 fromthe middle Eocene to the late Oligocene (¡-45 to 25 million years ago). Our results demonstrate that pCO2 ranged between 1000 to 1500 parts per million by volume in the middle to late Eocene, then decreased in several steps during the Oligocene, and reached modern levels by the latest Oligocene. The fall in pCO2 likely allowed for a critical expansion of ice sheets on Antarctica and promoted conditions that forced the onset of terrestrial C4 photosynthesis.

Additional note on the Eocene Period [66] [67] [68] [69]

Greenhouse gases, in particular carbon dioxide and methane, played a significant role during the Eocene in controlling the surface temperature. The end of the PETM was met with a very large sequestration of carbon dioxide in the form of methane clathrate, coal, and crude oil at the bottom of the Arctic Ocean, that reduced the atmospheric carbon dioxide. This event was similar in magnitude to the massive release of greenhouse gases at the beginning of the PETM, and it is hypothesized that the sequestration was mainly due to organic carbon burial and weathering of silicates. For the early Eocene there is much discussion on how much carbon dioxide is in the atmosphere. This is due to numerous proxies representing different atmospheric carbon dioxide content. For example, diverse geochemical and paleontological proxies indicate that at the maximum of global warmth the atmospheric carbon dioxide values were at 700 – 900 ppm while other proxies such as pedogenic (soil building) carbonate and marine boron isotopes indicate large changes of carbon dioxide of over 2,000 ppm over periods of time of less than 1 million years. Sources for this large influx of carbon dioxide could be attributed to volcanic out-gassing due to North Atlantic rifting or oxidation of methane stored in large reservoirs deposited from the PETM event in the sea floor or wetland environments. For contrast, today the carbon dioxide levels are at 390 ppm or .039%.

During the early Eocene, methane was another greenhouse gas that had a drastic effect on the climate. In comparison to carbon dioxide, methane has much higher consequences with regards to temperature emission as methane has 25 times more emission than carbon dioxide. The majority of the methane released to the atmosphere during this period of time would have been from wetlands, swamps, and forests. The atmospheric methane concentration today is 0.000179% or 1.79 ppmv.

---

[66] Bowen, J. G., and J. C. Zachos, 2010: Rapid carbon sequestration at the termination of the Palaeocene-Eocene Thermal Maximum. *Nature Geoscience*, **3**, 866-869.

[67] Pearson. P. N., and M. R. Palmer, 2000: Atmospheric carbon dioxide concentrations over the past 60 million years. *Nature*, **406**, 695-699.

[68] Royer. D. L. and Coauthors, 2001: Paleobotanical Evidence for Near Present-Day Levels of Atmospheric CO2 During Part of the Tertiary. *Science*, **292**, 2310-2313.

[69] Sloan, L. C., Walker, C. G., Moore Jr, T. C., Rea, D. K., and J. C. Zachos, 1992: Possible methane-induced polar warming in the early Eocene. *Nature*, **357**, 1129-1131.

Due to the warmer climate and sea level rise associated with the early Eocene, more wetlands, more forests, and more coal deposits would be available for methane release. Comparing the early Eocene production of methane to current levels of atmospheric methane, the early Eocene would be able to produce triple the amount of current methane production. The warm temperatures during the early Eocene could have increased methane production rates, and methane that is released into the atmosphere would in turn warm the troposphere, cool the stratosphere, and produce water vapor and carbon dioxide through oxidation. Biogenic production of methane produces carbon dioxide and water vapor along with the methane, as well as yielding infrared radiation. The breakdown of methane in an oxygen atmosphere produces carbon monoxide, water vapor and infrared radiation. The carbon monoxide is not stable so it eventually becomes carbon dioxide and in doing so releases yet more infrared radiation. Water vapor traps more infrared than does carbon dioxide.

As mentioned earlier, an issue arises, however, when trying to model the Eocene and reproduce the results that are found with the proxy data. Using all different ranges of greenhouse gases that occurred during the early Eocene, models were unable to produce the warming that was found at the poles and the reduced seasonality that occurs with winters at the poles being substantially warmer.

700,000 years ago to today

Let us now look at the temperature variations of the recent past, starting with 700,000 years ago. The temperature variation is best illustrated by the quantity of ice volume.

http://www.geocraft.com/WVFossils/ice_ages.html

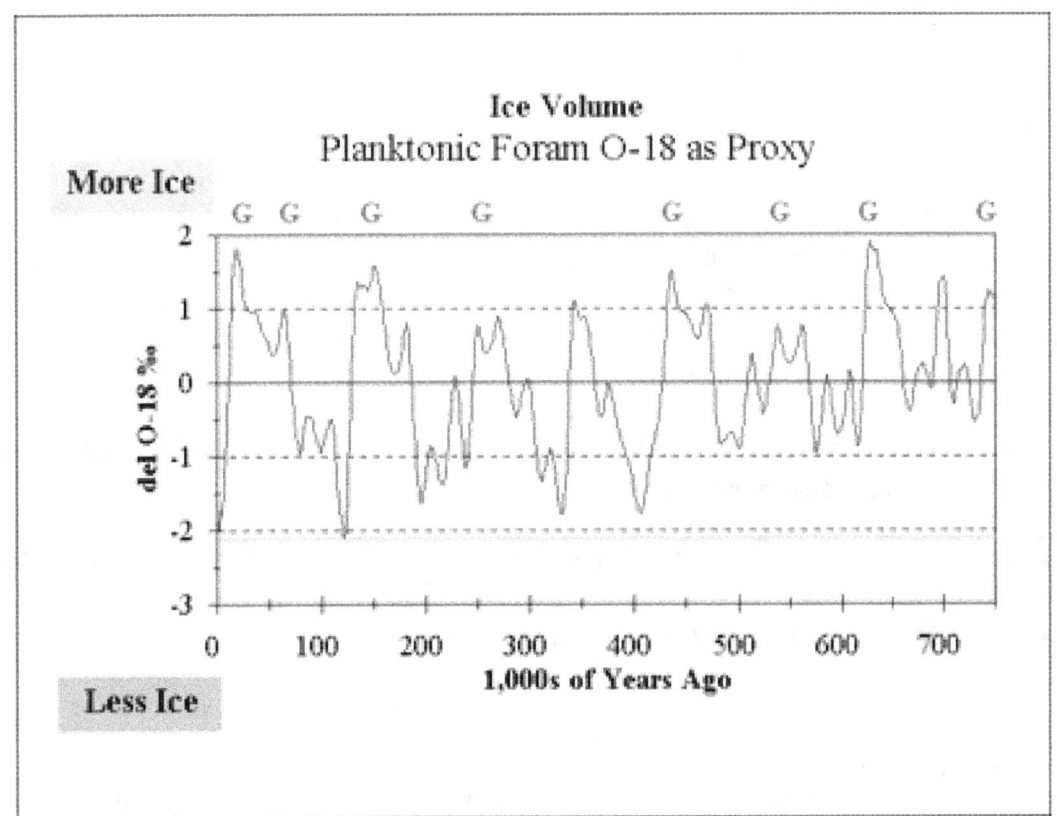

Observations

Earth's climate and the biosphere have been in constant flux, dominated by ice ages and glaciers for the past several million years. We are currently enjoying a temporary reprieve from the deep freeze.

Approximately every 100,000 years Earth's climate warms up temporarily. These warm periods, called interglacial periods, appear to last approximately 15,000 to 20,000 years before regressing back to a cold ice age climate. At year 18,000 and counting, our current interglacial vacation from the Ice Age is much nearer the end than the beginning.

Global warming started long before the "Industrial Revolution" and the invention of the internal combustion engine. Global warming began 18,000 years ago as the earth started warming its way out of the Pleistocene Ice Age-- a time when much of North America, Europe, and Asia lay buried beneath great sheets of glacial ice.

Global warming during Earth's current interglacial warm period has greatly altered our environment and the distribution and diversity of all life. For example:

- Approximately 15,000 years ago the earth had warmed sufficiently to halt the advance of glaciers, and sea levels worldwide began to rise.
- By 8,000 years ago the land bridge across the Bering Strait was drowned, cutting off the migration of men and animals to North America from Asia.
- Since the end of the Ice Age, Earth's temperature has risen approximately 16 degrees F and sea levels have risen a total of 300 feet! Forests have returned where once there was only ice.

Greenland (130,000 – 116,000 years ago)

Proponents of Anthropogenic Global Warming paint a picture of runaway global warming[70] in a hot world only getting hotter. Hot world in modern times, really? We are in an interglacial period. Therefore, let us examine how temperatures were like naturally, during the previous interglacial period before the last Ice age. As Greenland is often cited as an example of "tremendous" warming, let us review the climate of Greenland 130,000 to 116,000 years ago. Fortunately, extensive research has already been performed by scientists.

You will find summaries below on the findings of two of those studies, together with relevant sources of the research for additional information:

Willerslev, E.; et al. (2007). "Ancient biomolecules from deep ice cores reveal a forested southern Greenland".
Science 317 (5834): 111-4. doi:10.1126/science.1141758. PMID 17615355

---

[70] The following link is an example of the extreme hyperbole of runaway global warming regularly disseminated by proponents of AGW: http://www.zero-carbon-or-climate-catastrophe.org/runaway-heating.html

Scientists who probed 2 kilometers (1.2 mi) through a Greenland glacier to recover the oldest plant DNA on record said that the planet was far warmer hundreds of thousands of years ago than is generally believed. DNA of trees, plants and insects including butterflies and spiders from beneath the southern Greenland glacier was estimated to date to 450,000 to 900,000 years ago, according to the remnants retrieved from this long-vanished boreal forest. That view contrasts sharply with the prevailing one that a lush forest of this kind could not have existed in Greenland any later than 2.4 million years ago. These DNA samples suggest that the temperature probably reached 10 °C (50 °F) in the summer and -17 °C (1.4 °F) in the winter. They also indicate that during the last interglacial period, 130,000-116,000 years ago, **when temperatures were on average 5 °C (9 °F)** higher than now, the glaciers on Greenland did not completely melt away.

More details on the research can be found at these links:

- http://www.pubmedcentral.nih.gov/articlerender.fcgi?tool=pmcentrez&artid=2694912
- http://dx.doi.org/10.1126%2Fscience.1141758
- http://www.pubmedcentral.gov/articlerender.fcgi?tool=pmcentrez&artid=2694912
- http://www.ncbi.nlm.nih.gov/pubmed/17615355

Alley, 2000

Between 1989 and 1993, U.S. and European climate researchers drilled into the summit of Greenland's ice sheet, obtaining a pair of 3 km (1.9 mi) long ice cores. Analysis of the layering and chemical composition of the cores has provided a revolutionary new record of climate change in the Northern Hemisphere going back about 100,000 years, and illustrated that the world's weather and temperature have often shifted rapidly from one seemingly stable state to another, with worldwide consequences.

Conclusions

The reader is encouraged to take note that global temperature during the current interglacial period is 5 °C below the previous interglacial period.

The physical evidence from Greenland provides a clear confirmation of the graph presented in the previous section, in other words:

(i)   The climate has been embroiled in Ice Ages for most of the past several million years-

(ii)   There is periodic respite of warming every 100,000 years for only between 15,000 to 20,000 years.

(iii)   The last interglacial warming was 5°C higher than the current.

(iv)   One also wonders why there is no evidence of the release of massive deposits of subsea methane (called an oceanic Anoxic event) and related mass-extinctions during the last interglacial warming period when temperatures were 5°C higher than today. Yet, this does not prevent proponents of anthropogenic global warming from predicting such doom and gloom today.

(v)   From the Vostok Ice Core data, we also know that CO2 level peaked at around 290 ppm during the last interglacial period (see section The 800-year lag) a full 100 ppm or 26% below the current level of 390 ppm. In other words, temperature was 5°C above modern times but CO2 was at pre-industrial level.

Misinformation on "extreme" warming in Greenland disseminated by proponents of Anthropogenic Global Warming #1

The scientific research carried out by Alley in the early 2000s shows that temperature has been trending down in the past 10,000 years.[71]

- Even a thousand years ago, the temperature of Greenland was at -30.5 °C compared to -31.5 °C in modern times
- The Holocene Warming Period, not linked to any changes in CO2 concentration encountered several significant peaks in Greenland temperature, well above today's levels.

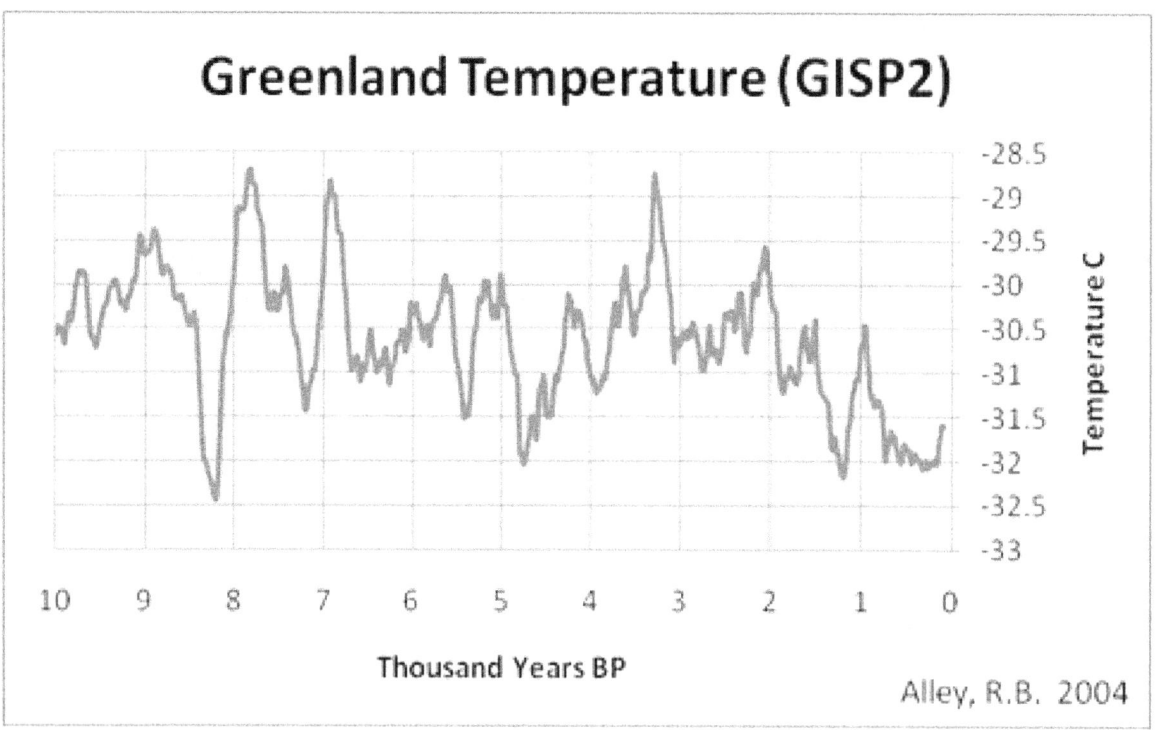

71 http://www.quadrant.org.au/img/content/May%202010/Kininmonth%20graph.gif

Misinformation on "extreme" warming in Greenland disseminated by proponents of Anthropogenic Global Warming #2

In September 2011, proponents of AGW, frustrated by increasing public disbelief in their claims, insidiously attempted to modify World Atlases to "demonstrate" dishonestly that 15% of the Greenland Icecap had disappeared.

http://www.australianclimatemadness.com/2011/09/times-world-atlas-falls-prey-to-climate-alarmism/

Times World Atlas falls prey to climate alarmism
Wednesday, 21 September 2011 8:40 am · 12 comments

Some things you really believe you can trust. The Times Comprehensive Atlas of the World, for example. Not anymore. Like so many grand old institutions, it has fallen prey to nonsensical claims that 15% of the Greenland ice sheet has disappeared in the last few years. So the cartographers meekly acquiesce, showing a massive retreat in the ice sheet in the latest edition (see image).

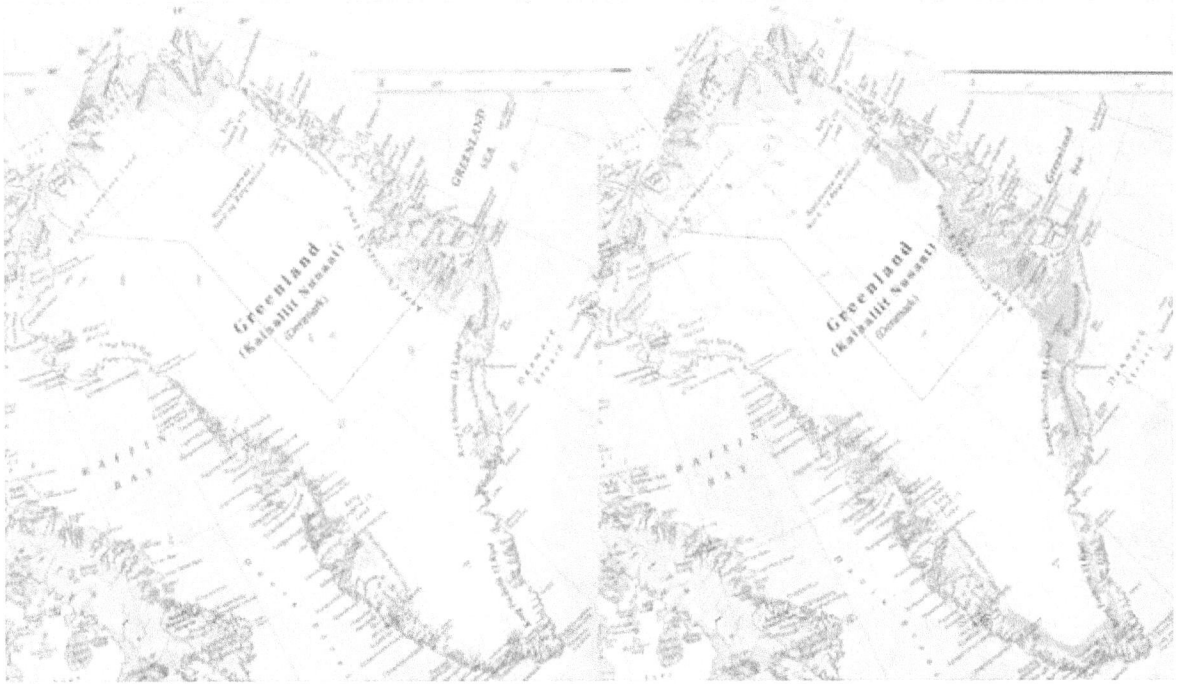

But having been pilloried in the press for the ridiculous claim (even by the BBC and Guardian), they've had to back down, as the Guardian reports:

The publishers of the Times Atlas were forced to admit on Tuesday that they were wrong to claim the Greenland ice pack had shrunk by 15%, as Arctic scientists rounded on the company for misinterpreting data and failing to consult them.

The humiliating climb down for HarperCollins – part of Rupert Murdoch's publishing empire – came after key sources of data on the Greenland ice denied that their research, cited by the Times Atlas, warranted the claims. Despite criticism of the claim by scientists, a spokeswoman for the atlas had, as recently as Monday, issued a robust defence of the claim, saying: "We are the best there is ... Our data shows that it has reduced by 15%. That's categorical."

But HarperCollins put out a statement on Tuesday saying: "For the launch of the latest edition of the atlas we issued a press release which unfortunately has been misleading with regard to the Greenland statistics. We came to these statistics by comparing the extent of the ice cap between the 10th and 13th editions of the atlas. The conclusion that was drawn from this, that 15% of Greenland's once permanent ice cover has had to be erased, was highlighted in the press release not in the atlas itself. This was done without consulting the scientific community and was incorrect. We apologise for this and will seek the advice of scientists on any future public statements." (source)

Maurizio Morabito has a theory. So the following series of events is consistent with the observations:

1. Times Atlas personnel read or listen from somewhere that the Greenland ice sheet is melting
2. They open the Wikipedia page on the Greenland ice sheet
3. As if by magic...that page contains a map of Greenland
4. Times Atlas personnel convert that map to the Times Atlas high-quality standard

Now where's the evidence for it? Where is it indeed, as Michael Corleone would have asked.

And furthermore, Hockey Schtick reports on a new paper that shows an ice sheet on the northern tip of Greenland has remained unchanged or grown slightly in the last few years:

Warmists tell us the effects of AGW should be most evident at the poles. A paper published today in the Journal of Geophysical Research closely examines the Flade Isblink Ice Cap at the northern tip of Greenland using data from two satellites from 2002-2008 and finds a slightly positive/near zero change in surface elevation and no change whatsoever in mass. However, according to the experts at The Times Comprehensive Atlas of the World, this entire ice cap has completely disappeared.

Another blow to alarmist credibility - and the Times Atlas - thanks to its desperation to advance an agenda by any means possible

Misinformation on the last Interglacial period disseminated by proponents of Anthropogenic Global Warming

Here is more panic generating misinformation, this time from Dr. James Hansen, NASA's Director of the Goddard Institute for Space Studies (GISS)

Big climate change could happen fast - and soon[72]

December 9th, 2011

New research from NASA into the Earth's paleoclimate history indicates we could be facing rapid climate change this century, including sea level rises of many meters.

And while international leaders have suggested a goal of limiting global warming to two degrees Celsius from pre-industrial times, Goddard Institute for Space Studies director James E Hansen says that even this would lead to drastic changes.

The Earth's average global surface temperature has already risen by 0.8 degrees Celsius since 1880, says Hansen, and is now increasing by more than 0.1 degree Celsius every decade.

At the current rate of fossil fuel burning, the concentration of carbon dioxide in the atmosphere will have doubled from pre-industrial times by the middle of this century, causing an eventual warming of several degrees, he says.

Hansen and his colleague Makiko Sato compared the climate of today, the Holocene, with previous similar interglacial epochs. By studying cores from both ice sheets and deep ocean sediments, they found that global mean temperatures during the Eemian period, which began about 130,000 years ago and lasted about 15,000 years, were less than one degree Celsius warmer than today.

If temperatures were to rise two degrees Celsius over pre-industrial times, global mean temperature would far exceed that of the Eemian, when sea level was four to six meters higher than today, says Hansen.

"The paleoclimate record reveals a more sensitive climate than thought, even as of a few years ago. Limiting human-caused warming to two degrees is not sufficient," he says. "It would be a prescription for disaster."

Two degrees Celsius of warming would make Earth much warmer than during the Eemian - indeed, similar to Pliocene-like conditions, when sea level was about 25 meters higher than today, says Hansen.

However, that sea level increase due to ice sheet loss would be expected to occur over centuries, and large uncertainties remain in predicting it accurately.

"We don't have a substantial cushion between today's climate and dangerous warming. Earth is poised to experience strong amplifying feedbacks in response to moderate additional global warming," says Hansen.

"Humans have overwhelmed the natural, slow changes that occur on geologic timescales."

---

[72] http://www.tgdaily.com/sustainability-features/60111-big-climate-change-could-happen-fast-and-soon

There are so many brazenly factually inaccurate misstatements in the above short article that it is hard to know where to begin.

1. As per research from Willerslev and Alley, temperature in the last interglacial period (referred to above as the Eemian period) were typically 5°C above that of modern times, while CO2 concentration was lower than today.

   - 5°C is a lot more than 1°C as minimized by the article

   - Dr. Hansen should also ask himself why the temperature rose that much during the Eemian period, without a significant increase in CO2 concentration, and provide an explanation on why this time is different.

2. As per the graph in the earlier section on the little Ice Age, global temperature has varied significantly in short periods of time on the geologic scale, independent of CO2 levels.

3. During the Carboniferous period, 360 to 330 million years ago, CO2 concentration averaged 800 ppm (twice that of today) but temperatures averaged 14°C (same as modern level). For more details, please see the section Paleo Temperatures.

   - A doubling of current CO2 concentration from pre-industrial times would mean 580 ppm which is still well below the 800 ppm of the Carboniferous period

   - Dr. Hansen should, therefore, provide a clear explanation why temperature during the Carboniferous period was not significantly warmer, and why he is so certain that an increase to 580 ppm, this time will result in catastrophic warming. The historical facts of Earth's climate proves otherwise.

4. As detailed in the ensuing section on the Holocene period, temperatures rose then without the impetus of additional CO2 in the atmosphere. Why? Then, during the little Ice Age which ended barely a century ago, temperature fell by around -0.8°C. Why?

   - Dr. Hansen should explain why the warming of the 20th century was not just a natural warming after the end of the little Ice Age.

5. Please refer to more detailed Paleo facts in this chapter History of Global Temperature. The historical record clearly demonstrates that the mean CO2 concentration display a correlation with the mean surface temperature. In addition, the significant climatic variations within each period indicate clearly that climate can vary widely from warm to cool with the same and often much higher level of CO2 concentration than we have today.

6. We shall check the facts on Dr. Hansen's assertions on sea levels in a later section.

In the July/August 2008 issue of Launch Magazine, here is how NASA Astronaut and Physicist Walter Cunningham described Dr. Hansen's tenure of enforcing climate pseudoscience at NASA:

> "Hansen is a political activist who spreads fear even when NASA's own data contradict him,"… "NASA should be at the forefront in the collection of scientific evidence and debunking the current hysteria over human-caused, or Anthropogenic Global Warming (AGW). Unfortunately, it's becoming just another agency caught up in the politics of global warming, or worse, politicized science."

Hansen uses his bureaucratic position as Director of NASA GISS, to pursue a political agenda. He inflated the issue of human induced global warming to a global fraud in 1988 testimony before a House and Senate committee when he said; "the greenhouse effect has been detected and it is changing our climate now" This shows either ignorance of climate science or a deliberate attempt to mislead or both. The phrasing suggests incorrectly the greenhouse effect is new. There is no evidence, except in the computer models, that it is causing current climate change. He capped this with another unsupportable statement that he was, "99 percent certain that the warming trend was not a natural variation but was caused by a buildup of carbon dioxide and other artificial gases in the atmosphere."

NASA GISS systematically "controls and "adjusts", temperature data, as do other organizations such as the CRU (Climate Research Unit). We shall go into this aspect in more detail in a later chapter, but below are a few examples from Hansen's present and past:

November 2011: Jimmy Works His Arctic Magic[73]

> Despite having no data north of 80N, Hansen has determined that it was very hot there in November. By fabricating a huge 4-8C anomaly at the North Pole, he is able to keep global temperatures (barely) rising this century, while HadCRUT shows global temperatures falling.
>
> He also did a bang up job warming Greenland well above measured temperatures. RSS showed almost all of Greenland cold, but Hansen's magic crayon did an impressive job of heating the place up.
>
> http://data.giss.nasa.gov/

(please see image on next page)

---

[73] http://www.real-science.com/jimmy-works-arctic-magic

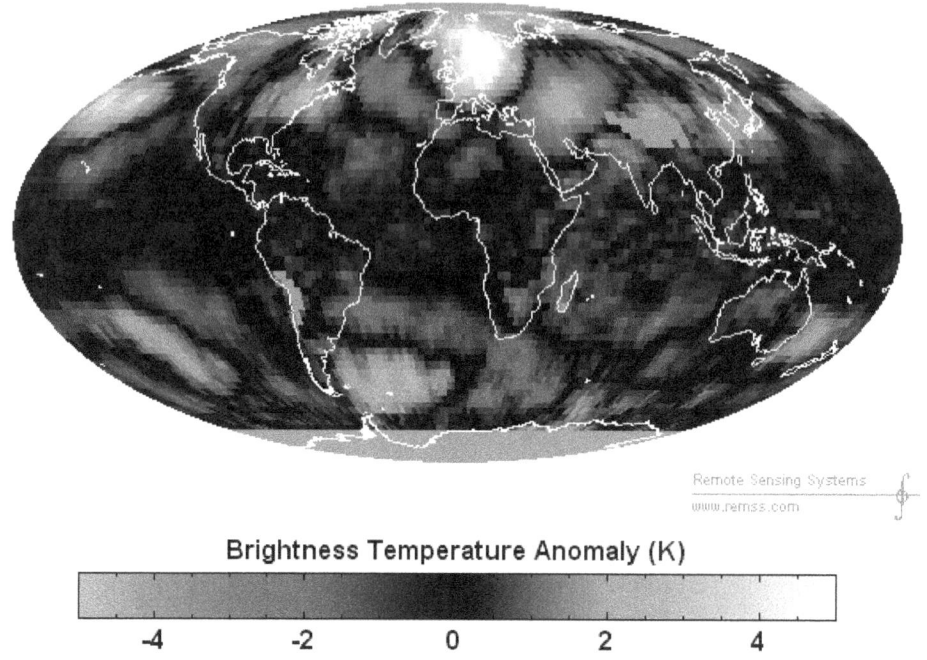

January 2009: GISS's Temperature trend revision

At the very end of the GISS update, under paragraph 4 in the next to last paragraph, GISS states that: "From climate models and empirical analyses this GHG forcing translates into a mean warming rate of 0.15C per decade".

In recent years, GISS temperature data has been increasingly at odds with satellite data. Therefore, GISS temperature data series again came under close scrutiny recently [Lubos Motl, et. al].

This revision was singularly noteworthy: the revised GISS GHG driven temperature trend was a whopping 25% lower than the IPCC's flawed and discredited prediction of 0.20C per decade.

Below is a GISS publication by Hansen in 1999:

Hansen et al. 1999[74]
Hansen, J., R. Ruedy, J. Glascoe, and Mki. Sato, 1999: GISS analysis
of surface temperature change. J. Geophys. Res., 104, 30997-31022,
doi:10.1029/1999JD900835.

We describe the current GISS analysis of surface temperature change
for the period 1880-1999 based primarily on meteorological station
measurements. The global surface temperature in 1998 was the warmest
in the period of instrumental data. The rate of temperature change
was higher in the past 25 years than at any previous time in the
period of instrumental data. The warmth of 1998 was too large and
pervasive to be fully accounted for by the recent El Nino. Despite
cooling in the first half of 1999, we suggest that the mean global
temperature, averaged over 2-3 years, has moved to a higher level,
analogous to the increase that occurred in the late 1970s. Warming in
the United States over the past 50 years has been smaller than in
most of the world, and over that period there was a slight cooling
trend in the eastern United States and the neighboring Atlantic
Ocean. The spatial and temporal patterns of the temperature change
suggest that more than one mechanism was involved in this regional
cooling. The cooling trend in the United States, which began after
the 1930s and is associated with ocean temperature change patterns,
began to reverse after 1979. We suggest that further warming in the
United States to a level rivaling the 1930s is likely in the next
decade, but reliable prediction requires better understanding of
decadal oscillations of ocean temperature

Hansen's statement, "Warming in the United States over the past 50 years has been smaller than in most of the world", is shockingly disingenuous. As documented in the section The 20th Century: Carbon Dioxide versus Global Temperature , based on NASA's data itself, there was a clear period of *global* cooling between 1950 and the late 1970s, not just in the United States. Furthermore, the reader will, hopefully, remember the statistical analysis performed in the same section, where an outlier below the lower control limit (LCL) was clearly established in 1976. How could any scientist with any integrity have ignored that outlier?

This brazen misrepresentation of data illustrates well the extent of mendacity in the field of climate pseudoscience.

---

[74] http://pubs.giss.nasa.gov/abs/ha03200f.html

The Salon interviewee and book author, Rob Reiss admits he somehow conflated 40 years with 20 years, and concedes that Dr. Hansen actually said 40 years for his prediction (back in 1988).

In 2012, 24 years later, we are not seeing anywhere the magnitude of sea level rise forecast by Hansen's pseudoscience. Actually 20 or 40 years does not make a difference. Per Dr. Hansen's prediction in 1988, even if we assume a 40 year time span, the sea level rise should have been about halfway up the side of Manhattan Island by now.

Anthony Watts first revealed this shoddy "science" to the world:

http://wattsupwiththat.com/2009/10/22/a-little-known-but-failed-20-year-old-climate-change-prediction-by-dr-james-hansen/

Here is Salon interviewee Rob Reiss:

http://www.salon.com/2001/10/23/weather/

Tuesday, Oct 23, 2001 7:41 PM UTC

Reiss spoke to Salon from his home in New York.

Extreme weather means more terrifying hurricanes and tornadoes and fires than we usually see. But what can we expect such conditions to do to our daily life?

While doing research 12 or 13 years ago, I met Jim Hansen, the scientist who in 1988 predicted the greenhouse effect before Congress. I went over to the window with him and looked out on Broadway in New York City and said, "If what you're saying about the greenhouse effect is true, is anything going to look different down there in 20 years?" He looked for a while and was quiet and didn't say anything for a couple seconds. Then he said, "Well, there will be more traffic." I, of course, didn't think he heard the question right. Then he explained, "The West Side Highway [which runs along the Hudson River] will be under water. And there will be tape across the windows across the street because of high winds. And the same birds won't be there. The trees in the median strip will change." Then he said, "There will be more police cars." Why? "Well, you know what happens to crime when the heat goes up."

And so far, over the last 10 years, we've had 10 of the hottest years on record.

Didn't he also say that restaurants would have signs in their windows that read, "Water by request only."

Under the greenhouse effect, extreme weather increases. Depending on where you are in terms of the hydrological cycle, you get more of whatever you're prone to get. New York can get droughts, the droughts can get more severe and you'll have signs in restaurants saying "Water by request only."

When did he say this will happen?

Within 20 or 30 years. And remember we had this conversation in 1988 or 1989.

Does he still believe these things?

Yes, he still believes everything. I talked to him a few months ago and he said he wouldn't change anything that he said then.

Here is the update from Dr. Hansen himself:

http://www.columbia.edu/~jeh1/mailings/2011/20110126_SingingInTheRain.pdf

Michaels also has the facts wrong about a 1988 interview of me by Bob Reiss, in which Reiss asked me to speculate on changes that might happen in New York City in 40 years assuming CO_2 doubled in amount. Michaels has it as 20 years, not 40 years, with no mention of doubled CO_2.

Reiss verified this fact to me, but he later sent the message: "I went back to my book and re-read the interview I had with you. I am embarrassed to say that although the book text is correct, in remembering our original conversation, during a casual phone interview with a Salon magazine reporter in 2001 I was off in years. What I asked you originally at your office window was for a prediction of what Broadway would look like in 40 years, not 20. But when I spoke to the Salon reporter 10 years later - probably because I'd been watching the predictions come true, I remembered it as a 20 year question." So give Michaels a pass on this one -- assume that he reads Salon, but he did not check the original source, Reiss' book.

So what exactly did Dr. Hansen predict in 1988?

As Anthony Watts explained in simple terms back in 2011:

Are the predictions coming true? Let's find out. Let's look at the tide gauge in New York and see what it says.

(Please see image on next page)

NEW YORK

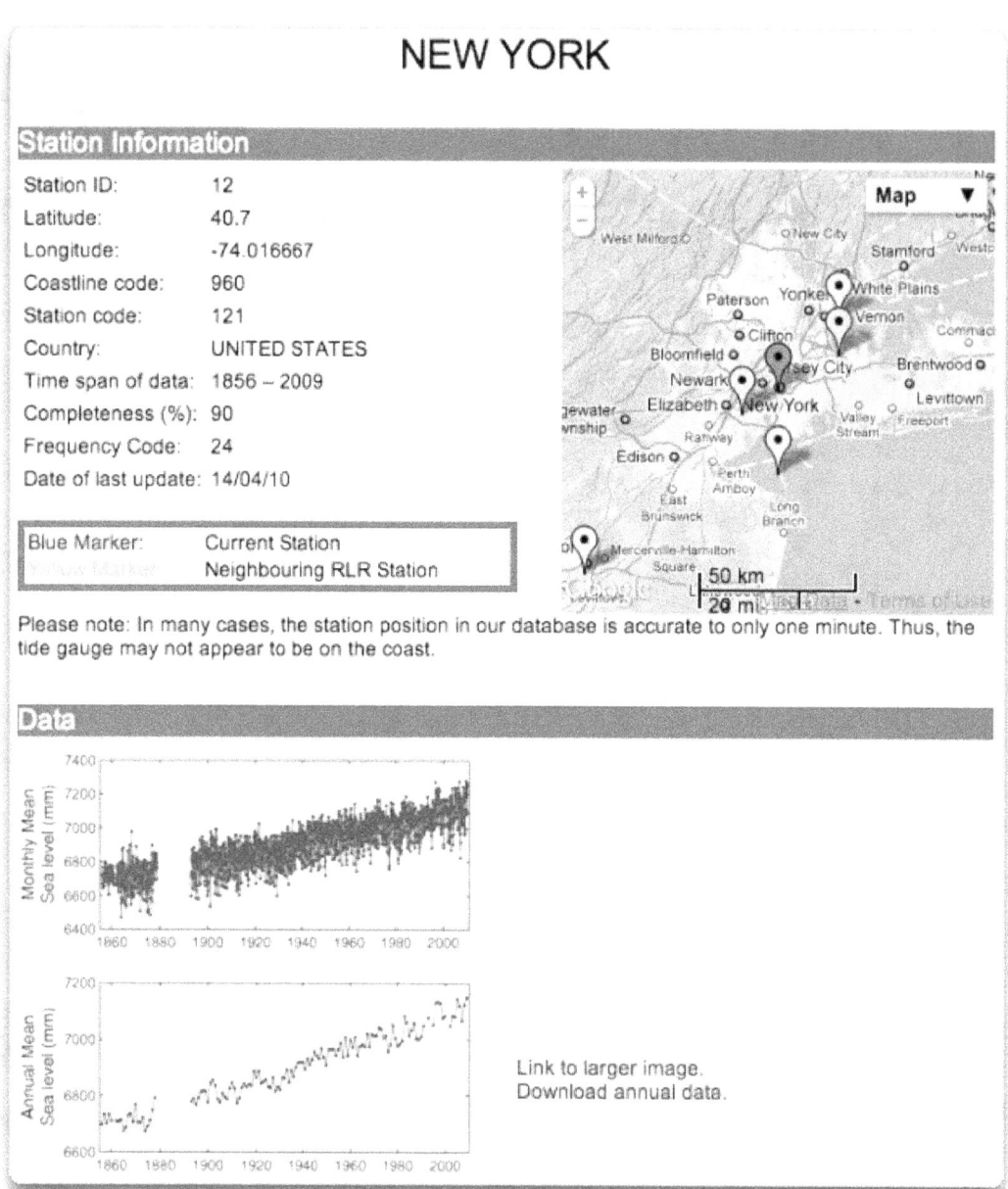

Here's the PSMSL page
http://www.psmsl.org/data/obtaining/stations/12.php

You can see the terrifying surge of acceleration in the sea level due to increasing GHGs in the 20th century [HUMOR]. Willis downloaded and plotted the data to see what the slope looked like, and then plotted a linear average line.

Here it is overlaid with the Colorado satellite data. Note the rate of rise is unchanged:

(Please see image on next page)

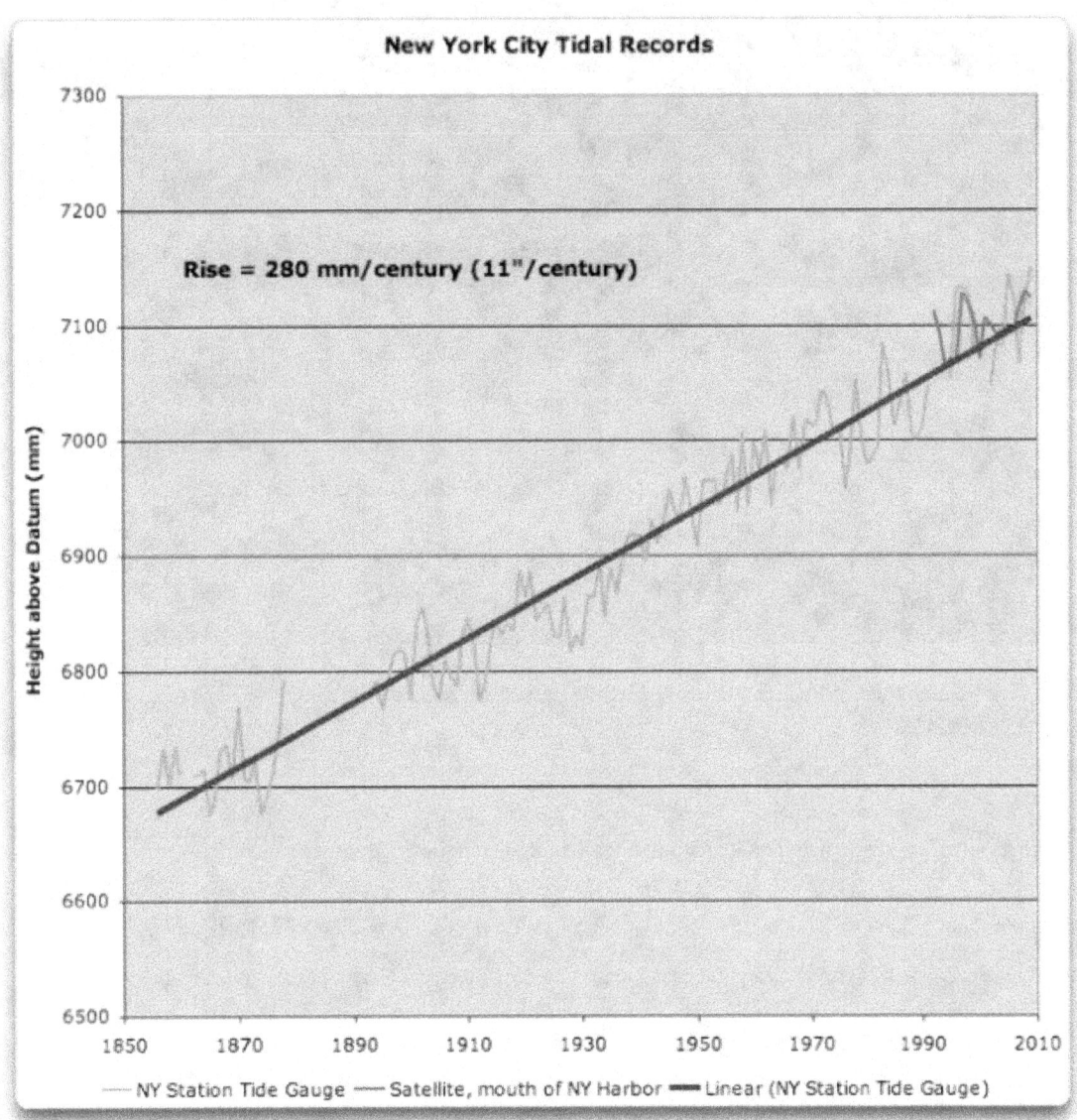

New York City Tidal Records

Rise = 280 mm/century (11"/century)

NY Station Tide Gauge ——— Satellite, mouth of NY Harbor ▬▬ Linear (NY Station Tide Gauge)

Add to that, the recent peer reviewed paper from the Journal of Coastal Research that said: "worldwide-temperature increase has not produced acceleration of global sea level over the past 100 years"

As of this update in March 2011, we're 23 years into his prediction of the West Side Highway being underwater. From what I can measure in Google Earth, Dr. Hansen would need at least a ten foot rise in forty years to make his prediction work. See this image below from Google Earth where I placed the pointe over the West Side Highway, near the famous landmark and museum, the USS Intrepid:

(Please see image on next page)

The lat/lon should you wish to check yourself is: 40.764572° - 73.998498°

Here's a ground level view (via a tourist photo) so you can see the vertical distance from the roadway to the sea level on that day and tide condition. Sure looks like at least 10 feet to me [HUMOR].

Note: We shall be looking at sea level science and pseudoscience in more detail in a later chapter.

Despite the clear disconnect with facts as conclusively demonstrated, whether or not Dr. Hansen actually believes in anthropogenic global warming is not for me to speculate. What is clear is that the Director of GISS has benefited handsomely from his crusade to "save the planet", including his very public row with the former President of the United States. As a consequence, Dr. Hansen has been showered with accolades, both of a symbolic and monetary nature from the high society of climate pseudoscience and benefactors such as George Soros who, at the Copenhagen Climate Summit of 2009, publicly called for $100 billion of IMF gold to be disbursed to "fight" climate change.

At the beginning of 2011, Chris Horner, a co-founder of The American Tradition Institute, filed a lawsuit against James Hansen, accusing him of receiving more than $1.2 million from the very environmental organizations whose agenda he advocated.

"Hansen's office appears to be somewhat of a rogue operation. It's clearly a taxpayer-funded global warming advocacy organization," said Horner, "The real issue here is, has Hansen been asking NASA in writing, in advance, for permission for these outside activities? We have reason to believe that has not been occurring."

The lawsuit cited the following awards lavished at Dr. Hansen in recognition of his "valiant" efforts to rescue humanity from ourselves:

- A shared $1 million prize from the Dan David Foundation for his "profound contribution to humanity." Hansen's cut ranged from $333,000 to $500,000," Horner said, adding that the precise amount is not known because Hansen's publicly available financial disclosure form only shows the prize was "an amount in excess of $5,000."

- The 2010 Blue Planet prize worth $550,000 from the Asahi Glass Foundation, which recognizes efforts to solve environmental issues.

- The Sophie Prize for his "political activism," worth $100,000. The Sophie Prize is meant to "inspire people working towards a sustainable future."

- Speaking fees totaling $48,164 from a range of mostly environmental organizations.

- A $15,000 participation fee, waived by the W.J. Clinton Foundation for its 2009 Waterkeeper Conference.

- $720,000 in legal advice and media consulting services provided by The George Soros Open Society Institute. Hansen said he did not take "direct" support from Soros but accepted "pro bono legal advice."

I have not followed, nor am I particularly interested in the outcome of the lawsuit. What is of interest to me is the really dismal lack of facts and scientific integrity behind a theory which continues to be pushed on the world's population as settled science. This has been demonstrated in numerous examples in this book, as well as many other cases in point that I have not documented. What is of most concern to me is that the veracity of this theory is still simply assumed to be accurate by policymakers around the world responsible for prioritizing the allocation of increasingly scarce public resources as well as the regulation of the private sector.

The most significant drought during the existence of human beings on this planet was not caused by a change in CO2 concentration

http://www.redorbit.com/news/science/1097887/drought_pushed_ancient_african_migration/

Drought Pushed Ancient African Migration
October 11, 2007

A University of Arizona researcher suspects that a monster drought is behind our ancestors' massive bug-out from Africa.

Lake Malawi, the third-largest lake in Africa and ninth-largest in the world, dropped nearly 2,000 feet during a mega-drought roughly 100,000 years ago, according to the UA's Andrew S. Cohen, the lead scientist on a report in the latest edition of the Proceedings of the National Academy of Sciences.

The research was funded by the National Science Foundation, the International Continental Drilling Program and the Smithsonian Institution.

Cohen says the now lush, forested tropical area — with rainfall like that of the Southeastern United States — probably looked something like Tucson during the drought. And beautiful, deep and

remarkably clear Lake Malawi would have been a salty, pea soup pond by comparison.

The massive droughts, thought to have occurred between 135,000 and 90,000 years ago would overlap the Out-of-Africa theory that said modern humans descend from a relatively small number of people living in Africa between 150,000 and 70,000 years ago.

Cohen theorizes that the drought and resulting desertification likely caused a crash in population. And the rebound of the lake coincides with

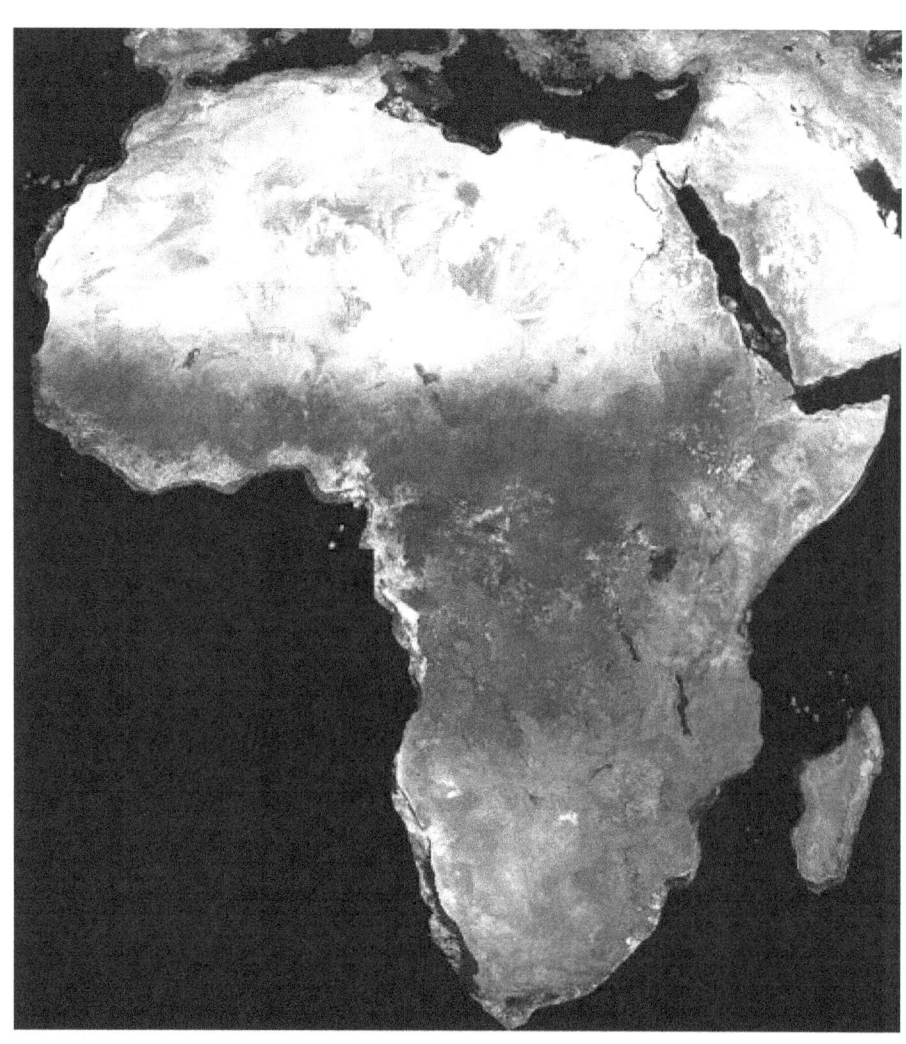

archaeological work that shows an increase in population and people leaving the continent.

Members of the Malawi Drilling Project used a ship to drop a deep sea-style rig 1,942 feet to the lake bottom and extract core samples 415 feet beneath the lake bottom. The core samples give a picture of conditions there over hundreds of thousands of years, Cohen said.

Carbon and other sample analysis — taken at 300-year intervals — provides a record of things that fell into or died in the lake. Besides remains of fish, plankton and invertebrates there was pollen from plants surrounding the lake and charcoal from fires.

In the case of fish bones, scales and teeth, chemical analysis of the fish parts in the core samples could tell whether the sample point was beneath shallow or deep water when the sample drifted down to the bottom.

"You are what you eat," said David Dettman, a member of the science team and the director of the UA's Environmental Isotope Laboratory.

As such, the ratio of carbon and nitrogen isotopes differs between fish that lived in shallow and deep water, Dettman said. (An isotope is a different form of an element, having a different number of neutrons.)

"It tells you about the food resources of these fish," says Dettman. "By the chemistry in the bones you can tell if they're fish feeding in the rocky shallows near the shore" or in the deep open water.

During some periods, the core sample taken from what is now the deepest part of the lake shows the lake was much shallower.

Like an "in box," the oldest deposits in the core sample were at the bottom of the stack — the farthest from the present lake bottom.

Cohen says it's completely different from drilling for oil.

Oil drillers, Cohen says, want to make a hole. The drilling scientists want the stuff that was in the hole (and truly hope they don't hit oil or gas).

"We have to get the mud back intact," says Cohen. "The upper 100 meters is full of water, squishy."

Cohen is already working on another Malawi drilling project, one that he hopes will be cheaper and easier. Coring on dry land, in former lake beds that have been elevated by geologic faults, should give them the same kind of climate and environmental records — but without the expense and difficulties of drilling in a 2,000-foot deep lake.

"We hope to get the same quality record," Cohen says, "but be able to drive a drill rig up to the (rock) outcropping."

But Lake Malawi itself is interesting for other reasons, too. It's a freshwater equivalent to the Galapagos Islands: an isolated place where scientists can study dramatic evolutionary changes in the lake's fish the way Darwin and later scientists studied the Galapagos' finch species.

Only four of Lake Malawi's 500 to 1,000 fish species are found anywhere else, says Peter Reinthal, a UA adjunct associate professor of ecology and evolutionary biology and curator of fish.

He compared them to Darwin's Galapagos finches, highly specialized fish that fill niches in the giant lake's food chain.

Some have made downright weird adaptations.

Many of Lake Malawi's fish are mouth brooders, incubating eggs in their mouth, even sheltering the hatchlings after birth to keep them out of danger. Nice adaptation, but it doesn't end there.

Other fish have adapted to take advantage of that trait.

"One (predator fish) has a head like a battering ram," Reinthal says. "They'll swim directly at a female with a mouthful of young and try to disgorge them."

Others, Reinthal said, feed on the scales of other fish, or scrape them off and mimic them.

—Contact reporter Dan Sorenson at 573-4185 or dsorenson@azstarnet.com.
Source: redOrbit (http://s.tt/16Bhd)

Misinformation disseminated by proponents of Anthropogenic Global Warming

In order to counter recent research revealing facts about Earth's past climate, proponents of Anthropogenic Global Warming have been spreading misinformation in the popular press as well as "scientific " journals that "humans have only been around for a few millenia and they will not be able to tolerate an extreme planet.", hence the climate of Earth's past is "irrelevant".

It is ironic to now see proponents of AGW partially singing the same song as Creationists that human existence is only a few thousand years old. The facts are, of course, different. Anatomically modern humans evolved from archaic Homo sapiens in Africa in the Middle Paleolithic, about 200,000 years ago. [75]

In other words, humans survived:

- the current interglacial period
- the last Ice Age
- the previous interglacial period preceding the last Ice Age when temperatures were 5°C higher than today while CO2 concentrations didn't exceed 300 ppm (see section
- Greenland (130,000 – 116,000 years ago)
- 70,000 years of the second-to-last Ice Age preceding the most recent Ice Age

---

[75] "Human Evolution by The Smithsonian Institution's Human Origins Program". Human Origins Initiative. Smithsonian Institution. http://www.mnh.si.edu/anthro/humanorigins/ha/sap.htm.

In addition, human beings have survived major natural changes in climate without the benefit of modern technology which we are accustomed to today.

It is truly amazing to see the lengths to which proponents of AGW are in denial about the facts and say anything in an act of sheer desperation to resuscitate their pet theory. And it is even more remarkable that many have subconsciously blinded themselves to the facts, so that they even believe, in all sincerity, in what they are saying.

Holocene Temperature Variations

The Holocene period, in other words the past 12,000 years, is a tiny speck in geologic time. http://www.ncdc.noaa.gov/paleo/data.html

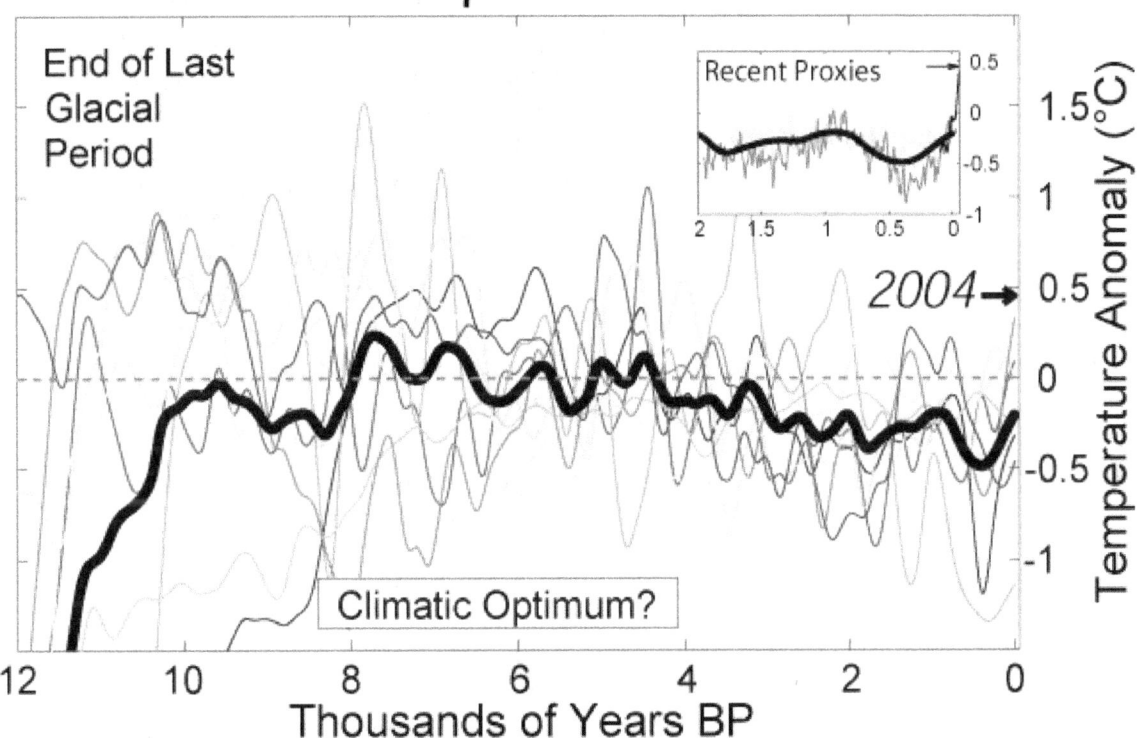

Observations

Period	Global Temperature Anomaly °C (0 = Modern times)	Comments
8,000 years ago	+0.3	• The 8000 year old temple of Stonehenge in England is off by 1/2 degree on alignment with the sun and moon- Recently, a second Stonehenge calendar was discovered by infrared satellite photos of the Sahara Desert

Period	Global Temperature Anomaly °C (0 = Modern times)	Comments
		on the far western borders of Egypt, once a lush grazing plain. • The farmers and herdsmen of ancient times needed an accurate calendar to time the seasons. What was driving climate change and the advancement of deserts long before the Industrial Revolution?
4,000 years ago	+0.25	
2,000 years ago	0	• Same temperature as modern times
1,000 years ago	0	• Same temperature as modern times • In the mid-2000s, there was a lot of hype from proponents of Anthropogenic Global Warming about the Northwest Passage being open to shipping, as if it was something completely unprecedented. In fact, the reason why it is named a Passage is because, a little over a millennium ago, the Vikings traversed this body of *water* in their rickety wooden boats; the Passage was definitely open to shipping at that time.
500 years ago	-0.5	• The little Ice Age, a sudden decline in global temperature, causes of which scientists are still investigating – to this day
2004	+0.4	• The most recent temperature figures are controversial, as was most notably highlighted by the Climategate "hide the decline" scandal as well as differences with other independent temperature measurements. Details are described in a later chapter • There is broad agreement that the rapid warming witnessed in the 1980s and 1990s has since plateaued since the early 2000s. Independent measurements by scientists such as Roy Spencer have been shown a slight decline at the end of the last decade.

Analysis by Don J. Easterbrook, Professor Emeritus at Western Washington University, Bellingham, WA [76]

Warmest years of the past century

- 1934 has long been considered the warmest year of the past century. A decade ago, the closest challenger appeared to be 1998, a super El Nino year, but it trailed 1934 by 0.54°C (0.97°F).
- Since then, NASA GISS has *adjusted* the U.S. data for 1934 downward and 1998 upward (see December 25, 2010 post by Ira Glickstein) in an attempt to make 1998 warmer than 1934 and seemingly erased the original rather large lead of 1934 over 1998.
- The last phases of the strong 2009-2010 El Nino in early 2010 made this year another possible contender for the warmest year of the century.
- However, December 2010 has been one of the coldest Decembers in a century in many parts of the world.

Does it really matter?

- Regardless of which year wins the temperature adjustment battle, how significant will that be? To answer that question, we need to look at a much longer time frame? Centuries and millennia...
- So where do the 1934/1998/2010 warm years rank in the long-term list of warm years?
- Of the past 10,500 years, 9,100 were warmer than 1934/1998/2010. **Thus, regardless of which year (1934, 1998, or 2010) turns out to be the warmest of the past century, that year will rank number 9,099 in the long-term list.**

Medieval Warm Period (0 – 1200 AD)

References

- http://nipccreport.org/articles/2011/dec/Neukometal2011b.gif

- Neukom, R., Luterbacher, J., Villalba, R., Kuttel, M., Frank, D., Jones, P.D., Grosjean, M., Wanner, H., Aravena, J.-C., Black, D.E., Christie, D.A., D'Arrigo, R., Lara, A., Morales, M., Soliz-Gamboa, C., Srur, A., Urritia, R. and von Gunten, L. 2011. Multiproxy: "Summer and winter surface air temperature field reconstructions for southern South America covering the past centuries. Climate Dynamics 37: 35-51."

Findings

The findings of Neukom et al. go a long way towards demonstrating that:

(1) The Medieval Warm Period was a global phenomenon that was comprised of **even warmer intervals than the warmest portion of the Current Warm Period**, and that,

(2) The greater warmth of the Medieval Warm Period occurred when there was far less CO2 in the air than there is nowadays, which facts clearly demonstrate that the

---

[76] http://wattsupwiththat.com/2010/12/28/2010%E2%80%94where-does-it-fit-in-the-warmest-year-list/

planet's current -- but not unprecedented -- **degree of warmth need not be CO2-induced.**

Working with 22 of the best climate proxies they could find that stretched far enough back in time, Neukom et al. (2011) reconstructed a mean austral summer (December-February) temperature history for the period AD 900-1995 for the terrestrial area of the planet located between 20°S and 55°S and between 30°W and 80°W -- a region they call Southern South America (SSA) -- noting that their results "represent the first seasonal sub-continental-scale climate field reconstructions of the Southern Hemisphere going so far back in time.

The international research team -- composed of scientists from Argentina, Chile, Germany, Switzerland, The Netherlands, the United Kingdom and the United States -- write that their summer temperature reconstruction suggests that "a warm period extended in SSA from 900 (or even earlier) to the mid-fourteenth century," which they describe as being temporally located "towards the end of the Medieval Climate Anomaly as concluded from Northern Hemisphere temperature reconstructions."

And as can be seen from the graph below, the warmest decade of this Medieval Warm Period was calculated by them to be AD 1079-1088, which as best as can be determined from their graph is about 0.17°C warmer than the peak warmth of the Current Warm Period."

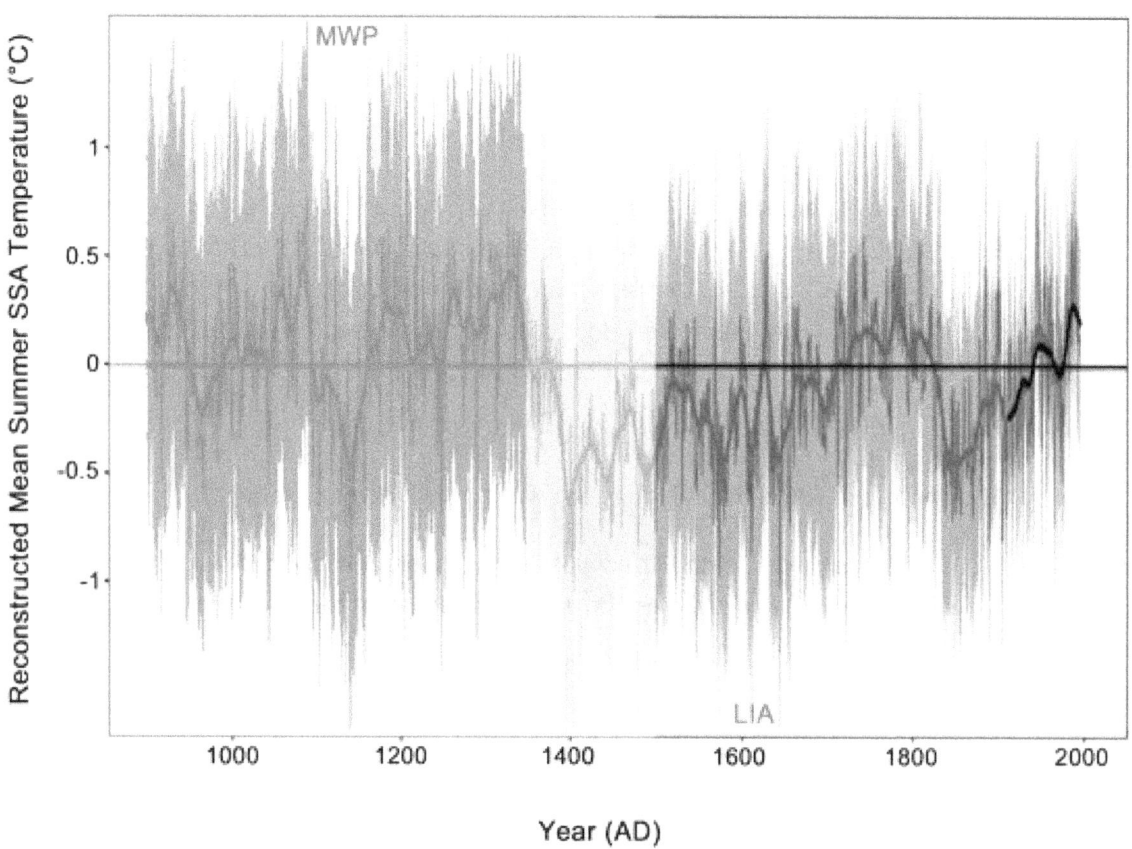

References
- http://jonova.s3.amazonaws.com/graphs/china/liu-2011-predictions-web.gif
- http://nextgrandminimum.wordpress.com/2011/12/07/amplitudes-rates-periodicities-and-causes-of-temperature-variations-in-the-past-2485-years-and-future-trends-over-the-central-eastern-tibetan-plateau-by-liu-et-al-2011-climate-science-ro/
- This is from a paper published in a China Science Bulletin:

Findings

Amplitudes, rates, periodicities and causes of temperature variations in the past 2485 years and future trends over the central-eastern Tibetan Plateau were analyzed.

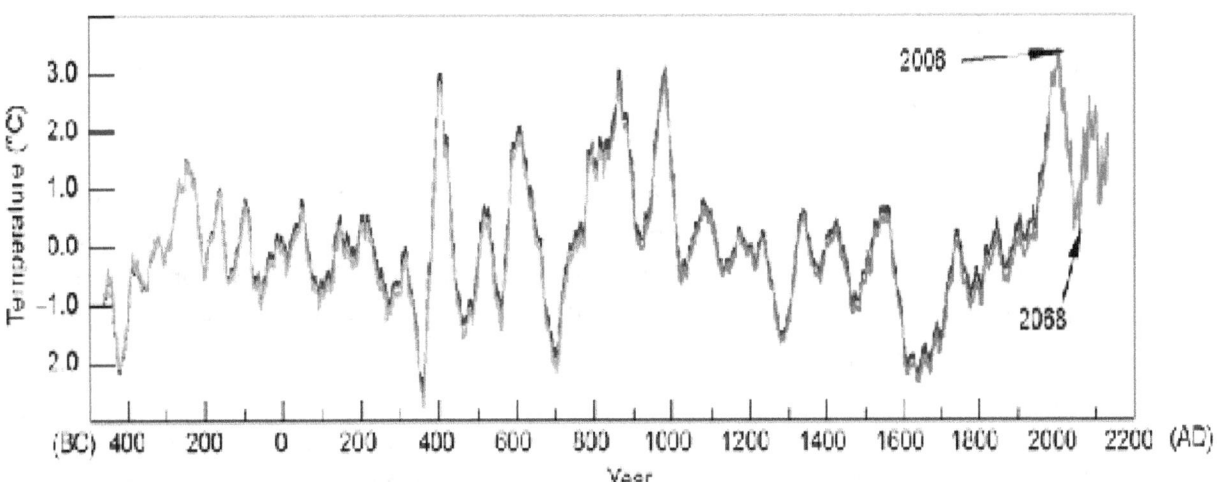

Abstract

The results showed that extreme climatic events on the Plateau, such as the Medieval Warm Period, Little Ice Age and 20th Century Warming appeared synchronously with those in other places worldwide.

The largest amplitude and rate of temperature change occurred during the Eastern Jin Event (343-425 AD), and not in the late 20th century. There were significant cycles of 1324 a, 800 a, 199 a, 110 a and 2-3 a in the 2485-year temperature series. The 1324 a, 800 a, 199 a and 110 a cycles are associated with solar activity, which greatly affects the Earth surface temperature.

The long-term trends (>1000 a) of temperature were controlled by the millennium-scale cycle, and amplitudes were dominated by multi-century cycles. Moreover, cold intervals corresponded to sunspot minimums. The prediction indicated that the temperature will decrease in the future until to 2068 AD and then increase again.

Below is how NOAA explains away the mid-Holocene period:

http://www.ncdc.noaa.gov/paleo/globalwarming/holocene.html

> In summary, the mid-Holocene, roughly 6,000 years ago, was generally warmer than today, but only in summer and only in the northern hemisphere. Moreover, we clearly know the cause of this natural warming, and know without doubt that this proven "astronomical" climate forcing mechanism cannot be responsible for the warming over the last 100 years.

Once again, this limited explanation is akin to looking at a glass half-full, while concealing the half-empty part of the glass. Unsurprisingly, the NOAA article fails to mention the following facts as well.

1. In the far southern hemisphere (e.g. New Zealand and Antarctica), the warmest period during the Holocene appears to have been roughly 8,000 to 10,500 years ago, immediately following the end of the last ice age[77][78]

2. "Only in summer": in recent years, have we not seen the re-appearance of extreme cold in winter, both in the Northern and Southern Hemispheres? Furthermore the statement of "warm winters" contradicts directly with the re-branded tale of "climate weirding" by proponents of Anthropogenic Global Warming, stipulating confidently that a warming planet results in extremes of both warm *and cold* temperatures.

3. The Holocene warming, first in the Southern, then in the Northern Hemisphere happened while CO2 concentration remained below 300 ppm. No alternate explanation has been provided, except a feeble attempt to minimize the warming itself.

What we observe here, once more, is pre-conceptual pseudoscience again at work. Logic is not essential. Contradictions are all right. Supporting facts are to be promoted, but any contrarian facts are to be concealed, minimized, rationalized or discredited. Anything goes as long as the foregone conclusion firmly remains that human CO2 emissions are causing climate change.

---

[77] Masson, V., Vimeux, F., Jouzel, J., Morgan, V., Delmotte, M., Ciais,P., Hammer, C., Johnsen, S., Lipenkov, V.Y., Mosley-Thompson, E.,Petit, J.-R., Steig, E.J., Stievenard,M., Vaikmae, R. (2000). "Holocene climate variability in Antarctica based on 11 ice-core isotopic records". Quaternary Research 54 (3): 348–358. doi:10.1006/qres.2000.2172.

[78] P.W. Williams, D.N.T. King, J.-X. Zhao K.D. Collerson (2004). "Speleothem master chronologies: combined Holocene ^{18}O and ^{13}C records from the North Island of New Zealand and their paleoenvironmental interpretation". *The Holocene* **14** (2): 194–208. doi:10.1191/0959683604hl676rp.

I would also request NOAA to reconcile its statement of "warm winters" with that of the Natural Resources Defense Council (NRDC) which has recently concluded that "extreme" hot and cold records were broken across the United States in 2011[79]

> Extreme weather records broken across the United States in 2011 - Though the year isn't yet over, 2011 proved to be rough when it came to extreme weather climate change is to blame. RECORDS FOR EXTREME HEAT AND EXTREME COLD WERE BROKEN IN ALL 50 U.S. STATES. The extremes at time have been disastrous, costing Americans an estimated overall $53 billion. In a conference Thursday, the Natural Resources Defense Council, an international nonprofit environmental organization, planned to release a map showing exactly how areas have been hit, including state-by-state analyses on weather extremes, record breakers, rainfall and snowfall.

> What's causing the changes? Perhaps climate change, according to the NRDC. A Special Report on Extreme Events from the Intergovernmental Panel on Climate Change has already concluded that the effects of climate change will intensify extreme heat, heavy precipitation, and maximum wind speeds of tropical storms.

So, aside from the fact that breaking a 150-year record in Earth's 4 billion years of existence may not necessarily be "extreme", which is correct? Warmer winters or hotter summers and colder winters? Depending on how the weather changed over the past two decades, proponents of Anthropogenic Global Warming have been trying to peddle both in parallel. This is clearly not science.

Science is based on developing a single theory based on structured observations and measurements, not two contradictory sets of predictions interchangeable based on how the weather evolves.

Personal Note

It is also ironic to see how NOAA has all of the funds it needs to pursue climate pseudoscience, but, when the NOAA administrator was recently asked by Congress about what the organization was doing to protect the West Coast of the United States from the massive debris of the 2011 Japan Earthquake making its way across the Pacific, Mr. David M. Kennedy pleaded "lack of funds" as the reason for doing very little about protecting the citizens and environment of the Western United States.[80]

> Begich, Chairman of the Subcommittee on Oceans, Atmosphere, Fisheries, and Coast Guard of the Senate Commerce Committee, presided over the hearing and expressed frustration to NOAA Assistant Administrator David Kennedy over the lack of a clean-up plan. Begich urged NOAA to shift its focus from monitoring the debris trajectory

---

[79] http://dscriber.com/dscriber/4045-climate-change-extreme-weather-records-broken-across-the-united-states-in-2011

[80] http://www.deltadiscovery.com/story/2012/05/16/statewide/begich-calls-for-faster-federal-response-to-tsunami-debris-calls-on-noaa-to-focus-on-clean-up-not-just-monitoring-asks-administration-for-45-million-for-effort/212.html

to clean-up efforts. Following the hearing, Lubchenco called Begich to discuss ways that NOAA can assist in the clean-up.

To fund the extensive and expensive clean-up, Begich sent a follow up letter to President Obama requesting $45 million to be made available this year and next to community groups to execute the debris clean-up

Temperatures in modern times

First there was runaway global warming....

At the turn of the millennium, when the warming period of the 1980s and 1990s reached its peak, when warming accelerated culminating in the spectacular heat of 1998, the credibility of the theory of anthropogenic global warming was also at its zenith. The climate science community, awash in billions of research dollars, was busy concocting nightmares of further acceleration in runaway global warming.

A panicked world population generally believed the climatologists and applauded them for their work, myself included. After all, with the decline of religion, especially in the Western world, scientists are often equated by people as the superhuman gods of modern times, their statements to be believed without question or further investigation.

Others were quick to ride the bandwagon. Al Gore (politician at large after the 2000 election) and Rajendra Pachauri (railway engineer who is the head of the IPCC) won the Nobel Peace Prize in recognition of their "selfless work for saving humanity".

The IPCC boldly predicted the end of snow in its Third Assessment Report, in the firm belief that the extraordinarily hot winter of 1998 would only get even worse in the next decade.

http://www.ipcc.ch/ipccreports/tar/wg2/569.htm

15.2.4.1.2.4. Ice Storms

Milder winter temperatures will decrease heavy snowstorms but could cause an increase in freezing rain if average daily temperatures fluctuate about the freezing point. It is difficult to predict where ice storms will occur and identify vulnerable populations. The ice storm of January 1998 (see Section 15.3.2.6) left 45 people dead and nearly 5 million people without heat or electricity in Ontario, Quebec, and New York (CDC, 1998; Francis and Hengeveld, 1998; Kerry et al., 1999). The storm had a huge impact on medical services and human health. Doctors' offices were forced to close, and a large number of surgeries were cancelled (Blair, 1998; Hamilton, 1998). One urban emergency department reported 327 injuries resulting from falls in a group of 257 patients (Smith et al., 1998b).

Dr. David Viner, senior research scientist at the CRU concurred:

http://www.independent.co.uk/environment/snowfalls-are-now-just-a-thing-of-the-past-724017.html

According to Dr David Viner, a senior research scientist at the climatic research unit (CRU) of the University of East Anglia,within a few years winter snowfall will become "a very rare and exciting event".

"Children just aren't going to know what snow is," he said

Al Gore was, of course, not going to be outshined.

It was November 7, 2006, and Los Angeles, California, had just set another record high temperature (97 degrees). That fact was not lost on Al Gore. "It shouldn't be this hot in November," he said. It was also Election Day, and the former U.S. vice president was on the campaign trail again. These days he is no longer a candidate, but he is still campaigning. This time, his battle is waged against global warming. He wants to alert people to the seriousness of this crisis, and with his movie, *An Inconvenient Truth*, a book of the same title, and his traveling slide show, Gore is trying to reach as many people as possible with his message.

Then, as usual, Mother Nature shattered human delusions once again. Cold winters returned in 2007-2008, followed by another in 2008-2009, then again in 2010-2011. And in 2011-2012, even if the continental United States has had a warm winter, followed by a very early start to summer, Eurasia, North Africa and Alaska have experienced epic cold not seen in many decades.

Climate scientists did know about the plateauing of global temperatures in advance. Hence, Phil Jones "completed Mike's Nature trick of adding in the real temps to each series for the last 20 years (ie, from 1981 onwards) and from 1961 to hide the decline." More about Climategate and the subsequent spin and attempted cover up will be described in a later chapter[81] [82].

Despite protestations of continued warming, credibility in the theory of Anthropogenic Global Warming began to subside quickly among the general population, not only in the United States as proponents often claim, but also in Europe and elsewhere in the world. A new tale had to be spun and quickly.

Then there was "Global Weirding"

Credit for interweaving the recently revived extreme cold weather into the tale of global warming can be attributed primarily to Thomas Friedman, journalist, columnist and author, and Dr. Jeffrey D. Sachs, Director of the Earth Institute of Columbia University.

Thomas Friedman renamed "Climate Change" to "Global Weirding" in his book: *Hot, Flat, and Crowded*

http://fora.tv/2008/09/23/Tom_Friedman_Hot_Flat_and_Crowded

---

[81] Briefly, here is the background:
- Of all the sample of tree rings taken over, one series was known to show a temperature decline this century
- All the other tree ring proxies (samples) moved with the temperature (showed a increase in temperature in the last century that matched the actual thermometer)
- Meaning it was felt there was some other error. So they arbitrarily excluded it. Such exclusion of outliers on a pre-conceptual "feeling" without extensive justification is statistical malpractice.
- Also refer to the following peer review: http://climateaudit.org/2011/12/02/peer-review-of-enhanced-hide-the-decline/#comment-314396.

[82] Since Climategate, proponents of AGW have attempted damage-control by saying that the "hide the decline" remark was taken "out of context", investigations "cleared" Jones, etc, without actually going into the specifics and without providing a factual explanation of why the discordant series was removed.

In February 2010, Friedman also started a new Op-Ed column in the New York Times named "Global Weirding Is Here"

http://www.nytimes.com/2010/02/17/opinion/17friedman.html?_r=1

"Of the festivals of nonsense that periodically overtake American politics, surely the silliest is the argument that because Washington is having a particularly snowy winter it proves that climate change is a hoax and, therefore, we need not bother with all this girly-man stuff like renewable energy, solar panels and carbon taxes. Just drill, baby, drill.

When you see lawmakers like Senator Jim DeMint of South Carolina tweeting that "it is going to keep snowing until Al Gore cries 'uncle,' " or news that the grandchildren of Senator James Inhofe of Oklahoma are building an igloo next to the Capitol with a big sign that says "Al Gore's New Home," you really wonder if we can have a serious discussion about the climate-energy issue anymore.

The climate-science community is not blameless. It knew it was up against formidable forces — from the oil and coal companies that finance the studies skeptical of climate change to conservatives who hate anything that will lead to more government regulations to the Chamber of Commerce that will resist any energy taxes. Therefore, climate experts can't leave themselves vulnerable by citing non-peer-reviewed research or failing to respond to legitimate questions, some of which happened with both the Climatic Research Unit at the University of East Anglia and the United Nations Intergovernmental Panel on Climate Change.

Although there remains a mountain of research from multiple institutions about the reality of climate change, the public has grown uneasy. What's real? In my view, the climate-science community should convene its top experts — from places like NASA, America's national laboratories, the Massachusetts Institute of Technology, Stanford, the California Institute of Technology and the U.K. Met Office Hadley Centre — and produce a simple 50-page report. They could call it "What We Know," summarizing everything we already know about climate change in language that a sixth grader could understand, with unimpeachable peer-reviewed footnotes.

At the same time, they should add a summary of all the errors and wild exaggerations made by the climate skeptics — and where they get their funding. It is time the climate scientists stopped just playing defense. The physicist Joseph Romm, a leading climate writer, is posting on his Web site, climateprogress.org, his own listing of the best scientific papers on every aspect of climate change for anyone who wants a quick summary now.

Here are the points I like to stress:

1) Avoid the term "global warming." I prefer the term "global weirding," because that is what actually happens as global temperatures rise and the climate changes. The weather gets weird. The hots are expected to get hotter, the wets wetter, the dries drier and the most violent storms more numerous.

The fact that it has snowed like crazy in Washington — while it has rained at the Winter Olympics in Canada, while Australia is having a record 13-year drought — is right in line with what every major study on climate change predicts: The weather will get weird; some areas will get more precipitation than ever; others will become drier than ever.

2) Historically, we know that the climate has warmed and cooled slowly, going from Ice Ages to warming periods, driven, in part, by changes in the earth's orbit and hence the amount of sunlight different parts of the earth get. What the current debate is about is whether humans — by emitting so much carbon and thickening the greenhouse-gas blanket around the earth so that it traps more heat — are now rapidly exacerbating nature's natural warming cycles to a degree that could lead to dangerous disruptions.

3) Those who favor taking action are saying: "Because the warming that humans are doing is irreversible and potentially catastrophic, let's buy some insurance — by investing in renewable energy, energy efficiency and mass transit — because this insurance will also actually make us richer and more secure." We will import less oil, invent and export more clean-tech products, send fewer dollars overseas to buy oil and, most importantly, diminish the dollars that are sustaining the worst petro-dictators in the world who indirectly fund terrorists and the schools that nurture them.

4) Even if climate change proves less catastrophic than some fear, in a world that is forecast to grow from 6.7 billion to 9.2 billion people between now and 2050, more and more of whom will live like Americans, demand for renewable energy and clean water is going to soar. It is obviously going to be the next great global industry.

China, of course, understands that, which is why it is investing heavily in clean-tech, efficiency and high-speed rail. It sees the future trends and is betting on them. Indeed, I suspect China is quietly laughing at us right now. And Iran, Russia, Venezuela and the whole OPEC gang are high-fiving each other. Nothing better serves their interests than to see Americans becoming confused about climate change, and, therefore, less inclined to move toward clean-tech and, therefore, more certain to remain addicted to oil. Yes, sir, it is morning in Saudi Arabia."

Both Friedman and Sachs outlined this amended storyline eloquently on mainstream news media such as CNN and journals such as Scientific American.

http://www.scientificamerican.com/article.cfm?id=the-deepening-crisis

"This year was ushered in by the phony "Climategate" controversy, which involved leaked e-mails of a British climate research unit; the political right wing depicted some ill-considered language in the messages as proof of a vast global plot. Independent reviews have since rejected the charges of scientific conspiracy, but the damage is done: the U.S. public once again swings toward disbelief in the basic science of human-induced climate change."

Popular science journals were quick to jump on to the re- engineered bandwagon. For example, Scientific American and real-science.com declared with all sincerity:

https://www.scientificamerican.com/article.cfm?id=ethanol-corn-climate
http://www.scientificamerican.com/podcast/episode.cfm?id=what-does-winter-weather-reveal-abo-10-02-11
http://www.real-science.com/record-early-snowfall-new-york

"Sadly, climate change won't save you from bundling up, or shoveling. Even in a much warmer world, there will still be colder than average winters."

It is interesting to note that different magazines and journals used the exact same wording. Google the above phrase and you will get over 200,000 hits.

Below is another example of the renovation of the tale of global warming from Scientific American:

"What Does Winter Weather Reveal about Global Warming? No single weather event proves or disproves the fundamental science of climate change, but extreme weather is what scientists expect from global warming. David Biello reports"

The mainstream media was not to be outdone by the science journals.

Below are more details:

http://www.mrc.org/node/27124

Media Use Crazy Weather to Hype Global Warming, Despite Admissions Weather Isn't Climate

"Last winter, as blizzard snowfalls piled up into several feet in the nation's capital, conservatives mocked global warming alarmists for trying to link weather incidents to global warming. But as summer heat waves, volcanoes and sinkholes have appeared recently, climate alarmists proved they missed the point.

A top Obama administration scientist attacked global warming skeptics during the winter by pointing out that "weather is not the same thing as climate." ABC's Bill Blakemore argued the same thing in order to defend the existence of manmade global warming on Jan. 8, 2010.

But Associated Press, USA Today, The New York Times and The Washington Post have all promoted a connection between the extreme heat and weather around the world this summer and global warming. One CNN host asked if the events were the "apocalypse" or global warming. The Huffington Post proposed naming hurricanes and other disasters after climate change "deniers."

"Floods, fires, melting ice and feverish heat: From smoke-choked Moscow to water-soaked Iowa and the High Arctic, the planet seems to be having a midsummer breakdown. It's not just a portent of things to come, scientists say, but a sign of troubling climate change already under way," the AP wrote, sounding more like Al Gore than an objective news agency.

AP cited the World Meteorological Organization, NASA and the Intergovernmental Panel on Climate Change (IPCC) saying that "extremes" were expected in a warming scenario. But its report didn't include any other viewpoints or propose other possible reasons for the weather events. And it failed to point out the scandals connected to IPCC, NASA and the warming movement as a whole.

The 2009 ClimateGate scandal and subsequent scandals undermined the very credibility of the climate alarmist movement, but were underreported by the network news media.

AP left out meteorologists who explained some of those events based on jet stream activity. According to New Scientist magazine, the jet stream is being blocked right now and has consequently slowed down. Meteorologists say that the jet stream's slower movements are responsible for the deadly fires in Russia, the floods in Pakistan and other rare weather events. "The unusual weather in the US and Canada last month also has a similar case," New Scientist wrote.

Discover Magazine expounded on the New Scientist article saying "this happens from time to time, and it sets the stage for extreme conditions when weather systems hover over the same area."

Despite other explanations and viewpoints, The New York Times also linked weather to climate saying, "the collective answer of the scientific community [whether global warming is causing more weather extremes]" is "probably."

Like the Times, many news outlets promoted the connection between warming and weather, but were careful to briefly note that individual weather events cannot be proven to have been caused by global warming. Out of the Times' 1,302 word article, only 113 words were used to offer a caveat saying it is difficult to link "specific weather events" to climate change and to quote a NASA scientist who admitted he hasn't "proved it" yet.

Semantics aside, those mainstream stories were nearly as biased in their coverage as blatantly left-wing websites like the Huffington Post.

Huffington Post argued that "global weirding" incidents such as landslides, sinkholes and volcanoes are "consistent" with global warming.

The site interviewed David Orr, a professor of environmental studies and politics at Oberlin College, who said, "you ask is this evidence of climate destabilization, the only scientific answer you can give is: It is consistent with what we can expect." The complete list of "weird" stuff was heat waves, floods, landslides, wildfires, ice islands, sinkholes, volcanoes, dead fish and oyster herpes."

Dead fish and oyster herpes? Huffington Post said, "These are certainly stories to be filed under weird: Although climate change can't necessarily be held responsible, some scientists are suggesting it as the instigator of strange ocean occurrences."

The fact is that the alarmists and the news media will find someone to support claims that just about everything is correlated to man-made global warming. MSNBC host Dylan Ratigan even claimed that Snowpocalypse (the nickname for the blizzard activity on parts of the East Coast) was consistent with global warming. "

.... (Please refer to the link http://www.mrc.org/node/27124 for the rest of the article)

****

Opposing Views.com summarizes the situation well

http://www.opposingviews.com/i/east-coast-blizzard-is-also-a-sign-of-global-warming-say-alarmists

"The real problem is with both claims taken together: last year's relative lack of snow was cited as proof of global warming, and this year's over-abundance of snow is also cited as proof of global warming. A theory that is considered confirmed by whatever happens, no matter what happens, is not a scientific theory at all. It's like Petr Beckmann's example of an "inherently irrefutable" claim, which is that there is a second moon orbiting the earth which has zero mass and becomes transparent when illuminated. That theory is entirely consistent with such an object never being observed — which means it can never be disproved no matter how false it is.

So it is with global warming. If there is lower-than-average snow it's due to global warming ("too warm for snow to form") and if there is higher-than-average snow it's due to global warming ("more moisture there, more snow here"), and if snowfall is average, the two cancel out. If northern Europe as a less severe winter, it's due to global warming making winters less cold; if northern Europe as a more severe winter, it's due to global warming interfering with the Gulf Stream.

No matter what happens, it's "proof" of global warming. It's the theory that can never be disproved no matter how false it is. Which is to say, it's not really a scientific theory at all.

Does anyone remember the clear contradictions with "Global Warming" advocated at the turn of the millennium?

There are, evidently, some clear contradictions with the message of the IPCC and Al Gore's "Inconvenient Truth" at the beginning of the decade.

But who is going to remember anything that was said ten years earlier, especially as the mainstream media is still generally supporting the Anthropogenic Global Warming scenario. A few media organizations which were not, such as Fox News could easily be labeled as fringe, rightwing and in the pocket of the large oil companies.

What is not so noble is the strategy to also taint anyone or any organization attempting to investigate or document factual discrepancies with the theory of Anthropogenic Global Warming as also fringe, rightwing and in the pocket of the large oil companies.

The Daily Telegraph has been one of the few mainstream media outlets to take on the now "Global Weirding" lobby and has, as a consequence be subject to the above accusations. The below article very factually points out the contradictions between "Global Warming" and "Global Weirding"

http://blogs.telegraph.co.uk/news/jamesdelingpole/100148381/global-weirding-the-new-big-lie/

Blog version as of March 31st, 2012

"They're calling it Global Weirding now, as I suppose, inevitably they were bound to do in the end. Well "global warming" stopped working in 1989 when the globe stopped warming. Climate change was always a bit of a non-starter because climate does change regardless of whether or not we all drive 4 x 4s, or buy carbon offsets or listen to Stephen Fry and Ron Weasley's injunction to take our holidays in England this year. And Global Climate Disruption, as some pillock tried to christen it, was never going to catch on because, well, it's just too blatantly contrived and desperate isn't it?

So Global Weirding it is. The concept was popularised last week in a characteristically dire and parti pris BBC Horizon documentary which purported to have lots of new evidence (or 'hearsay' as it would more likely have been termed in a court of law) showing that our weather is getting more extreme – weirder. It seems to have been broadcast to coincide with a new IPCC report which has been excitedly written up in newspapers like the Guardian and the Detroit Free Press as evidence that we are heading towards climate disaster.

Global warming is leading to such severe storms, droughts and heatwaves that nations should prepare for an unprecedented onslaught of deadly and costly weather disasters, an international panel of scientists has said.

The greatest danger is in highly populated, poorer regions, but no corner of the globe is immune. The document, by the Intergovernmental Panel on Climate Change, forecasts stronger tropical cyclones and more frequent heatwaves, deluges and droughts, and blames man-made climate change, population shifts and poverty.

But this is pretty much the exact opposite of what the IPCC report actually says. As Roger Pielke Jr has noted, the report is a far cry from the IPCC's usual slipshod, scaremongering standards:

Kudos to the IPCC — they have gotten the issue just about right, where "right" means that the report accurately reflects the academic literature on this topic. Over time good science will win out over the rest — sometimes it just takes a little while.

A few quotable quotes from the report (from Chapter 4):

"There is medium evidence and high agreement that long-term trends in normalized losses have not been attributed to natural or anthropogenic climate change"

"The statement about the absence of trends in impacts attributable to natural or anthropogenic climate change holds for tropical and extratropical storms and tornados"

"The absence of an attributable climate change signal in losses also holds for flood losses"

The report even takes care of tying up a loose end that has allowed some commentators to avoid the scientific literature:

"Some authors suggest that a (natural or anthropogenic) climate change signal can be found in the records of disaster losses (e.g., Mills, 2005; Höppe and Grimm, 2009), but their work is in the nature of reviews and commentary rather than empirical research."

So what this IPCC report is saying is that WE DO NOT KNOW if there's an anthropogenic signal in extreme weather patterns, and that there does not seem to be a trend towards increased extreme weather events such as tornados and tropical storms. Yet the liberal MSM is reporting the opposite. How come?

Well here's the weird part. The misinformation comes from the IPCC's summary of its own report (available here) which has been regurgitated, in classic churnalism style, by all the usual lazy MSM suspects. (H/T Katabasis)

It begins:

"Evidence suggests that climate change has led to changes in climate extremes such as heat waves, record high temperatures and, in many regions, heavy precipitation in the past half century, the Intergovernmental Panel on Climate Change said today."

The IPCC, of course, has form in this regard. With its four Assessment Reports, its Summaries for Policymakers have been notably more extreme and confident in CAGW than the reports themselves warrant. So yet again what we have here is the work of serious-minded, neutral scientists (yes, they do still exist) being twisted for political purposes by activists.

As we know, the great global warming alarmism Ponzi scheme is looking extremely vulnerable at the moment. Global warming has stopped. There's a growing public backlash against eco-taxes, ugly flickery lightbulbs, higher energy bills, bat chomping eco-crucifixes and all the other paraphernalia of the environmental religion. And unfortunately, as we saw in '44 and '45, what these kind of people do when they get backed into a corner is not surrender but get nastier and more devious.

We've seen this recently in the Fakegate affair. And in Leo Hickman of the Guardian's contemptible "expose" of one of the hitherto anonymous donors of the Global Warming Policy Foundation. And in the Planet Under Pressure comedy conference staged last week by comedy organisations including the Royal Society, mainly in order to try to breathe new life into the stagnant, green-tinged corpse of climate alarmism.

One of the speakers at Planet Under Pressure claimed – in apparent seriousness – that climate scepticism was an illness that needed to be treated.

Scepticism regarding the need for immediate and massive action against carbon emissions is a sickness of societies and individuals which needs to be "treated", according to an Oregon-based professor of "sociology and environmental studies". Professor Kari Norgaard compares the struggle against climate scepticism to that against racism and slavery in the US South.

Prof Norgaard holds a B.S. in biology and a master's and PhD in sociology.

"Over the past ten years I have published and taught in the areas of environmental sociology, gender and environment, race and environment, climate change, sociology of culture, social movements and sociology of emotions," she says.

As Paul Joseph Watson notes at Prison Planet:

The effort to re-brand legitimate scientific dissent as a mental disorder that requires pharmacological or psychological treatment is a frightening glimpse into the Brave New World society climate change alarmists see themselves as ruling over.

Due to the fact that skepticism towards man-made global warming is running at an all time high, and with good reason, rather than admit they have lost the debate, climate change alarmists are instead advocating that their ideological opponents simply be drugged or brainwashed into compliance."

Will there be further re-branding of the tale?

Who knows? The unusually warm winter and early start of summer in the continental United States in 2011-2012, even if not replicated elsewhere on the globe, may give pause for thought for the likes of Friedman and Sachs. Perhaps a new re-writing of the script back to "Global Warming" is in the works, consistent with the latest weather? Only time will tell.

What are the facts? Is the current weather really getting warmer and/or more extreme compared to the past?

Firstly, one should consider the previous extremes of weather that occurred
- Recently during the Little Ice Age. See Section Hypothesis #1: Sunspots are disappearing (Don J. Easterbrook, H. Svensmark and others))
- Significant natural climate changes that occurred during the Holocene and previous periods of Earth's history dating back to half a billion years ago. See Chapter History of Global Temperature

The current weather difficulties pale in comparison to past troubles.

Let us now also examine some facts right in modern times - the 20th century and the beginning of the 21st century. We should bear in mind that a hundred years is a very, very short period in the geologic timescale. But even within this limited period, there are clear differences between what is hyped in the media and the reality.

<u>Global Weather & Climate Extremes</u> World Meteorological Organization

Continent	Highest Temperature Recorded during past century	Lowest Temperature Recorded during past century
Africa	57.8 °C (136.0 °F) 'Aziziya, Libya 13 September 1922.	−23.9 °C (−11.0 °F) Ifrane, Morocco 11 February 1935
Antarctica	15 °C (59 °F) Vanda Station 5 January 1974 −89.2 °C (−128.6 °F)	Vostok Station 21 July 1983
Asia	55 °C (131 °F) Mitraba, Kuwait 15 June 2010C	−67.8 °C (−90.0 °F) Measured Verkhoyansk, Siberia, Russia (then in the Russian Empire) 5 February 1892 −71.2 °C (−96.2 °F) Extrapolated Oymyakon, Siberia, Russia (then in the Soviet Union) 26 January 1926[3]
Europe	48.0 °C (118.4 °F) Athens, Greece (and Elefsina, Greece) 10 July 1977 D	−58.1 °C (−72.6 °F) Ust-Shchuger, Russia 31 December 1978
North America	56.7 °C (134.1 °F) Death Valley, California, U.S.A. 10 July 1913	−63 °C (−81.4 °F) Snag, Yukon, Canada 3 February 1947 −66.1 °C (−87.0 °F) North Ice, Greenland 9 January 1954
Oceania (Australia)	50.7 °C (123.3 °F) Oodnadatta, South Australia, Australia 2 January 1960	−25.6 °C (−14.1 °F) Ranfurly, New Zealand 18 July 1903
South America	48.9 °C (120.0 °F) Rivadavia, Salta Province, Argentina 11 December 1905	Argentina 11 December 1905 −32.8 °C (−27.0 °F) Sarmiento, Argentina 1 June 1907

Observations

- Of course, both warm and cold records are broken *locally* in many locations every summer and every winter respectively. However, the absolute extremes of temperature within the past century have almost all been set many decades ago with only one exception.

- A cautionary note again: the above table only documents temperature extremes measured by human instruments in the past 100 years, which is 0.0000025% of Earth's geologic history. When the entire record of Earth's temperature history based on research with flora and fauna and other proxies is taken into account, the variation in temperature extremes on Earth are much, much wider.

Cooling between 1880 and 1910
http://www.ncdc.noaa.gov/img/climate/research/2008/ann/global-jan-nov-error-bar-pg.gif

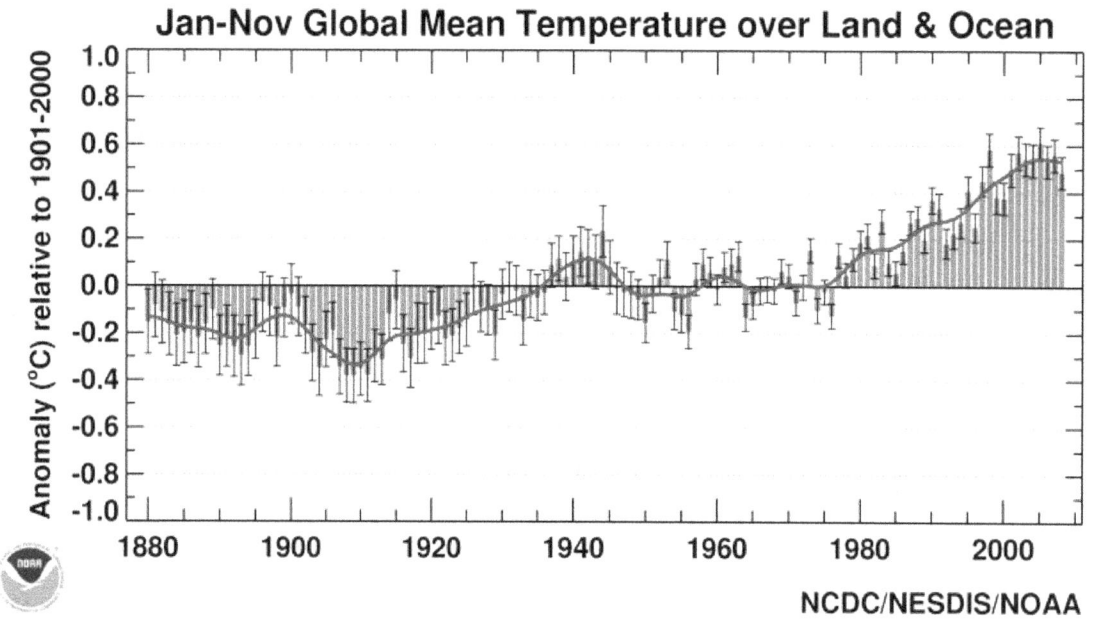

Observations
- The Temperature anomaly dipped from -0.1 °C to -0.35 °C between 1880 and 1910
- CO2 concentration went up from 284.3 ppm in 1832 to 299.7 ppm in 1910.
- Regional increases of CO2 concentration in rapidly industrializing parts of the world, such as Europe and North America were, obviously much higher, although exact figures are unavailable.

1936 North American Heat Wave

http://www.crh.noaa.gov/arx/events/heatwave36.php

http://www.ncdc.noaa.gov/oa/climate/research/cag3/cag3.html

The 1936 North American heat wave was the most severe heat wave in the modern history of North America. It took place in the middle of the Great Depression and Dust Bowl of the 1930s, and caused catastrophic human suffering and an enormous economic toll. The death toll exceeded 5,000, and huge numbers of crops were destroyed by the heat and lack of moisture. Many state and city record high temperatures set during the 1936 heat wave still stand to this day.

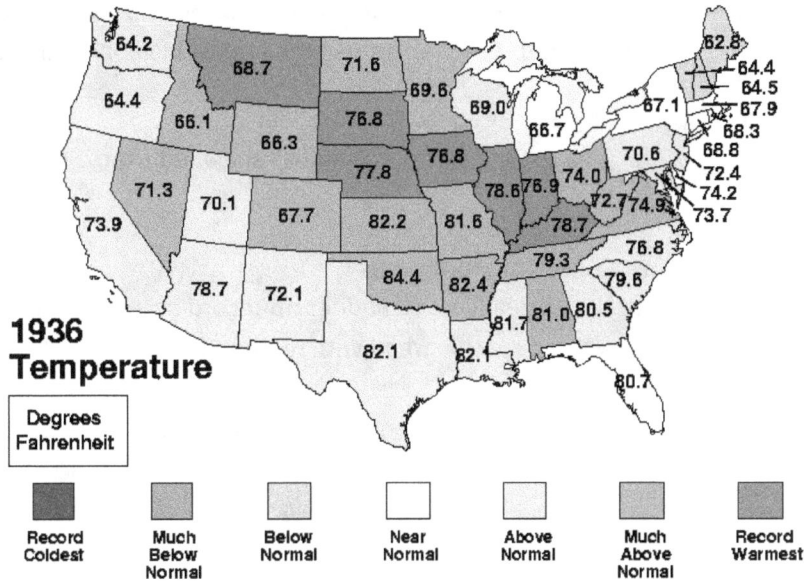

The heat wave followed one of the coldest winters on record.

Fortunately, for Americans living in the 1930s, neither "Global Warming" nor "Global Weirding" was invented at that time. They had enough to worry about with the Great Depression.

Tornadoes

Much has been hyped about the recent spate of tornado activity in the United States in 2011 and 2012. What do the facts tell us?

Most tornadoes in single 24-hour outbreak [83]

> The April 25–28, 2011 tornado outbreak is the most prolific tornado outbreak in US history. It produced approximately 358 tornadoes, with 206 of those in a single 24-hour period. 335 deaths occurred in that same 24-hour time period of which 322 were tornado related.

> The outbreak has also helped smash the record for most tornadoes in the month of April with 770 tornadoes, more than double the prior record (267 in April 1974).

> The overall record for a single month was 542 in May 2003, which was also broken.

but....

Largest outbreak in the fall [84]

> Most tornado outbreaks occur in the spring, but there is a secondary peak of tornado activity in the fall. In 1992, 95 tornadoes broke out in 41 hours of continuous tornado activity from

[83] "April 2011 tornado information". NOAA. http://www.noaanews.noaa.gov/april_2011_tornado_information.html

[84] Grazulis, Thomas P. (July 1993). Significant Tornadoes 1680–1991: A Chronology and Analysis of Events. St. Johnsbury, VT: The Tornado Project of Environmental Films. ISBN 1-879362-03-1

November 21 to 23. Many other very large outbreaks have occurred in the fall, especially in November and early December

Longest continuous outbreak [85]

Under most definitions, the November 1992 tornado outbreak is also the longest continuous tornado outbreak, and among the largest in geographic scope, as well.

Most tornadoes spawned from a hurricane

The greatest number of tornadoes spawned from a hurricane is 117 from Hurricane Ivan in 2004

10 deadliest American tornadoes [86]

9 of the 10 deadliest tornadoes on record occurred many decades or more than a century ago with no specific trend with relation to time period. The deadliest was in 1925. See table below.

Rank	Name (location)	Date	Deaths
	10 deadliest American tornadoes		
1	"Tri-State" (Missouri, Illinois and Indiana)	March 18, 1925	695
2	Natchez, Mississippi	May 7, 1840	317
3	St. Louis, Missouri and East St. Louis, Illinois	May 27, 1896	255
4	Tupelo, Mississippi	April 5, 1936	216
5	Gainesville, Georgia	April 5, 1936	203
6	Woodward, Oklahoma	April 9, 1947	181
7	Joplin, Missouri	May 22, 2011	161
8	Amite, Louisiana and Purvis, Mississippi	April 24, 1908	143
9	New Richmond, Wisconsin	June 12, 1899	117
10	Flint, Michigan	June 8, 1953	116

Again a cautionary note: the above table only documents tornadoes recorded in the past 150 years, which is 0.00000375% of Earth's geologic history. Extreme tornadoes which may have occurred in ages past are not taken into consideration, given that no data is available.

[85] Grazulis, Thomas P. (July 1993). Significant Tornadoes 1680–1991: A Chronology and Analysis of Events. St. Johnsbury, VT: The Tornado Project of Environmental Films. ISBN 1-879362-03-1
[86] http://www.spc.noaa.gov/faq/tornado/killers.html

V.   Examples of cold weather in recent years

In this chapter, I shall not mention "extreme" hot weather events in contemporary times. The reader can easily refer to television programs and articles in both the pseudoscience and mainstream media have been, and continue to be cherry-picked and puffed-up. I recommend the *Huffington Post*, if the reader wishes to be enlightened with a comprehensive list of recent blistering weather events, such as the 2010 Russian heat wave, the 2011 Texan heat wave, the very mild winter in 2011 in the Eastern and Central United States, the early cherry blossom in Washington, D.C. on 2012, to name but a few sizzling "excesses" of Mother Nature. By no means am I denying that these hot weather events happened. But the showering of record hot weather with hyperbole, while at the same time ignoring, concealing, minimizing and/or explaining away record cold weather, is tantamount to a de-facto exaggeration and distortion of the true situation, derived from a predetermined conclusion. The cold weather events listed below are intended to bring balance into the discussion, not to deny the heat waves that have occurred recently, and not to even remotely suggest global cooling.

However, the resurgence of "extreme" cold weather since 2007 is, once again, indicative of how little humanity really knows about the multitude of factors that determine both local and global climate. Most of what is published today is subject to change, due to new research, new facts uncovered from the past and/or observed in the future. Everyone, from the foremost "expert" in climate pseudoscience to the peasant grandmother in Central Asia predicting the weather, is guessing in the wind.

At the turn of the century, Dr David Viner, a senior research scientist at the climatic research unit (CRU) of the University of East Anglia, stated confidently that within a few years winter snowfall will become "a very rare and exciting event". "Children just aren't going to know what snow is," he said.

The Independent newspaper in the United Kingdom was quick to add overstatement to this global warming fire: [87]

```
Britain's winter ends tomorrow with further indications of a striking
environmental change: snow is starting to disappear from our lives.

Sledges, snowmen, snowballs and the excitement of waking to find that the
stuff has settled outside are all a rapidly diminishing part of Britain's
culture, as warmer winters - which scientists are attributing to global
climate change - produce not only fewer white Christmases, but fewer white
Januaries and Februaries.
```

As mentioned earlier, it did not take long for the IPCC to "scientifically" confirm the replacement of snow with ice storms[88]

---

[87] http://www.independent.co.uk/environment/snowfalls-are-now-just-a-thing-of-the-past-724017.html

[88] http://www.ipcc.ch/ipccreports/tar/wg2/569.htm

Later in the decade, when Mother Nature augmented, instead of eliminated snow, Richard Somerville, lead IPCC author, did not blink an eye when he stated on the American Television channel, ABC, that climate scientists had been predicting for "decades" that there would be more snow. [89]

```
Q: for some decades climate scientists have been predicting that heavy
snowfalls in the winter would be more frequent.

A: That's right, in fact .....
```

Unfortunately for Mr. Somerville, the age of the Internet ensures easier transparency of the past record than it was ever possible in the past.

A final word before we go into the examples of cold weather. Humanity should be very careful not to equate "record" with "extreme". From the perspective of the Earth's 4 billion year history, both record hot and cold weather are, by no means, extreme. The Earth is constantly telling us "been there, done that!" At best, we have climate records for a century and a half, which is 0.00000375% of the entire period of existence of the Earth. Furthermore, as clearly depicted in the Chapter History of Global Temperature, today's Earth is a very cold, frigid "Ice House" compared to the much higher temperatures in most of Earth's history, including the world of the dinosaurs when this planet was teeming with life.

As mentioned earlier, the following sections are not meant to deny the heat waves glorified in the mainstream and pseudoscience press, but rather to provide a mirror-image of facts in order to achieve balance. Consider these sections as a reverse-engineering of the *Huffington Post*, the misnamed *thinkprogress.org* and other hot weather cherry-picking publications. Mainstream and pseudoscience publications such as TIME Magazine hyped up the mild winter in the Eastern and Central United States with the absurd headline: "The Year That Winter Forgot: Is It Climate Change?", under *TIME Science*[90]. This is a good example of how pervasive bad pseudoscience has overrun much of what people are offered to read in the media.

June 2012

While there was much hype in the popular press about the hot summer in the United States, Americans swayed by the flurry of climate mania would have done well to have reminded themselves that the entire expanse of the continental United States, large as it may seem, is but a small fraction of the overall surface area of the planet. On the other side of the pond, cool conditions lingered on right up to the summer solstice – June 20th in this leap year.

The Thames Diamond Jubilee Pageant took place on June 3rd, a parade of more than 1,000 ships and boats on the Tideway of the River Thames in London as part of the celebrations to commemorate sixty years of Elizabeth II. The Queen, Prince Philip and other members of the Royal Family were aboard vessels that took part in the parade. Right in the middle of the festivities, the all-seeing television cameras could no longer see the Queen waving from the *Spirit of Chartwell* , the boat carrying the royals. For a brief suspense-filled moment, reporters were wondering what happened!

---

[89] http://abcnews.go.com/

[90] http://www.time.com/time/health/article/0,8599,2104040,00.html?xid=gonewsedit

Where was the Queen? Had anything untoward occurred to Her Majesty? Sadly for the ever sensationalist media, there was no breaking bad news to report.

As an octogenarian, the Queen was well accustomed to the damp and cool summers of the United Kingdom. Yet she had not anticipated the ferocity of the cold wind on the Thames in "globally warmed" 2012. A short while after the Queen's "disappearance", she magically resurfaced on-deck. The Queen had merely popped below-deck to put on some warm clothing and a scarf.

Further inland, in the heartland of Eurasia, the weather was not much hotter. In southwest Russia, near the coast of the Black Sea, cool and wet conditions persisted, culminating in floods in early July. 172 people died during the floods, and Russian police said the floods damaged the homes of nearly 13,000 people. Peddlers of climate fraud lost no time in attributing the watery conditions to global warming. In 2010, the same people had cited the extreme heat and drought in Russia as ominous "signs" of a warming planet, and unashamedly engaged in wild predictions of such warm and dry conditions "surely" getting even more extreme in the future. Only two years later, Mother Nature had proven them completely wrong, yet that did not even make the tiniest of dents in the faith of proponents of Anthropogenic Global Warming. Amnesia is a good attribute to nurture when one is in the business of climate pre-conceptualism.

On the other extremity of the planet, Antarctica was experiencing one of the coldest winters on record in this "warming planet[91]. On June 12th, 2012, temperature at Vostok Station dropped to -81 °C, below the freezing point of carbon dioxide. Fortunately this did not mean that dry ice (frozen

---

[91] https://stevengoddard.wordpress.com/2012/06/12/antarctic-temperature-drops-below-the-freezing-point-of-co2/

carbon dioxide) accumulated, because the low partial pressure of CO_2 in the air meant that frozen CO_2 molecules sublimated at an average rate as fast as they could freeze.

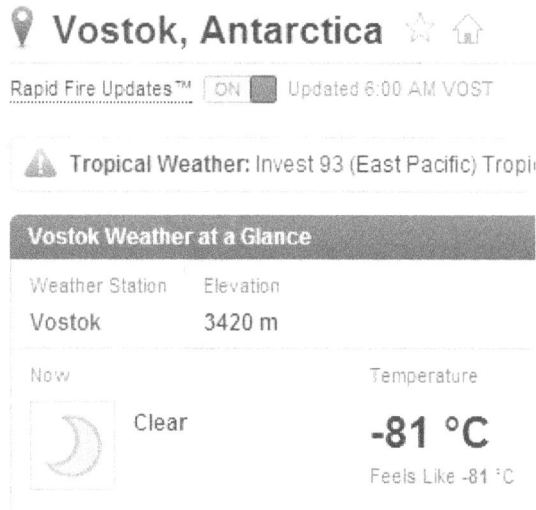

Extreme cold temperatures, even by Antarctic standards, persisted throughout the Southern Hemisphere winter, right until the end of July. The average temperature in Vostok during June and July is around -65°C[92], meaning that temperatures had dipped more than 15°C below average this winter. In fact, the temperatures experienced in Vostok in June and July of 2012 came very close to the lowest temperature ever recorded on this planet by human instruments: -89.6°C at Vostok station on July 21st, 1983. Strange for an ever warming planet, isn't it? Incidentally, the highest temperature ever recorded at the South Pole (Amundsen-Scott Station) was −13.6°C on the 27th of December, 1978[93].

I would surmise that this is the first time the reader has even heard of near record cold temperatures in Antarctica in 2012. On the other hand, I am also sure that the reader has been bombarded with reports of icebergs melting. This is not surprising, demonstrating once again the near-universal bias of the mainstream and pseudoscience media: hide and/or explain away the cold and exaggerate the warmth.

May 2012

I am allergic to pollen. It is often said that people who are allergic to pollen are the first to notice the beginning of spring. After all, itchy eyes, uncontrollable tears, a running nose and fits of runaway sneezing are not easy to ignore. Most of us await the beginning of spring with joy. However, those of us who are affected by pollen anticipate the end of winter with some trepidation.

It is true that in the late 1990s and early 2000s, winters became warmer and shorter. At the turn of the century, I remember being surprised when I felt the start of spring as early as end of January; this helped reinforce my then strong belief in Anthropogenic Global Warming. The trend reversed around 2007 when colder and longer winters returned in many parts of the world. Pollen attacks on allergic people shifted back to March or even later.

---

[92] http://www.world-climates.com/city-climate-vostok-antarctic-antarctica/
[93] http://wiki.answers.com/Q/What_is_the_coldest_temperature_ever_recorded_at_the_South_Pole

Mother Nature still has the propensity to continually surprise us. In the winter of 2011-2012, she decided to spare the Eastern and Central United States with a mild winter, but the rest of the Northern Hemisphere was not so lucky.

May 26th, 2012

The winter may have been mild in a good part of the continental United States, but, as usual, Mother Nature, and not the climate change proponents had the last word, even in North America.

- A late May storm in Billings Montana set 2 snowfall records.[94]

The storm that dropped snow in south-central Montana on Friday also set a couple of weather records.

The Billings area received 2.2 inches of snow as of midnight Friday, around the time it stopped snowing, said Tom Humphrey, forecaster for the Billings office of the National Weather Service. That was a record snowfall for May 25, Humphrey said.

"That broke the old record of 0.3 inches in 1975," he said.

A second record that was set has to do with snow depth. The NWS measures snow depth at 6 a.m., and on Saturday, that depth measured 1 inch, Humphrey said.

"That's the latest record of 1 inch this late in May," he said.

The previous record was set in 1983, when on May 13 it snowed 10 inches, Humphrey said. All of that snow had melted by the next day.

Humphrey called the storm "a very effective precipitation producer, when you factor in rain and snow." Most of the snow fell along the east slopes of the Absaroka/Beartooth Mountains, with the town of Red Lodge getting 5 to 10 inches.

---

[94] http://billingsgazette.com/news/local/potent-storm-sets-snowfall-records-for-billings/arti%20cle_7151b56e-3947-5f5a-9e08-7bfae17e6599.html?comment_form=true

The Bull Mountains south of Roundup got 6 inches, and the area south
of Livingston got 2 to 4 inches.

Snow is not uncommon in May, Humphrey said, but it's more likely to
fall in the first half of the month. When it does fall, it tends to
melt quickly because of the warmth of the ground.

Humphrey said an even bigger storm system was expected to move into
the Billings area late Saturday afternoon, bringing mainly rain, and
continuing through the night and into Sunday. The wet weather will
likely taper off by Sunday night.

"Hopefully we'll be able to squeeze one dry day for the Memorial Day
weekend," Humphrey said.

The cooler weather pattern is expected to continue, with temperatures
in the low to mid 50s on Sunday and in the low 60s on Monday. Normal
temperatures for this time of the year are in the low 70s, he said.

- And in Ontario, 80% of Ontario apples and other tree fruits were wiped out by an unusual
  surge of cold during much of May. Ontario was not alone in its misery. In Kentucky, for
  example, heart nuts were destroyed as frost hit during bloom.

http://www.theobserver.ca/2012/05/26/80-of-ontario-apples-other-tree-fruits-wiped-out

The state of John Zekveld's orchard is about as bad as he first feared
after severe frost struck several weeks ago.

The owner of Zekveld's Garden Market and orchard at Reece's Corners,
like apple and other tree fruit growers across a large chunk of
eastern North America, was predicting his crop could be all but wiped
out.

But, he expected to know better after a few weeks. That time has
passed and the news is not good.

"When you drive through the orchard it's a pretty sight, as far as the
trees looking healthy," Zekveld said

"But there's nothing hanging on them."

Almost a month after the killer frost, early assessments of severe damage to this year's crop appear to be holding, although a few trees have some apples growing on them, he said.

It's estimated 80% of the apple crop has been lost, said Kelly Ciceran, general manager of the Ontario Apple Growers, an organization representing about 215 of the province's apple orchards.

Last season's Ontario apple crop had an estimated value of $65 million. The organization is planning a survey in June to collect firm data on what's left of this year's crop, Ciceran said.

A couple of weeks of summer weather in March brought fruit trees back to life early, opening the door to frost damage later in the spring.

Peter Geerts, who owns Birnam Orchards near Arkona and sits on the board of the Ontario Apple Growers, said the damage is Ontario-wide and even into Michigan and New York.

"Normally, when you get a frost like that it's fairly localized," he said. "But the whole Great Lakes basin has been hurt."

In his 25 years of growing apples, Geerts said, "I've never seen it like this."

Often, frost can cause apples that survive to be misshapen or spotted, he said.

"Every time is different so you won't really see that for another three weeks yet . . . once that apple starts to grow."

Either way, it will be a tough year for apple growers. "We'll survive though," Geerts said.

He's like about 65% growers who buy crop insurance. "That helps quite a bit, but it's like any insurance," he said. "You're better off not to have to use it."

The damage comes at a time when Ontario's apple industry was improving, with better prices and major retailers onside with selling Ontario-grown produce, Geerts said.

The hope now, he said, is that growers can pick up where they left off next season "and carry on business as usual, I hope."

In Zekveld's orchard, the Empire and McIntosh trees are empty and there are a few Golden and Jonagold apples appearing.

"There seems to be quite a few Gala," he added.

"I talked to another farmer and he's got no Gala but he's got McIntosh, some anyways."

There are a few peaches in the tops of some trees in that section of the orchard.

"The same with pears," Zekveld said. "No plums, a couple of cherries on the trees but nothing that's going to provide us with a whole lot to sell."

But the strawberry crop looks good, and Zekveld said he was able to
pick a few flats already to sell in his market. "That's a bonus.

Zekveld grows and sells about 30 different fruit and vegetable crops,
but apples are still one of the largest parts of the business.

"I'm feeling a little bit better now that I have some Gala on there,"
he said, "but I have to wait and see if those apples mature properly."

paul.morden@sunmedia.ca

May 25th, 2012

- Despite conditions being colder than average in most of Australia during the Southern Hemisphere summer, the summer heat wave in Perth – the hottest January in the city in 34 years was much hyped by proponents of Anthropogenic Global Warming. Now the city and surroundings have shivered through their coldest May night *in 98 years* with the temperature dropping to 1.3C, while Jandakot dipped below zero.[95] By contrast, the reader will be hard-pressed to find mention of the sudden reversal of conditions in Perth by the global mainstream and pseudoscience media

Perth's coldest May night in 98 years, but it's -3.4C at Wandering

Bureau of Meteorology spokesman, Neil Bennett, said Perth's overnight
low equalled the record temperature set on May 11, 1914 at 7am this
morning.

Mr Bennett said the chill was caused by a "very large" high pressure
system sitting on the south coast of WA which produced clear skies,
which meant sunny days but a lack of cloud cover to slow the cooling
down overnight.

"With no clouds and light winds, the cooling down is going off at its
maximum rate," Mr Bennett said.

Jandakot recorded -0.6C at 5.49am, the only spot in the metropolitan
area to fall below zero, while several centres in the South West also
dipped below zero.

Early risers in Jandakot and Bibra Lake had to scrape the ice off car
windscreens before they left for work and lawns were covered in a
layer of frost..

May 20th, 2012

- Players of Chelsea, the soccer team that won the European Championships, returned to the United Kingdom to be cheered by fans in overcoats → even the daytime temperature was close to freezing

---

[95] http://www.perthnow.com.au/news/western-australia/yep-03-at-jandakot-that-really-is-cold/comments-e6frg13u-1226366379911

- Still winter in Northern Europe, barely a month before the Summer Solstice
- For example, there was heavy snow all over Germany[96]

Personal note

I live in Western Europe. On the morning of May 16th, I woke up late, looked out of the window and thought, what a gorgeous sunny day. Surely, the hot May Sun would warm up the day to 25°C or higher. Then I made the mistake of opening the window. It was COLD! I checked the weather on my smart phone and the high temperature was going to be just 12°C! In mid-May! It was barely a month away from the summer solstice. Overnight the temperature was in the lower single digits (°C). The cold air from the Arctic had managed to vanquish the hot May Sun.

April 2012

End–April 2012

Despite the early cherry blossom in Washington, D.C., a large snowstorm hit the East Coast of the United States. Once again, Mother Nature tried to teach a lesson to the guessers in the wind who had been picking early cherries in D.C. for every cent of hype they were able to earn.

And in Europe, after a short thaw at the end of March, winter had returned with a vengeance and lingered on, nearly uninterrupted until mid-May. It snowed during Easter, even in the lower elevations. A plethora of fresh new snow kept the Alps white well past the spring-by date.

In China, a mixture of cold air and dust caused havoc.

---

[96] http://www.walesonline.co.uk/news/uk-news/2012/05/16/world-in-pictures-may-16-2012-cows-in-germany-lady-gaga-in-taiwan-and-sacha-baron-cohen-in-cannes-gallery-100252-30982834/

April 11th; 2012

Amarillo, Texas, one of the normally warmer places in the US, was hit by an ice storm which dropped 3 to 4 feet of hail on the area. While the IPCC did predict ice storms replacing snowstorms in winter (December - February) in the northern latitudes, pseudoscience prophecy never mentioned ice replacing rain in sub-tropical locations in late spring.

April10th, 2012

Winter did not let up across much of Europe. The wind was not just cold; it was chilling middle-of-winter cold. In the morning, I attempted to go out in a sweater, but had to rush back home and don my winter blazer again!

In Central Europe, the Alps were again full of snow, right down to 3,000 feet or even lower elevations. The only potential good news was for skiers who could look forward to extending an already phenomenal ski season. Unfortunately, many ski stations, especially in lower altitudes, had already closed, not due to lack of snow, but in anticipation of a spring that did not arrive as expected.

Easter Weekend, 2012

Up north where the glaciers are supposedly melting faster than a chocolate Easter Bunny on a Florida beach in August, Anchorage, Alaska celebrated Easter with a heavy snowfall. Anchorage had been whacked hard by 133 inches (3.4 meters) of snow in the winter of 2011-2012. Season-long below average temperatures, broke a previous record set sixty years earlier. There will be more about the extraordinary winter in Alaska later.

And in Europe, cold and snow persisted across a good part of the continent. There was white Easter in the Rhiental and Central Europe in general. Snow delays hampered aircraft in airports in Berlin and Munich. Snow also blemished the weekend in Eastern Europe and Scandinavia. There were snowstorms even in Portugal, the southern extremity of Europe, next to North Africa! Tourists escaping to western Iberia for some sun were visibly disappointed.

Berliners were wondering where the elusive early spring as predicted by global warming was, when their tulips were destroyed by frost at the exact same time they were supposed to be blooming.

In South Asia, a landslide in Kashmir killed over 100 Pakistani soldiers, more than all of the recent border skirmishes with India. Despite the northern section of the Indian Subcontinent being around the Tropic of Cancer (22 ½ degrees North), it continued to snow heavily into March and even the beginning of April. This had not happened in over forty years.

Summary for April 2012

In summary, in 2012, spring was delayed, yet again, in many parts of the Northern Hemisphere, with the notable exception of the Eastern and Central United States.

March 2012

At the beginning of March, it was still winter in sub-tropical Turkey and the Middle East, with sub-freezing temperatures affecting much of the region. On the positive side, the heavy snow thwarted a ground and helicopter assault by the Syrian Army against innocent civilians, albeit only for a few days.

In Europe, although many areas experienced a short-lived thaw in temperatures, in Moscow, it was so cold during the inauguration of the Russian President in Moscow that even cold-hardened strongman Putin was tearing up in his victory speech.

In North America, although the Eastern and Central United States were reaping the benefits of an unusually mild winter, which even boosted the economy and spiked employment numbers, the Western United States and Alaska were not so lucky. On March 18th, severe winter storms gripped Arizona, Nevada and other Western States.

Late January to Late February 2012

A month and a half earlier, a snowless Christmas and a mild New Year in Europe led many proponents of Anthropogenic Global Warming to prophesize a soft, short, winter and an early start to spring. The predictions of the IPCC's Third Assessment Report almost a decade ago were coming true, they hailed with wholehearted glee. Then, in late January, once again, Mother Nature decided to teach the prophets a very harsh lesson. The epic cold spell of 2012 in much of Eurasia and North Africa, in other words, most of the Northern Hemisphere, was the longest and coldest in sixty years. Even the well-remembered cold February of 1985 now paled in comparison. In much of Eurasia, including locations as far south as Milan, temperatures descended to -15°C or lower and stayed at these sub-freezing lows for weeks. Factoring in wind-chill, Real Feel temperatures descended to between -20 and -25°C, even in cities such as Milan. It felt like much of Europe and Northern Asia had been transplanted to a camp at the North Pole. As far south as the Sahara, the sand turned white.

This was not extreme cold weather for just a few days. Much of Eurasia remained enveloped in an Arctic grip for nearly a month without a single respite of warmth in-between. A deep freeze stretched from the British Isles to Italy to North Africa to the Middle East to the Balkans to Scandinavia to Russia to Siberia to Japan to China to Northern India, the likes of which, both in terms of breadth and depth had not been experienced in a very, very long time. The ground was completely frozen for weeks. Snow that fell in the early part of the cold spell turned into ice that remained in place for weeks without melting. With every new snowfall, the thickness of the ice only grew.

Even cities well accustomed to cold winters struggled to keep roads and railways clear, and airports running. Even people well accustomed to cold ventured out less and less in the cold Arctic conditions. It was so cold in Western Europe, that from the moment somebody stepped outside, their hands and years froze, eyes watered, and one felt like being in the deep freeze compartment of one's refrigerator with the added discomfort of a wind howling at thirty or more miles per hour. Skiing became impossible, not due to lack of snow, of which there were plenty, but due to fear of frostbite.

When temperatures eventually returned to normal conditions, i.e. around the freezing point, many people had the sensation that summer had arrived. Average winter temperatures felt "warm" compared to the Arctic deep freeze that had gripped the continent for weeks.

Some highlights[97]

Region	Highlight
Eurasia	Summary In much of Europe and Northern Asia, Real Feel temperatures descended to -20 and -25°C or lower. Water from lakes large enough not to freeze started to evaporate in a bizarre natural phenomenon, given that the air temperature was typically 15 - 20°C below the temperature of the water. Many children and adults alike gasped in awe, as they had never seen such a thing in their entire lives. Water washing up from large lakes on to the shore froze instantly and stayed frozen for weeks creating large ice decorations, the likes of which were not seen by many people during their entire lifetimes. Masses of lake water washed up and frozen on the shores of Lake Geneva, Switzerland causes havoc[98]  Personal Note In mid-February, I visited a little village on the Bodensee on the Swiss/German/Austrian border. To my amazement, I saw a thick layer of ice covering the lakeside, beach, pier, etc.

---

[97] The facts highlighted here have been collected from http://globaldisasterwatch.blogspot.com and a variety of other sources (listed individually) as well as personal experience. Some precaution needs to be taken however: the associated commentary often retains a global warming bias. For example the aforementioned site states the facts of the cold weather, but with the occasional comment of "extreme", a possible "between the lines" implication of "climate-weirding" caused by a "warming" planet. Even more significantly, the site overstates the mild winter in the Eastern and Central United States, a relatively small proportion of the overall surface area of the Northern Hemisphere, quoting one-sided hyperbole from TIME magazine such as: "The Year That Winter Forgot: Is It Climate Change? - 2012 is shaping up to be the year that winter forgot in the U.S. December and the first week of January have seen atypically mild temperatures throughout much of the country". (http://globaldisasterwatch.blogspot.com/2012/02/friday-february-3-2012.html).
This kind of predisposition is typical in the reporting of weather by the mainstream and pseudoscience media.

[98] http://www.parismatch.com/Actu-Match/Monde/Photos/fevrier-2012/De-glace-375933/

Region	Highlight
	First, I thought that it was heavy snow frozen into ice. But then, I noticed that birds – seagulls, ducks, etc were busy eating the ice. A truly bizarre scene…. I asked the locals. They informed me that lake water had frozen over and accumulated on the shore for weeks of extreme cold between -10 and -20 °C. The birds were attracted by the rich plankton in the frozen lake water. One man in his thirties said that he had lived in the village all of his life and never saw such a thing! I told the locals provocatively that climate scientists tell us that the world is warming. There was a chorus of laughter, and several profanities were expressed in German which are best not shared in this book. February 10th, 2012 • The bitter winter cold sweeping across Asia and Europe is not just miserable for commuters. The inclement weather is causing widespread disruption to transport services, power supplies - and a surge in sales of winter jackets. Meteorologists warn that the subzero temperatures are expected to continue into next week. Analysts say, it could also be the winter of discontent for the global economy. Experts say, the most direct impact from the severe cold is on global food prices.
Europe	February 17th, 2012 • Europe continues to do battle against extreme weather, as plunging temperatures and heavy snowfall sweep through large parts of the continent. The cold snap that has killed 480 people to date – about a quarter in eastern Europe, and many of them homeless – shows no signs of stopping, while hundreds of Eastern European villages remain cut off because of cold and snow. February 12th, 2012 • Regions across Europe sustained another deluge of extreme cold weather. • Winter weather conditions wreaked havoc across much of Europe Friday, with travel delays, school and public office closures and deaths. Central and Eastern Europe got the worst of the recent weather. Death tolls rose into the hundreds as countries struggled to deal with bitter temperatures. February 5th, 2012 • Europe cold snap death toll has risen to more than 260 with hundreds having to be rescued after a ferry caught in a snow storm hit a breakwater off Italy. February 3rd, 2012 • The cold snap has claimed 164 lives, as countries from Ukraine to Italy struggle with temperatures that have plunged to record lows.

Region	Highlight

February 2nd, 2012

- Heavy snow has caused disruption across Europe, carpeting much of Italy to the south and Turkey to the east.

January 31st, 2012

- Deaths in Ukraine and Poland in freezing Europe weather

Western Europe

Summary

The river Seine froze, and authorities in Paris had to resort to the use of an *ice-breaker* to try to keep the river navigable.[99]

In Milan, Italy, temperatures descended to -21°C in the first week of February.

Rome, Italy was blanketed in snow at the beginning of February, only the third time in the past fifteen years; the Colloseum was closed in order to prevent damage.

In the United Kingdom, temperatures dipped to -10°C or lower, well below the average. Half of the flights to Heathrow Airport were cancelled.

In Northern Germany, despite the Arctic temperatures, the shining sun melted the top layer of snow, resulting in horizontal freezing "rain" which threatened shipping in the Baltic.

In Berlin, Germany, temperatures dipped to -18°C (without the effect of wind chill added), compared to a long-term average of -6°C.

In the ski competitions in Bavaria, Southern Germany, top ski athletes from all over the world had to don four ski masks to protect themselves from frostbite. Temperature at the higher elevations was down to as low as -45°C without taking into account wind chill. Even with protection, the athletes were openly nervous about the Arctic conditions when interviewed by World Sport (CNN International).

In my own home, the inside temperature was so low that I needed to don a sweater all of the time, even with central heating thermostat set to

---

[99] http://actualite.lachainemeteo.com/actualite-meteo/2012-02-11-13h32/paris---le-brise-glace-de-la-capitale-entre-en-scene---15687.php

Region	Highlight
	maximum.
	Near Zurich, Switzerland, the low-lying mountain (1,200 meters) near the village I use as a base for ski vacations was completely frozen. In all of the years that I have visited this village, I have never seen the mountain like it was this year.
	More than 260 people died from the polar weather in Western Europe alone, in the first week of February alone.
	Over the entire 25 or more continuous days of deep freeze without respite, more than 500 people died in Western Europe from the bitter chill and thousands more ended up in hospital with severe frostbite.

February 19th, 2012

- Italy - Extreme Weather in Rome Damages Colosseum. Severe snow and cold in Rome have taken a toll on one of its architectural wonders, the Colosseum. The Colosseum had to shut its doors to tourists after bits of the massive structure crumbled and fell. The loosening of plaster masonry and stone was attributed to ice forming on the walls. The damage is a result of what is known as the "freeze-thaw cycle." The cold wave was called exceptional. "Maybe every 30 years it gets this cold, but it's very rare." Most nights this month have been subfreezing in the city. Moreover, there have been two outbursts of snow that have left an accumulation of wet snow on parts of the Colosseum.

February 15th, 2012

- Britain has endured its coldest night of the winter so far with temperatures plummeting close to -18C in some areas - part of a record-breaking cold snap.

February 12th, 2012

Region	Highlight

- Rome was blanketed with a rare dusting of snow for the second time in a week.

February 10th, 2012

- Ice to bring danger to roads in fresh wave of severe weather. Ice warnings have been issued for northern parts of Britain as forecasters warned of more extreme weather and urged motorists to take extra car on slippery roads. A potentially treacherous mixture of snow, sleet and rain was set to sweep across much of the north on Thursday night, causing fresh chaos for the country's transport networks. Forecasters warned travelling conditions would remain "difficult" across much of Britain with ice potentially creating havoc for drivers as many parts still recover from last weekend's snowfalls. Northern areas particulary at risk included Cumbria, Northumberland, the Pennines and into Scotland. Further snow was expected on Thursday, with up to four inches falling across much of southern Britain. Forecasters warned temperatures in southern areas will struggle to remain above freezing throughout the week and into the weekend, with the mercury in some areas plunging overnight to as low as -13C. The north, meanwhile, will remain significantly warmer with temperatures as high as 8C.

February 5th, 2012

- Britain braced for 'severe weather' as 15cm of snow is forecast - The UK was on red alert as experts urged people to make plans to deal with a bout of 'severe weather'. According to the Met Office, up to 15cm of snow is predicted to fall and temperatures will plummet to an icy -11 C.

February 3rd, 2012

- Even London is braced for snow as Britain shivers - Forecasters warned that extreme cold will grip the country over the next few day with the possibility of snow even in London and the south. The Met Office has upgraded its cold weather alert to level three, which means 'severe'.
- In Italy, weather experts say it is the coldest week in 27 years
- Heavy snow has caused widespread disruption in northern and central Italy. More than 600 passengers were trapped on an unheated train in the Apennine mountains for seven hours on Wednesday night, when the brakes and electrical cables froze.
- Rescuers in Germany were unable to save an elderly woman after she had gone swimming in the frozen waters of a gravel pit in Lower Saxony. Reports said she had often swum in the lake.

February 4th, 2012

- Romans bewildered by their city's first big snowfall in 26 years used government-issued shovels to clear sidewalks and piazzas, and kitchen utensils to clear windshields on Saturday, Feb. 4, 2012. The snow was as deep as 8 inches in some

Region	Highlight
	neighborhoods, according to the Associated Press reports, and shut down tourist sites such as the Colosseum.[100] 
	February 2nd, 2012
	• In central Italy, heavy goods lorries were barred from motorways and several top-flight football matches have fallen victim to the wintry conditions.
	• Snowfalls were recorded as far south as southern Italy and Corsica, where at least 20cm of snow covered the centre of the Mediterranean island.
	• Italian rail services were reduced because of the wintry conditions.
	• German media reported that ice and sub-zero temperatures had led to the deaths of two women: a pedestrian froze after falling into a drainage ditch and a driver was killed when she lost control of her car on an icy road.
Eastern Europe	Summary
	Much of the river Danube froze.
	In much of Eastern Europe / the Balkans, more than 3 meters (9 feet) of snow buried entire houses.
	Heaviest accumulation of snow in more than six decades; over 10,000 people were cut off from the rest of civilization.[101] 
	February 21st, 2012
	• Serbia - Boats sink in Belgrade as thaw causes Danube ice chaos. A rapid thaw has brought chaos to the River Danube in the Serbian

---

[100] http://www.sacbee.com/2012/02/04/4239146/rome-in-the-snow.html
[101] http://www.bbc.co.uk/news/world-europe-16830034

capital Belgrade, where ice damaged boats, pontoons and floating restaurants. The thick ice covered one of Europe's busiest waterways during the recent freeze, but began to break up on Sunday as temperatures rose.

- Debris was scattered among the breaking ice for hundreds of meters along the river, and several floating restaurants, barges and boats were beached on river banks after the ice snapped anchor lines. Belgrade emergency services said there was no ice risk to bridges and other infrastructure in the city, and there was no threat of flooding. The Danube flows 2,860 kilometers (1,777 miles) through nine countries and is vital for transport, power and industry. It has been almost entirely frozen from Austria to the Black Sea. Ice more than 30cm (11in) thick in places broke up over the weekend as temperatures rose.

- At least 20 people have died from the cold in Serbia in recent weeks and economists say damage from the cold snap may cost Serbia as much as 500 million Euros (£415m; $660m). Some 3,300 people remain stranded by snow and ice in rural areas, where they can only be reached by helicopter.

February 17th, 2012

- The Serbian government declared a state of emergency on Feb. 5, due to heavy snowfall and very low temperatures. Before the declaration, 37 municipalities in Serbia had declared emergencies due to impassable roads and challenges in providing food supplies, medicine and electricity to remote areas. Heavy snowfall and extremely low temperatures (-28 degrees Centigrade in some parts) led to 13 deaths, with an estimated 70,000 people in remote villages severely affected. The government advised to keep all schools closed until February 17. The energy suppliers issued a public appeal to conserve energy to prevent any restrictions as extremely low temperatures are seriously jeopardizing the energy system and supply of energy and fuel. The power company announced that it can meet the present level of demand for a week longer. Freezing weather is expected to continue until the end of February, and the government is already preparing for the big melt and possible floods.

- Starting from Jan. 23, Moldova has experienced extremely cold weather conditions, with temperatures averaging between -12 and -16°C. The recent forecast issued by the national hydro-meteorological service indicates that the cold temperature will stay at low levels, reaching possibly -27°C. Assisting severely affected localities in the northern part of the country is difficult because snow measures are so high; at least 12 villages do not have electricity. The most affected are people living alone, especially the elderly and families with many children. The Ministry of Education reports that 248 schools are closed in the country due to cold conditions and high levels of snow, most of which are located in northern cities. The government has mobilized and has been providing assistance to the most vulnerable people in the country; 17 tents have been installed by Civil Protection and Emergency Situations Service around the country for heating, provision

Region	Highlight
	of snacks and hot drinks, which serve about 2,200 to 2,400 persons daily. There is a lack of warm clothing to distribute to the most vulnerable, and lack of fuel to ensure functioning of the support tents for the coming two weeks.

February 13th, 2012

- Kosovo - A young girl has been pulled alive from a house hours after it was hit by an avalanche which killed at least nine people in southern Kosovo. She was found buried under 10 metres (33 ft) of snow after officers heard her voice and a mobile phone ringing. Rescuers are still looking for one person after the avalanche hit a remote mountain village on the border with Albania and Macedonia on Saturday. Several homes were destroyed. Only two were said to be occupied at the time.
- NATO peacekeepers deployed in Kosovo had been called in to help rescue operations, but were unable to land a helicopter in the blizzard. Those who were found dead in the village of Restelica after Saturday's avalanche were two brothers and their families. "No bigger tragedy has ever struck this region." The cold snap which started in Europe in late January has left dozens dead in the Balkans.

February 10th, 2012

- Hungary Extreme Winter Claims Lives as Cold Weather Set to Stay - Extreme cold weather claimed more than five lives in Hungary over the weekend as some schools were closed and settlements temporarily sealed off. Southeastern regions of the country were blanketed with as much as 60 centimeters (24 inches) of snow.

February 5th, 2012

- Extreme cold weather in Ukraine causes 101 deaths - Of the Ukrainians who have died since the cold weather hit Jan. 27, 64 were found frozen on the streets, 11 died in hospitals and 26 in their homes. It was so cold there that some 1500 swans, sea gulls and ducks froze to the ice in a small harbor near Ukraine's Black Sea port of Odessa, forcing emergency workers to use ships to break up the surface and free the birds. The weeklong cold snap — Eastern Europe's WORST IN DECADES — is causing power outages, frozen water pipes and the widespread closure of schools, nurseries, airports and bus routes.

February 3rd, 2012

- Serbia snow strands thousands. Heavy snow has left at least 11,000 villagers cut off in remote areas of Serbia amid the European cold snap. At least six people have died in Serbia, with emergency services expressing concern for the health of the sick and the elderly in particular.
- Emergency services in Serbia have described the situation, close to the country's south-western borders with Kosovo and Montenegro, as very serious. In places, the snow has reached a depth of 2m (6ft 6in). Fourteen municipalities are affected. Helicopters have helped move several people to safety, and food and medicines have been airlifted to isolated areas.

Region	Highlight

- Snow began falling in Serbia on 7 January and has hardly stopped since. Serbian media say further snow is expected in the coming days
- Temperatures are below -30C (-22F) in parts of Europe and 63 people have died in Ukraine and 29 in Poland.
- Ukraine has seen the highest number of fatalities, many of them homeless. Over a 24-hour period, as many as 20 people died. Food shortages have been reported in the capital, Kiev, because trucks have been unable to transport supplies.

February 2nd, 2012

- The freeze that has swept south through the continent has caused at least 80 deaths, mainly in Ukraine and Poland.
- Temperatures were so low that some areas in Romania along the shores of the Black Sea froze.
- Ukrainian officials reported that the number of deaths attributed to the freeze had risen to 43, with 13 people falling victim to hypothermia in the past 24 hours.
- Villages were cut off in Bosnia where temperatures fell to -10C
- In Bosnia and Serbia helicopters were used to airlift supplies to villages cut off by drifting snow.
- Several towns and cities in Bulgaria saw record lows, with -29C reported in Kneja in the north for the second day running. For much of the country an "orange" alert was in place, warning of dangerously low temperatures.
- Seven more deaths were reported in Poland, bringing to more than 20 the number who have fallen victim to the cold snap.

February 1st, 2012

- Police in Poland say at least ten people froze to death as the weather worsened over night after what had been until now, a mild winter. There are warnings the country can expect 30 degrees below zero at the end of the week.
- In the Czech Republic the railways suffered as the extreme weather buckled tracks and engines broke down.
- Romania, central Serbia and Bulgaria are all in the grip of the big freeze with Ukraine reporting over 18 deaths.
- In Romania an unlikely crew came to the rescue of 300 stray dogs being held in kennels just outside Bucharest. Volunteers from a local prison worked for several hours to dig out alleyways at the shelter to help the freezing animals
- Bulgaria faced another day of record low temperatures on Tuesday, January 31, with a "Code Orange" weather warning in force for the whole country and with the severe weather having claimed two lives the previous day.

January 31st, 2012

- Emergency shelters have been set up in the Bulgarian capital Sofia after days of freezing weather
- At least 18 people have died in Ukraine and 10 in Poland after heavy

Region	Highlight
	snow fall and a sudden drop in temperatures across east Europe. • Three deaths were also reported in Serbia and one in Bulgaria. • Ukrainian officials said nearly 500 people had sought treatment for frostbite and hypothermia in just three days. • And over that time, more than 17,000 people had sought refuge in some 1,500 shelters. Temperatures have plunged to -16C (3F) during the day and -23C (-10F) at night. • Temperatures dropped in Poland last Friday from just below freezing to -26C (-15F). Polish forecasters have warned that temperatures could fall further during the week, to below -20C during the day and -30C at night. • In Serbia, police reported that the snowy conditions had led to the deaths of a woman and two elderly men. Two other men, in their 70s, are believed to be missing in the south of the country. • The freezing conditions also claimed a life in neighboring Bulgaria. • Emergency shelters offering food and heat are being set up in the Bulgarian capital Sofia and the Czech capital Prague.
Russia & Scandinavia	Summary In Russia and the Scandinavian Peninsula, temperatures tumbled even lower to -30°C feeling like -40°C with wind chill. The unusually cold spell was too much even for Russians, who are usually accustomed to bitter winters. Schoolchildren were obliged to keep their winter gear on during classes, as indoor temperatures were 9°C or lower, even with central heating operating at full capacity. Russia's GazProm was obliged to cut the natural gas flow to Western Europe, not due to any commercial dispute, but because, for the first time in many years, the company did not have enough to cater to both domestic needs and the requirements of their customers to the West. February 13th, 2012 Russia cold leave balance of 215 victims. The low temperatures, unusual even for these latitudes, carried five thousand 546 people to go to clinics and hospitals, including 154 children, 52 of whom were hospitalized. [102]  • Cold broke records in some localities on the outskirts of Moscow,

[102] http://www.nixguy.com/russia-cold-leave-balance-of-215-victims.html

Region	Highlight
	where temperatures dropped to 36 degrees Celsius below zero, advises the Meteorological Centre (CM) of Russia.
	• "There were 36 degrees below zero, the lowest temperatures in these locations for a February 13 in the history of meteorological observations, which dates back to 1879 in Istra and Klin", said a spokesman for the CM.
	• In Moscow, today the minimum temperature does not beat the record of 1911, 29.3 degrees below zero, but came very close: 28.7 degrees below zero.
	• According to the CM, in the Russian capital and surrounding regions temperatures in the last week have been on average 7 degrees Celsius lower than the norm.
	• Almost 60 regions of Russia, especially in the South of the country, suffering from unusually low temperatures since early February.
	• According to meteorologists, not he noted in Russia a month of February so cold in several decades.
	February 3rd, 2012
	• The coldest temperatures have been recorded in Russia and Kazakhstan. Snow is piled high in parts of the country. In the Urals and Siberia, the temperature fell to -40C (-40F) while in the capital of Kazakhstan, Astana, the wind-chill factor meant the real temperature was down to -52C, even though the air temperature was -35C. In southern Russia, cars and lorries became stuck in snow drifts between Novorossiisk and Krasnodar.
	February 1st, 2012
	• It is not just humans having to battle the cold – in southern Russia severe frosts have caused the deaths of thousands of fish after river levels which were already low completely froze. Transport, communications and power lines are all being hit. Several stranded lorry drivers faced a night in their cabs as roads became impassable.
Arctic	One of the popular conjectures of proponents of AGW is that much of the planets' excess heat circulates via ocean and air to the poles causing the magnified warming seen at the poles. Well, February 2012 was different: Barrow, Alaska: -26°C, Hammerfest, Norway: -18°C, Murmansk, Russia: -26°C, Kiruna, Sweden: -32 to -45°C (all temps without Real Feel which would be 5-10°C colder). More about the epic cold Alaskan winter in a later section.
Greece, Turkey & Middle-East	Summary It was snowing in Greece, and Turkey had remained blanketed in snow since early January. **February 12th, 2012** • As Europe battles a deep freeze, record low temperatures and heavy snow are making life even more miserable for more than 140,000 Turkish quake survivors still living in tents or temporary homes four months after the nation's devastating earthquake. The cold

Region	Highlight

snap, which began in Europe in late January, has left some families in Turkey's quake relief centers trying to stay warm by using coal stoves or electric heaters, and watching their drinking water freeze overnight.

February 9th, 2012
Greece and Bulgaria battle flooding as Europe struggles with extreme weather - Europeans across the continent have been battling more than a week of extreme weather, with thousands still trapped by snow in remote, mountain villages in the Balkans; hundreds - most of them homeless - dead after temperatures hit as low as minus 33.

February 3rd, 2012
- Heavy snow hit Turkey, with 50 centimeters falling in Istanbul on Wednesday. An avalanche in the south-east of the country killed a woman in her home. Another avalanche blocked a main road connecting the provinces of Bitlis and Diyarbakir.

February 2nd, 2012
- School closures were reported in northern Greece, where temperatures of -16C (3F) were recorded.

February 1st, 2012
- 20 centimeters of snow fell in Turkey, the most seen in 50 years

January 24th, 2012
- United Arab Emirates - Country continues to shiver as Met department records sub-zero temperatures at Jebel Jais area of Ras Al Khaimah. The UAE will wake up to the 'chilling' news that the cold snap that has enveloped the country will continue through the week.
- Unaccustomed to the chill, UAE residents are putting up a brave front as a cold wave continues to spread across the length and breadth of the country, with extreme weather conditions sweeping mountainous regions and the country recording its lowest temperatures for the year yesterday. Minimum temperatures are set to dip to 0°C in the UAE's mountainous regions while internal areas could see temperatures dipping to a nippy 5°C and coastal areas could see minimum temperatures of 13°C.
- Rough weather has prompted a warning to beachgoers across the UAE to refrain from venturing into the sea. The UAE's met department has predicted that the rough weather will continue until Friday. Strong winds have been buffeting Dubai since early morning.

Region	Highlight

North Africa   It was snowing in the Sahara, in Libya, all over North Africa[103]

Northern Asia

Summary
Polar temps continued in Northern Asia, for example -20°C in Japan and northern areas of China during a good part of February.

February 3rd, 2012
- The streets of Seoul, South Korea are covered in some of the heaviest snow on record. The snow caused traffic jams all over the city which deployed more than 200 police officers to control traffic. Streets will turn to ice as temperatures drop below freezing.

South Asia   Extreme cold temperatures in the tropical latitudes of Northern India, 8-15°C below normal in many places, afflicted a population not accustomed to such chilly conditions. In some instances, the cold even extended further south.

February 12th, 2012
- India - The extremely low temperatures in the city of Mumbai are the latest indicator of climatic fluctuations in Mumbai. While Mumbai had a minimum of 14.2 degrees Celsius on Wednesday, this dipped to 8.8 degrees on Thursday.

January 23rd, 2012
- Sri Lanka - Nuwara Eliya torn between extreme weather patterns. Warm days and freezing nights play havoc with the lives of girls, boys, fruits, flowers and veggies. The UNUSUAL drop in temperature in Nuwara Eliya has had widespread repercussions, with people finding it difficult to cope with the extreme weather, damage to crops and drop

---

[103] Link

Region	Highlight
	in water levels. Last Monday (16), the temperature in Nuwara Eliya dropped to 2.7 Celsius accompanied by ground frost. In 2009, the temperature dropped to 2.6 Celsius. Similar temperatures were experienced in 1929 and 1953.

Typical Explanations from proponents of Anthropogenic Global Warming

Below are some of the explanations improvised by proponents of AGW, ever since cold winters returned since the winter of 2007-2008.

Explanation	Translation
Climate "Weirding"	• All "extreme" weather – hot, cold, dry, wet – are signs of a warming planet • In particular, very cold weather is as much a symptom of a warming planet as is very hot weather • Any temperature, storm, etc., not witnessed during a maximum of 150 years of records in a world that has existed for 4 billion years is "extreme".
Blizzards and heavy snow accumulations cannot occur in a cooling planet	• A cooling planet has a less energetic atmosphere, therefore, does not have the energy for more heavy snowstorms
Surface weather is nothing more than a delayed and very misleading reflection of all the circulations above the friction layer	• Global temperature is not an indication of heat or cold. It can be colder but the planet is still warming. • Here is Kevin Trenberth of the IPCC with this variant[104]: Kevin Trenberth, a lead author of the chapter of the IPCC report that deals with the observed temperature changes, said he accepted there were problems with the global thermometer record but these had been accounted for in the final report. "It's not just temperature rises that tell us the world is warming," he said. "We also have physical changes like the fact that sea levels have risen around five inches since 1972, the Arctic icecap has declined by 40% and snow cover in the northern hemisphere has declined."
The world is not just the Northern Hemisphere	• It was summer in the Southern Hemisphere and, therefore, warm • Perth, Western Australia has sweltered through its hottest January in 34 years [105] • Of course, no mention should be made about the fact that Sydney, Eastern Australia has recorded its soggiest January in 11 years and the

---

[104] http://www.timesonline.co.uk/tol/news/environment/article7026317.ece
[105] http://globaldisasterwatch.blogspot.com/2012/02/wednesday-february-1-2012.html

Explanation	Translation
	coldest one in 12 years.
Cherry-picking	• To cherry pick the mild winter in the Eastern and Central United States and declare 2012 as the year that winter forgot is OK • To cherry pick the heat wave in Perth, Australia is OK • But to cite the extreme cold in Eurasia – the majority of the surface area of the Northern Hemisphere, the very mild summer in Eastern Australia, in fact to cite any instance of unusually cold weather is blasphemy, and must be discredited as cherry picking.

The fact that there is some contradiction within and between the various rationalizations is irrelevant.

It is likely that proponents of AGW would have found suitable explanations, had they been alive during the last Ice Age.

All roads lead to global warming

Humor

Here is TIME Magazine's "Science" section, confidently stating on January 9th, 2012 with the eye-catching headline: **The Year That Winter Forgot: Is It Climate Change?**

Europe was included in this "brilliant" analysis. The below paragraph once again illustrates the tremendous lack of science in climate pseudoscience. [106]

> "Nor is the unseasonable warmth confined to the U.S.; Europe has had mild temperatures so far as well. When cold goes missing, snow does too and it's been an unusually green (or brown) winter."

Even on January 9th, this statement was inaccurate. Although there was a slightly late start to winter, by the end of December, the Alps already had one of the best ski seasons ever. There will be more about this in a later section.

There are numerous examples of unscientific prophesy replacing predictions based on solid science. This book documents only some such cases. Being repeatedly proven wrong with predictions has, of course, not prevented the mainstream and pseudoscience media from brazenly writing new guesses in the wind. After all, the aim is to convince people into accepting global warming – what proponents of AGW call "educating" the public. The lack of factual relevance appears unimportant.

Some other notable cold weather events in February 2012

February 20th, 2012

Western United States - Deadly avalanche near ski resort in Washington State - 4 dead: A police spokesperson said the skiers were "very experienced" and well prepared. Rescuers were dispatched

---

[106] http://www.time.com/time/health/article/0,8599,2104040,00.html?xid=gonewsedit

to an out-of-bounds area near Stevens Pass ski resort after reports around noon local time (20:00 GMT). Three were killed. Eight skiers declared missing were later accounted for. The resort is located in the Cascade Mountains north-east of Seattle. The Northwest Weather and Avalanche Center warned of a high avalanche danger above 5,000 ft in the Stevens Pass area with a considerable danger at lower levels. The centre said there had been heavy snowfall over the last few days. Another person was killed in a separate incident when a snowboarder went over a cliff at Alpental ski area east of Seattle.

December 2011 to mid-January 2012[107]

The TIME Magazine "science" article, *The Year That Winter Forgot: Is It Climate Change: "it's been an unusually green (or brown) winter"* was written on January 9th, 2012. As the following section will demonstrate conclusively, without cherry picking mild temperatures in certain parts of the United States (a tiny fraction of the surface area of the world) and gentle conditions in Europe for a *few days* around Christmas and New Year, such an assertion had no basis in fact, even before the Eurasian deep freeze in which started in late January 2012 and lasted weeks. Winter did not forget the rest of the Northern Hemisphere. Moreover, parts of the Southern Hemisphere experienced a cooler than normal summer. In addition, aside from the unusually blistering conditions in and around Perth, Western Australia, there were no other extraordinary heat waves of note reported from the Southern Hemisphere.

Region	Highlight
Europe: The Alps	Summary Let us begin with the Alps. It is true that the ski season failed to start on-time due to a very dry November 2011. Temperatures in November were mainly average, although in some instances above average. But there was no precipitation. Both the locals in Switzerland and Austria, as well as prospective or in-place ski vacationers were disappointed. Straight on cue, proponents of Anthropogenic Global Warming prematurely cried climate wolf. Unsurprisingly, Mother Nature did not disappoint ski enthusiasts for long. Starting from the weekend of December 2nd, copious snow replenished not only the Alps, but also blanketed the lower elevations in much of Central and parts of Northern Europe as well. For example, all of Switzerland was snow-white and remained so for a couple of weeks. A thaw set in just before Christmas, but by that time, the mountains were full of fresh snow – ideal ski conditions. Even during the thaw, overnight temperature without the benefit of sunlight remained at or below the freezing point. The thaw only lasted until the new year after which the cold and snow resumed with a vengeance.

---

[107] The facts highlighted here have been collected from http://globaldisasterwatch.blogspot.com and a variety of other sources (listed individually) as well as personal experience. Some precaution needs to be taken however: the associated commentary often retains a global warming bias. For example the aforementioned site highlights the following article:" Big climate change could happen fast - and soon - New research from NASA into the Earth's paleoclimate history indicates we could be facing rapid climate change this century" (Link). See the section, Misinformation on the last Interglacial period disseminated by proponents of Anthropogenic Global Warming, for the factual inaccuracies in this article and the underlying statement by NASA.

This kind of predisposition is typical in the reporting by the mainstream and pseudoscience media.

By the time of the Davos summit in late January, just before the epic deep freeze in Eurasia, a meteorologist on CNN International proclaimed: "The snow pack in the Alps is unbelievable, we are talking about the heaviest snowfalls in many, many years."

Even before February, Europe already had a fantastic ski season, one of the best ever, as the local media did not hesitate to point out. Here is TSR (Television Suisse Romande) in late December[108]. I have taken the liberty to translate some of the key sentiments from French to English:

> "Les stations de montagne affichent un large sourire **(Big Smile)**.... les touristes ont afflué en nombre **(Many Tourists)** .... ne sont pas trop regardants sur leurs depenses **(Free Spending Tourists)** – A Leysin, on ne savait plus où mettre les gens **(TOO MANY PEOPLE, didn't know where to put them)**"

It appears that, this time, both the Alpine ski resorts and foreign tourists did not buy into the scaremongering by proponents of AGW back in November. Climate doomsayers had already panicked the local population in the early 2000s, by rashly forecasting the end of snow in the Alps and urging a switch to summer-time livelihoods   Although there were few years of less than ideal snow in the early part of the last decade, even resulting in the cancellations of ski competitions, the predictions of calamity were proven completely wrong in the latter half of the decade, with some of the best ski conditions ever since the winter of 2007-2008, and continuing into this decade. In the winter of 2011-2012, reasonable locals and tourists took the dry November in stride and subsequently prospered from a wonderful ski season once again.

January 24th, 2012
Snow Blankets World Economic Forum in Davos. Here is Ted Kemp, a CNBC Reporter[109]

> I talked to the local workman who was tasked with clearing the snow from the roof of the reinforced tent outside Congress Centre that is my new office. When was the last time, I asked, that this much has come down?
>
> He raised his brows and tilted his head back, registering his exasperation.
> "Oh, I don't know. Seven years? Eight years?"
>
> Then he climbed back onto the tent.

---

[108] http://www.tsr.ch/info/suisse/3687902-les-stations-de-montagne-affichent-le-sourire.html
[109] http://www.cnbc.com/id/46112639/Snow_Blankets_World_Economic_Forum_in_Davos

Region	Highlight

January 15th, 2012
The central mainland area in Europe has recently suffered several meters of snowfall. Austria has been particularly badly affected.

January 8th, 2012
- 1.7 metres of SNOW in Langen, Austria in one evening
- 3 meters of SNOW in Western Austria in 3 days. Skiers trapped

Here is the Daily Mail about snowed-in British skiers stranded in resorts across Europe after record snow dump [110]

> Austrian weather service ZAMG said some places had not seen snow 'so deep' in more than 30 years, adding that the village of Nauders last saw a similar depth of 47in in 1951. And in Germany the 9,718ft summit of the Zugspitze, which only had 7.5in of snow six weeks ago, now has 150in.

And below is globaldisasterwatch.blogspot.com:
- Austria hit by heavy snow - heavier than usual snowfall and high winds have caused chaos on roads and railways in many areas of Austria. Part of a major railway route has been shut down in the west of the country, and some villages and tourist resorts have been cut off.
- Up to 1.2 meters (4 feet) of fresh snow has been recorded in some areas since Thursday.
- The authorities say the probability of avalanches is extremely high and widespread. Nearly 2,000 homes have been without power. On Friday, around 15,000 tourists and locals were snowed in at ski resorts on the Arlberg Mountain. The roads there are now open again, but others in the region remain shut. More heavy snow is predicted over the next few days.
- Cyclone "Andrea" rages in western Austria - The effects of cyclone "Andrea" were felt around Austria Thursday. Parts of the country have been left without power, fallen trees cover the streets and more than 100 call outs for the fire brigade have been recorded by the emergency services.
- The regions of Vorarlberg and Tyrol were particularly badly affected.
- Western Austria in general and Bregenzerwald in Vorarlberg were hard hit by

[110] http://www.dailymail.co.uk/travel/article-2085131/Snow-Alps-Freak-snowfall-traps-1-000-British-skiers-Austrian-Alps.html

Region	Highlight

the low pressure front on Thursday afternoon. The majority of the incidents that took place were a result of fallen branches and trees on roads. By the evening both Walgau and Montafon began to suffer with railway lines also being badly affected. A fire also broke out in Alberschwende in Bregenzerwald as a result of a lightning strike, with parts of a metal garage blowing away altogether in Feldkirch, blocking a street.

- The storm continued to cause havoc in Dornbirn, Austria, where several cable cars came to a standstill in the strong winds. Wind speeds of 108 kilometers per hour (km/h) were measured, with similarly high speeds of 95km/h recorded in Feldkirch.

United Kingdom

January 15th, 2012

The thaw of Christmas/New Year did not last long in the United Kingdom.

- An anti-cyclone replaced storms and brought bitter cold. An intense high pressure moved slap-bang over the British Isles on the 12th.
- The anti-cyclone moved away the intense storms experienced after Christmas and New Year and replaced conditions with bitter cold nights and striking sunsets. The weather phenomenon has been described as acting like a "boulder in the scree" by BBC weather experts.
- Storms continue to blow around the west coast of Ireland and north of the Scottish mainland.
- The atmospheric conditions which engulfed the British Isles on Thursday night imposed clear blue skies, bitterly cold starry nights and hardly any wind. Night frosts greeted most areas in Britain with thick fog lingering in the countryside.
- There was an 80% probability of severe cold weather from midnight on Saturday and 0900 on Tuesday in parts of England. The combination of low sun, blue skies and red evening skies is producing some stunning sunsets.

Christmas 2011/ New Year 2012

Most of December was frosty in the United Kingdom, definitely not a green winter as proclaimed by TIME Magazine. But there was a thaw around Christmas and New Year. There were a couple of weeks of rain, high winds and exceptionally mild temperatures. During this period, even nighttime lows hovered above the average levels of 6C (43F) in London and 4C (39F) in central Scotland with some areas in southern England recording 14c (57f) on New Years Eve, warmer than the north African state of Morocco.

December 17th, 2011 [111]

- Britain is bracing for a weekend of wintry weather, with ice warnings in place for the whole country until Monday.
- The Met Office has issued a severe weather warning advising people to be aware of snow and icy conditions across the UK.
- Motorists have been warned to take care in the treacherous conditions as *sub-zero temperatures* continue over the weekend.
- On Friday an elderly man was killed in a four-car pile-up on the Formby bypass in Merseyside, where sleet fell earlier in the day.
- The Highways Agency said it was well prepared for the cold weather, with a fleet of 500 state-of-the-art winter vehicles on stand-by and enough salt to deal with severe conditions: "Road users are reminded to plan for their

---

[111] http://uk.news.yahoo.com/britain-braces-icy-weekend-072914486.html

Region	Highlight

journeys before setting out, checking the forecast, road conditions and leaving extra time if travel conditions are poor, or to delay their journeys if the weather becomes severe," a spokesman said. "They are also advised to be prepared and carry warm clothing and an emergency pack, which includes food and water, boots, de-icer, a torch, and a shovel in case of snow."

- After a frosty and, in places, icy start to Saturday, some areas will have scattered rain, sleet and snow showers.
- Elsewhere there may be some repeats of winter wonderland scenes that many parts of the country saw on Friday.
- There was 6cm of snow recorded in Glasgow, 2cm in High Wycombe and Buckinghamshire, and 4cm in parts of Northern Ireland.
- *Sky News weather presenter Jo Wheeler said there will no escape from the wintery conditions wherever you are.* "What's happening is we've got this very cold northwesterly flow, which has got a lot of showers on it, and the snow line is right down through the country," she said. "So basically, anything that falls in that is likely to be sleet or snow - it's going to be wintery whichever way you look at it. "Add to that the fact that we've got some very low temperatures. This morning many central parts of Britain are well below freezing so we're always looking at this ice risk, and that's really what we're concerned with this weekend and into Monday. Of course this time last year [coldest December in a century], and the year before, we were knee deep in snow so we've been quite lucky to get away with things so far. But certainly it's going to be a cold weekend and with lots of people out shopping for Christmas, I think the slippy pavements and roads are going to be a concern."

December 14th, 2011
Snow and strong winds disrupt travel in Scotland - Snow and high winds have disrupted road and rail travel across Scotland.
- Gusts of up to 70mph swept across the central belt and forecasters warned of heavy snow on high ground.
- In the Highlands, seven vehicles were involved in a collision which closed the A9 at Dalwhinnie for almost seven hours. Rail services were severely disrupted, with one passenger being injured after a train travelling from Wick struck a fallen tree at Inverness.
- The Met Office upgraded its weather alert from yellow to amber for heavy

Region	Highlight
	snow across some parts of the country. Amber warnings cover Central, Tayside, Fife, Strathclyde, the south-west of Scotland, Lothian and Borders, Highlands and Western Isles.
	• The Grampian area is on yellow, while southern and western Scotland could also experience very strong winds at times. The A83 at Rest And Be Thankful was closed for emergency repairs because of the severe weather and remained closed overnight.
	• In the Shawlands area on the south side of Glasgow, the A77 Kilmarnock Road was closed in both directions because of an unsafe building. The fire service were dealing with a satellite dish which was hanging from a tenement.
Sweden	January 8th, 2012
	• A new low pressure area with strong winds and precipitation was on the way toward Sweden. It passed over central Sweden on Wednesday and bringing rain and snow over most of the country, as well as gale force winds in the south and along the coast.
	• "It is a very strong low pressure area, which can be best described as an 'atmospheric bomb' as the pressure drops so suddenly. We're talking of gale force winds up to 35-40 metres per second."
	• Many are still without power after storm "Dagmar" which hit on Dec. 26. Sweden has not had winds as strong as those recorded during recent weeks for years. "Some of our stations measured the strongest winds in at least 15 years." Chaos was reportedly left in its wake, with rail traffic at a standstill, hundreds of thousands of households without power and fallen trees blocking many roads.
	• In some areas of Jämtland county, in central Sweden, Dagmar reached hurricane strength. Between Christmas and New Year the weather calmed down slightly in the wake of storm Dagmar but by then northern parts of the country had suffered extensive forest damage caused by the strong winds.
	• The National Board of Forestry (Skogsstyrelsen) has been taking stock of damages done to Swedish woods over the Christmas period and while work continues it has to do so with more bad weather en route for Sweden. The Swedish National Railway is also still working at establishing the extent of the damages done to its network in the aftermath of the storms.
Middle East	Throughout much of January, unusually cold conditions dominated much of the Middle East. In particular, Syria was so cold that anti-government forces were as much concerned about people dying from the cold as they were from the bullets and artillery of President Assad's army[112]

---

[112] http://www.photoblog.com/lccsyria/2012/01/25/january-25th-2012.html

Region	Highlight

Region	Highlight
Turkey	**Summary** In the winter of 2011-2012, unusually cold weather and record snowfall blanketed Turkey for a couple of months – definitely not a green winter for this Mediterranean country in sub-tropical latitudes.

January 19th, 2012 [113]

Cold weather and snowfall hits Turkey

- Many schools and roads are shut down throughout Turkey, and meteorology officials warned of ice and frost, while hundreds of villages have been cut off due to excessive snowfall. 19th January 2012
- Heavy snowfall and storms disrupted life throughout Turkey.

January 9th, 2012 [114]

[113] http://www.demotix.com/news/1009985/cold-weather-and-snowfall-hits-turkey-duzce
[114] http://www.todayszaman.com/news-268082-more-blustery-cold-weather-forecast-for-turkey.html

Region	Highlight

- As Turkey faces the onslaught of what promises to be a stormy winter, the General Directorate of Meteorology forecast on Sunday more cold, rainy and snowy weather across the country this week.
- Southern and western Turkey can expect more rainfall and central and eastern Turkey will likely see continued snowfall this week, adding to a winter season that began with unseasonably cold temperatures and snowfall back in October.
- The provinces of Erzincan, Tunceli, Bingöl, Elazığ and Adıyaman can expect a mix of heavy rain and snow. Residents of these provinces, as well as of Antalya and Anamur, have been warned to take precautions against possible floods and probable transportation delays.
- After severe winds and rainfall hit İzmir and Antalya hard on the weekend, meteorologists said the southern Mediterranean shores and central Anatolia will experience similar stormy weather, with winds reaching up to 60 kilometers per hour. The temperature in eastern Anatolia's Kars province dropped to *minus 12 degrees Celsius* Monday night.
- Meanwhile, the provinces of İstanbul, Kocaeli, Giresun and Sakarya can count on continued showers and thunderstorms.

Asia:
India

Summary

Throughout much of the winter, temperatures in Northern India, and occasionally even further south, temperatures were 5 to 8°C below normal. Many millions of Indians were shivering in a usually warm tropical land. Most of India at the lower elevations typically has very mild "winters", equivalent or only slightly colder than the summers of Northern Europe. In the Himalayas, this was no normal winter wither. There was so much snow that Kashmir was literally cut off from the rest of India – a really abnormal situation. Predictions by proponents of AGW of a warming Bay of Bengal expanding and causing abnormal flooding are, once again, probably premature panicking about the climate wolf.

January 18th, 2012

Cold wave in the Northern and Central India.

- After keeping its date with the Sun for a few days, Bangalore's weather has swung to extreme cold conditions.
- Monday saw *the coldest morning in the state in decades*.
- The harsh cold weather *broke records* in many places, with Madikeri registering its *lowest in 132 years* at 4.8 degrees Celsius, Mysore's coldest day

Region	Highlight

in 120 years at 7.7 degrees Celsius and Bangalore's coldest day of January in the past 19 years with minimum temperature touching 12 degrees Celsius. The outskirts of Bangalore like HAL airport area touched below 10 degrees Celsius and GKVK was freezing at 8.4 degrees Celsius. Maximum temperature across the state remained around 26 degrees Celsius. Bijapur recorded *the lowest ever minimum temperature recorded* at 8.4 degrees Celsius.

- "The severe cold wave from the North and North-East, combined with clear skies, has appreciably brought down the minimum temperatures across the state. Also the shorter day time and longer nights have reduced the radiation on Earth's surface." "The severity of cold also depends on the air-moss strip passing through the co-ordinates of Karnataka. This strip keeps shifting and the area which falls under this strip gets affected by the cold wave. Also the prolonged north-east monsoon has contributed to the cold spell."

January 2nd, 2012 [115]
Heavy rain, dense fog and cold weather disrupts New Year travel plans in northern India. Even the dogs and monkeys are feeling the cold

December 12th, 2011
- The death toll in a cold wave sweeping through northern India has risen to 39. The majority of deaths were in the state of Uttar Pradesh. Punjab and Haryana are among the other northern states badly hit.
- Most deaths take place among the homeless and the elderly. Last week, the country's Supreme Court ordered states to provide adequate night shelters for the homeless during the winter. "You should not allow even a single person to die this winter from the freezing cold."
- Heavy fog and a cold wind have disrupted life across northern India with sub-zero temperatures in Indian-administered Kashmir. The capital, Delhi, is also in the throes of a cold snap, with temperatures dipping to 2.3C and fog disrupting flight schedules.
- The cold wave has forced schools to shut in the state of Bihar until 25 December. An official from the meteorological department said the cold weather would continue for a number of days.

---

[115] http://www.aljazeera.com/weather/2012/01/201212105534101207.html

Region	Highlight
China	**Summary** China had an unusually cold December and January – definitely not a green winter
Hong Kong & South East China	In Hong Kong, the average low in January is 14°C. Low temperature averaged 5°C in January 2012. On Lantau island, the cable car was blocked by the cold.

January 2012 [116]

- Under the dominance of the winter monsoon, January 2012 was colder than usual. The mean temperature of the month was 15.1 degrees, 1.2 degrees below the normal figure of 16.3 degrees. The month was also wetter and gloomier than usual. The monthly rainfall of 42.1 millimetres was 17.4 millimetres above normal. The monthly total duration of bright sunshine was 86.0 hours, a deficit of 40 percent against to the normal figure of 143.0 hours.

- Under the influence of a continental airstream, the weather in Hong Kong was mainly fine and dry for the first three days of the month. An intense surge of the winter monsoon reached the coastal areas of Guangdong during the night on 3 January and brought cold weather with rain patches to the territory from 4 to 6 January. The weather remained cloudy and cool on 7 January. With clouds thinning out, it became generally fine with some haze for the ensuing three days.

- Affected by a broad rain bearing cloud band over southern China, local weather turned rainy from 11 to 13 January. While the winter monsoon moderated in the next two days, the rain continued with mist and fog patches. The visibility at Waglan Island once fell to 100 metres on the morning of 15 January. Meanwhile, a cold front formed over inland Guangdong on 15 January and moved across the coastal areas the next morning bringing cooler weather to Hong Kong on 16 January.

- With the prevalence of a relatively mild easterly airstream, there was plenty of sunshine and the temperature rose gradually from 17 to 19 January. The temperature at the Observatory reached a maximum of 22.4 degrees on 19 January, the highest of the month. Local weather became cloudy with some coastal fog when the easterly airstream freshened for the next two days.

- An intense winter monsoon reached the south China coast on the morning of 22 January and the weather became progressively colder with rain patches from 22 to 26 January. The temperature at the Observatory fell to a minimum of 7.4 degrees on the morning of 25 January, the lowest during the Lunar New Year holidays since 1996. With the winter monsoon over the south China coastal areas being generally replaced by a humid easterly airstream, local weather was misty with temperatures rising gradually on 27 and 28 January. Affected by a weak replenishment of the winter monsoon, the weather became slightly cooler for the next two days. A dry continental airstream brought mainly fine and dry weather to the territory on the last day of the month.

- There was no tropical cyclone over the South China Sea and the western North Pacific in the month.

December 2011 [117]

---

[116] http://www.weather.gov.hk/wxinfo/pastwx/mws201201.htm
[117] http://www.weather.gov.hk/wxinfo/pastwx/mws201112.htm

Region	Highlight
	<ul><li>Attributed to the frequent replenishment of the winter monsoon, December 2011 was colder than usual with a monthly mean temperature of 16.9 degrees, 0.9 degrees below the normal figure of 17.8 degrees. There were six cold days (daily minimum temperature at 12.0 degrees or below) in the month, about two days more than normal. The prevalence of continental airstream also brought drier than usual weather to the territory in December 2011. The total rainfall recorded in the month was 2.8 millimeters, less than a tenth of the monthly normal. The annual rainfall for 2011 was 1476.7 millimetres, a deficit of about 38 percent compared to the annual normal of 2382.7 millimeters.</li><li>A cold front crossed the coast of Guangdong on the morning of the first day in the month and the associated intense winter monsoon brought significantly cooler weather with sunny periods to Hong Kong on that day. Dominated by the winter monsoon, it remained rather cool, fine and dry for the ensuing three days. Affected by a fresh easterly airstream, the weather became cloudy with rain patches on 5 and 6 December. With the moderation of the easterly airstream, there were sunny intervals on 7 December.</li><li>An intense winter monsoon reached the south China coast on the morning of 8 December and brought significantly cooler weather to the territory on 8 and 9 December. Affected by the intense winter monsoon, it was cold, fine and very dry for the next three days. The temperatures at the Hong Kong Observatory fell to a minimum of 9.6 degrees on 11 December, the lowest of the month. While the winter monsoon moderated gradually on 13 December, daytime temperatures rose to above 20 degrees generally from 13 to 15 December.</li><li>With the arrival of a replenishment of the winter monsoon at southern China, it remained generally fine and dry from 16 to 19 December. Affected by a broad band of clouds, the weather became mainly cloudy for the ensuing two days.</li><li>Another replenishment of the winter monsoon reached southern China on the morning of 22 December and brought fine and dry weather to the territory on 22 and 23 December. With the strengthening of the winter monsoon, local weather was cold and very dry on the next two days. With the persistence of the winter monsoon, it remained cool and dry with localized haze for the rest of the month.</li><li>One tropical cyclone occurred over the South China Sea and the western North Pacific in the month.</li></ul>
Northern China	January 31st 2012 [118] Cold wave maintains grip over north China<ul><li>China's meteorological authority said Tuesday that the country's northern regions remain in the grip of severe cold, with minimum temperatures plunging to *minus 40 degrees Celsius*.</li><li>Minimum temperatures in the county of Mohe in Heilongjiang province and some parts of Inner Mongolia autonomous region have stayed below minus 40 degrees Celsius for more than a week, according to a statement posted on the National Meteorological Center's website.</li><li>On Sunday, minimum temperatures in the city of Manzhouli in Inner Mongolia reached a record low of *minus 44.9 degrees Celsius*.</li></ul>

---

[118] http://english.cntv.cn/20120131/117104.shtml

Region	Highlight

- The center forecast a fresh cold wave will hit the country over the next three days, bringing heavy winds and causing temperatures in the northern regions to drop by at least 4 to 6 degrees Celsius.
- The center suggested people keep warm when doing outdoor activities to prevent respiratory disease.

January19th 2012
Cold front to bring large-scale rain & snow to China
- China's National Meteorological Bureau says a cold front is expected to bring large-scale rain and snowfall to many parts of China.
- Dense fog is also set to blanket parts of central and eastern China, and the provinces of Fujian and Yunnan, with visibility of less than 1,000 meters. Snow is predicted to hit most areas in north China, while the south will experience light to moderate rain.
- The snowfall is expected to stop by the eve of the Chinese New Year. Relevant departments have been suggested to closely monitor weather changes, and take precautionary measures to minimize impact on transportation during the Spring Festival.

Taiwan

January 5th 2012 [119]
- The ongoing cold front that has brought temperatures in northern and northeastern Taiwan to lower than 9 degrees Celsius will not subside until tomorrow morning, the Central Weather Bureau (CWB) announced yesterday. The lowest temperature in a non-mountainous area in the country yesterday, 8.7 degrees Celsius, was recorded in New Taipei's Tamsui yesterday morning, the CWB said.
- The CWB said the temperature in the northern part of Taiwan could become even colder today. The cold front will continue to affect northern part of the country until Friday when the temperature will begin to slightly rise then.
- The CWB also issued special warning for heavy rain along the north coast, Keelung, Yilan, Taipei and New Taipei City mountainous regions.
- Residents in coastal areas, Hengchun Peninsula in southern Pingtung County, and offshore island county of Penghu, Orchid Islands, and Green Island should also be prepared for strong wind, the CWB added.

[119] http://www.chinapost.com.tw/taiwan/national/national-news/2012/01/05/328037/Temperatures-below.htm

Region	Highlight
Japan	Japan had an unusually cold winter, even resulting in an increase in household spending in December and January. In just one example of cold weather, three meters (9 feet) of snow blanketed the nation at the end of January. *Japan Dec Household Spending Up Unexpectedly On Cold Weather* [120] • Japan's average household spending unexpectedly rose in December thanks largely to cold weather, which boosted demand for winter clothing, the Ministry of Internal Affairs and Communications showed Tuesday. • Household spending stood at a real Y327,949 in December, up 0.5% from a year earlier, marking the first year-on-year rise in 10 months. • The December reading was much better than the median forecast of a 0.1% dip in a Market News International survey of economists. • In December, outlays on clothing and footwear rose 7.6% in real terms while spending on housing (repair, maintenance and rents) increased 7.0%. • In nominal terms, household spending rose 0.3% y/y in December. • The average real income of salaried workers' households was unchanged from a year earlier in December at Y893,427, following a fourth straight y/y drop. • Real disposable income in the average salaried workers' household fell 1.0% in December from a year before to Y749,449, the fifth consecutive y/y drop.
North America: Canada	January 22nd 2012 • Extreme and wacky weather continues as RECORD LOWS hit parts of Canada. Coast-to-coast, Canadians continue to deal with 'weird and wacky' weather. Freezing rain, severe wind chills and mild temperatures persist across the country, with temperatures dropping in the West and winter storm warnings hitting much of the East. • Bitterly cold weather has helped Alberta has set a new record for electricity demand. On Monday around the supper hour, demand peaked at an all-time high of 10,609 megawatts, which beat the previous record set Sunday. • On Wednesday, *record lows* hit Alberta as arctic air moved east. Edmonton faced a severe wind chill of about -43 C while Calgary expected a chill of -35 C. Bitterly cold air combined with a westerly wind of 15 to 30 km/h will produce wind chill values in the -40 to -49 C range in southern Saskatchewan throughout Wednesday and into Thursday. Winnipeg also got a blast of winter as temperatures were set to hit -36 C. At these extreme values, frostbite on exposed skin can occur in less than 10 minutes.
Continental United States	January 18th 2012 • U.S. north-west braces for *record snowfall* - A potentially historic winter storm is forecast to dump heavy snow across the Pacific Northwest today, probably wreaking travel havoc in areas not used to so much of the white stuff. States in the US Pacific Northwest are bracing for one of the worst snowstorms the region has seen in a generation. • A westerly storm is expected to engulf the state of Washington, bringing up to 2ft (61cm) of snow. Mountainous areas already hit by a weekend storm will be hardest hit, with areas of Oregon also seeing deep snow. Seattle officials fear

---

[120] https://mninews.deutsche-boerse.com/index.php/japan-dec-household-spending-unexpectedly-cold-weather?q=content/japan-dec-household-spending-unexpectedly-cold-weather

the storm could bring the heaviest snow at the city's airport since 1985.

- Travel could become dangerous or impossible in the region.
- The NWS described the upcoming snow storm as a "classic overrunning scenario" seeing an approaching warm front drawing cool air down from British Columbia, across the border in Canada. Most of the Washington lowlands will receive 5-10in (13-25cm) of snow overnight on Tuesday and into today - equivalent to the city's *annual snowfall in one day*. Other forecasters suggested the snow levels could be even higher.
- Weather officials in Canada were keeping their eye on the storm front. Vancouver has already seen snow and ice, with concerns that Wednesday's heavy snow could head over the border and into British Columbia.

January 9th 2012 [121]

Record snowfall in Western Texas:

- 19.5 inches of snow has fallen during the 2011-2012 season at Midland, Texas as of January 9th 2012. This breaks the old record of 13.9 inches set in 1946-47.
- A record daily snowfall of 10.6 inches was recorded on Monday, January 9th 2012 at Midland. This breaks the old record of 5.9 inches set in 1955. This is also now the greatest daily snowfall on record...breaking the old greatest daily snowfall on record of 9.8 inches set on December 11, 1998.

January 4th 2012 [122]

Extreme cold weather hit most of Florida at the beginning of January 2012

- A layer of protective ice covers a new strawerry blossom in a field Wednesday, Jan. 4, 2012, in Dover, Fla. Farmers spray a coating of water over their plants to help keep them from freeze damage. They had learned from previous cold spells in 2009 and 2010. Temperatures in central Florida dipped into the 20's overnight.

- Icicles cling to oranges in a small grove just after sunrise Wednesday, Jan. 4, 2012, in Seffner, Fla. Temperatures in central Florida dipped below freezing into the 20's degrees Fahrenheit overnight. Farmers spray water on their crops to help keep them around 32 degrees, protecting them from possible freeze damage.

---

[121] http://reasonabledoubtclimate.wordpress.com/2012/01/
[122] http://photoblog.msnbc.msn.com/_news/2012/01/04/9947895-frozen-fruit-and-flowers-in-florida-as-cold-weather-hits?lite

Region	Highlight

January 2nd 2012

- Frigid air blasting over the Great Lakes blew in the season's first major lake effect snowstorm on Monday, blocking visibility and causing massive pileups on icy roads from Michigan to Kentucky. As much as 2 feet of snow was expected to fall on upstate New York by Tuesday as the storm moves eastward from Michigan, where over 1 foot of snow fell by Monday afternoon.
- "You can see all of the snow showing up from the upper Peninsula of Michigan through western New York state, all the way through western Virginia and Kentucky. It's this west-northwest flow over the lakes that's causing this lake effect."

December 22nd 2012

- A family has been rescued from a car that had been buried in a snowdrift for almost two days on a rural highway in the US state of New Mexico.
- Rescuers had to dig through 1.2 meters of ice and snow to free the family, whose four-wheel drive got stuck on a highway when a blizzard moved through the area on Tuesday.
- The parents and their five-year-old daughter were clinging to each other and lethargic when they were found at 2.45am local time Wednesday. State police say they got a distress call and launched a search for the family yesterday.
- They were among 32 vehicles state police and guardsmen rescued from the storm, but they were the only ones who police say needed medical attention. The family, from Texas, was recovering in hospital today

December 21st 2012

- A massive winter storm blamed for at least six deaths made travel nearly impossible in parts of the central United States. "Blizzard conditions that caused fatal accidents and rendered highways impassable in five states crawled deeper into the Great Plains early Tuesday (local time).
- Hotels filled up quickly along major roadways from eastern New Mexico to Kansas, and nearly 100 rescue calls came in from motorists in the Texas Panhandle." Snow drifts reached three meters in parts of Colorado after strong winds whipped at the 38 centimeters of snow had fallen since the storm began on Monday. Parts of New Mexico were blanketed by 61 centimeters of snow while Kansas got up to 30 centimeters by overnight. While the heaviest snowfall had mostly ended by midday on Tuesday (local time), blizzard

Region	Highlight

conditions continued in many areas as strong winds of up to 80km/h whipped up heavy white flakes.

December 19th 2012

- U.S. Southwest, plains face blizzard warnings in big storm - The panhandles of Texas and Oklahoma were placed under a blizzard warning on Sunday as the Southwest and plains states braced for two days of bitter cold, heavy snow, rain and high winds.
- The storm is expected to produce up to 16 inches of snow and wind gusting to 50 mph in the first major snowstorm of the winter for Texas and Oklahoma. A blizzard watch was also in effect until Tuesday for parts of Colorado, New Mexico and Kansas, with high winds and up to a more than a foot of snow expected across the region. The storm was expected to edge into the mountains of Arizona, New Mexico and Colorado before heading east on Sunday night or this morning. By tonight, conditions further east in the Texas and Oklahoma panhandles are expected to have deteriorated so much that officials warned motorists to stay off the roads. The mix of rain and snow will move into the Chicago and Detroit areas on Tuesday, forecasters said. In Guymon, in the Oklahoma Panhandle, emergency management officials met on Sunday to go over storm preparations. With wind gusts up to 50 mph predicted, blowing snow could cause problems. "It's unbelievable. Right now it's 65 degrees. Tomorrow we're expecting to have our snow boots on."

Alaska

Summary

One of the popular conjectures of proponents of Anthropogenic Global Warming is that much of the planets' excess heat circulates via ocean and air to the poles causing the magnified warming seen at the poles. Well, this winter, Alaska definitely did not comply.

Many part of Alaska experienced some of the coldest temperatures on record, so cold that the Bering Sea froze for an extended period and cut off parts of Alaska from fuel supplies. Icebreakers, most notably a Russian tanker, became jammed in unprecedented ice and towns were at risk of running out of supplies.

Even for Alaskans hardened to extreme cold, this was a very difficult winter.

Once again, proponents of AGW were caught prematurely crying out global warming wolf back in 2010 when Alaska experienced a few months of warmer temperatures.

January 31st, 2012 [123]

Posted at 12:56 PM ET, 01/31/2012
Coldest January on record for parts of Alaska
By Jason Samenow (Washington Post)

A huge, persistent pool of cold air (indicated by low pressure at high altitudes in

---

[123] http://www.washingtonpost.com/blogs/capital-weather-gang/post/coldest-january-on-record-for-parts-of-alaska/2012/01/31/gIQAQVyIfQ_blog.html

Region	Highlight

the image above) sat over Alaska for much of January (whereas warm air flooded areas to the southeast, including the U.S.) (NOAA). *Mind-boggling, historic cold* has gripped the Last Frontier in recent weeks. Nome, Galena, and Bettles - in Alaska's west and west interior - are all likely to have their coldest Januarys on record the National Weather Service reported today. It will likely be the 5th coldest January on record in Fairbanks, with a hard-to-imagine average temperature of -26.7. Anchorage is likely to log its 4th coldest January.

Consider some of the following chilling facts:
- Using satellite data, the University of Wisconsin detected surface temperatures as cold as -73 below zero around the town of Arctic Village in northeast Alaska
- Low temperatures in the 60 to 65 below zero range have chilled the towns of Galena, Fort Yukon and Huslia since last Friday; the coldest recorded temperature was -65 at both Ft. Yukon and Galena
- Fairbanks hit -50 on January 28, and -51 on January 29, the first -50 degree readings there since 2006
- Fairbanks dropped to 40 below on 16 different days during January, the greatest number since 1971 (hat tip: Jim Cantore)
- The average low in Ft. Yukon, 145 miles northeast of Fairbanks, has been -35 (Source: Our Amazing Planet)
- The average temperature in Anchorage has been just 2.7 F in January compared to an average of 15. Only three other years have been colder (Source: MSNBC). More info from National Weather Service.

In addition to cold, snow socked Prince William Sound earlier in the month, paralyzing Valdez and Cordova. Anchorage has already received 92" of snow this winter, compared to an average of 74.5" for an entire season.

The cold, snowy pattern has arisen from a persistent storm track through the Gulf of Alaska, bringing an onslaught of snow events to the coastal part of the state and record-shattering cold in the interior.

This pattern, known as the +EPO (characterized by low pressure over Alaska, the Gulf of Alaska, and the Bering Strait) has actually helped keep Arctic cold out of the lower 48. As of today, a paltry 19% of the Lower 48 had snow on the ground compared to 42% last year on January 31.

There are some signs this pattern will, at least temporarily, break down. As the NWS Office in Fairbanks wrote earlier today:

As the month of February begins a large scale pattern change will get underway...but it will still be several days before the pattern change is complete.

(Note: An apparent reading of -79 below zero reading at Jim River Maintenance camp - very close to the all-time U.S. record low temperature of -80 in Prospect Creek from 1971 - was deemed bogus by the National Weather Service. The observing station was not up to standard and the reading may have been caused by a failing battery it said. For more, see detailed analysis from the WeatherMatrix blog[124])

---

[124] http://www.accuweather.com/en/weather-blogs/weathermatrix/was-us-record-low-temperature-of-80-

Region	Highlight

January 16th, 2012

Ultra-Harsh Alaska Winter Prompting Fuel Shortages:

- Living in Alaska's outer reaches is challenging enough, given the isolation and weather extremes, but at least three remote communities also have experienced weather-related late deliveries of fuel so crucial to their survival during an especially bitter winter.

- The iced-in town of Nome and the northwest Inupiat Eskimo villages of Noatak and Kobuk faced fuel shortages that illustrate the vulnerability of relying solely on deliveries by sea or air, potentially subjecting communities to the mercy of the elements. The villages, which just received their fuel, are especially vulnerable, unable to afford more additional storage tanks for gasoline and heating oil, which can run as high as $10 a gallon.

- Compounding a problem with no easy answers, temperatures dipping as low as minus 60 degrees Fahrenheit over the past few weeks means air deliveries are delayed at the same time people are consuming more fuel more quickly. Some people in both villages also use wood-burning stoves for supplemental heat, but diesel is the critical commodity.

- Nome missed its pre-winter delivery of fuel by barge when a huge storm swept western Alaska. In a high-profile journey, a Coast Guard icebreaker has cut a path in thick sea ice for a Russian tanker delivering 1.3 million gallons of fuel to the community of 3,500. Without a fuel delivery, Nome would likely run out of certain petroleum products before the end of winter and a barge delivery becomes possible in late spring.

- Until recently, the situation was much more dire for the smaller communities of Noatak and Kobuk, located farther north above the Arctic Circle, where relentless extreme cold prevented fuel deliveries by plane until this week. Before the new supply of fuel arrived in Noatak, the village store borrowed some heating oil from the village water and sewer plant. But filling the store's two 23,000-gallon tanks has diverted any potential crisis. "We're good for another month and a half."

- Residents in Kobuk also were highly relieved by an air shipment of heating oil that arrived Wednesday in the village of 150 people about 175 miles to the east. It's been too cold for people to use their snowmobiles much, so gasoline isn't as much of a concern. Running low on the diesel used to warm homes was another matter.

- In Noatak, residents once had fuel shipped by barge on the Noatak River, but that has long been impossible since the river shifted and became shallow there. Two years ago, residents began tapping into another source of fuel, thanks to the Red Dog zinc mine 40 miles to the northeast. The mine in 2009 began a program to sell gasoline and diesel to Noatak and another close neighbor, the village of Kivalina. The latest Red Dog fuel day for Noatak took place on the day the village store ran out of diesel. So villagers formed a convoy of about 30 snowmobiles and freight sleds, and headed out in weather marked by temperatures of 47 below and, for the first 10 miles, dense fog.

January 15th, 2012

Record-breaking snowfall in Valdez, Alaska and it's not stopping.

broken-in-alaska/60951

Region	Highlight

- Media coverage of the snowfall in the Prince William Sound city describes it as "Valdez Snowmageddon". As of Jan. 5th - 246.6″ of snow had fallen in Valdez. That's over 20 feet. The snowstorms continued. "Valdez usually has 151.8 inches of snow by Jan. 12. As of 2 p.m. Thursday, Valdez had seen a total of nearly 321.8 inches this season. That's more than 14 feet above normal." The old timers here are all puzzled by how EARLY these dumps (big short- spanned torrents of snow) are. It usually hits in February.
- "The record-breaking winter pushed the local elementary and high school past their legal roof snow-load limits of 90 pounds per square foot." On Jan. 13, the schools were forced to close "for the first time in recent memory" due to fears of roofs collapsing. In terms of what's in store weather-wise for Valdez residents in the next few days, "We are still having a little snow and rain, but it's gonna get cold (as the jet[stream dips south) and everything is gonna freeze solid. Then single digits for a few days." More snow was in store for this weekend

January 12th, 2012 [125]

Anchorage on pace for snowiest winter on record. Alaska's Anchorage and snow-weary Cordova hit with what they already have a lot of deep snow

- Heavy snow fell in Alaska's largest city Thursday, adding to what already has been the snowiest period for Anchorage since records have been kept.
- The National Weather Service predicted a snowfall of 8-16 inches, with the city's upper Hillside neighborhoods expected to get the bulk of the snow.
- It began snowing shortly before midnight. The heaviest snow was expected between 3 a.m. and noon.
- "It's snowing pretty good right now," forecaster Christian Cassell said at 4:30 a.m. Thursday at the Anchorage weather service office. The snow had been intermittent but the rate was increasing. "We still expect a decent amount of snow," Cassell said.
- The weather service counts a snow year from July to June. From July 1 through Tuesday, Anchorage has received 81.3 inches of snow. Meteorologist Shaun Baines said that makes it the snowiest period for Anchorage since records have been kept.
- If the pace keeps up through the last snows in either April or May, Anchorage is on track to have the snowiest winter ever, surpassing the previous record of 132.8 inches in 1954-55, Baines said.
- About 150 miles to the southeast of Anchorage, the Prince William Sound community of Cordova has already been buried under 172 inches of snow since Nov. 1 and is trying to dig out from recent storms.

---

[125] http://finance.yahoo.com/news/anchorage-pace-snowiest-winter-record-144823024.html

Region	Highlight
	<ul><li>Another 4 to 7 inches could fall Thursday, Baines said.</li><li>City officials in Cordova, a picturesque fishing community, already have learned that a regular shovel just won't cut it when you're digging out from nearly 15 feet of snow.</li><li>There were plenty of standard shovels around town. But what they needed was a larger version with a scoop that can push a cubic foot of snow or better at a time. "That's what's missing in Alaska," city spokesman Tim Joyce said Wednesday. Not anymore. "We will be shipping 72 shovels to Alaska by plane (Thursday) to help," said Genevieve Gagne, product manager at the shovel's maker, Quebec, Canada-based Garant. The new shovels cost about $50 each, and the city is paying for them with its emergency funds. The Yukon ergo sleigh shovels, with a 26-inch scoop, have a huge advantage over regular shovels. "Trying to lift snow all day with those is pretty backbreaking," Joyce said. "We have the National Guard right now using the standard shovel, and they're getting pretty trashed everyday — not the shovels but the Guardsmen themselves," he said.</li><li>Temperatures warmed Wednesday, when residents awoke to standing water because of stopped-up drains. The rain also made the existing snow heavier.</li><li>The warmer temperatures — about 35 degrees midday Wednesday — brought another hazard to the Prince William Sound community of 2,200 people: avalanche danger.</li><li>There's one road leading out, and it was closed though it could be opened for emergency vehicles.</li><li>The city also is warning people not to stand under the eaves of their houses to clear snow off the roof for fear the snow will come down on them. "There's a real high potential that if it does slide, they'd be buried," he said.</li><li>So far, four commercial buildings and two homes have been damaged from snow accumulation on roofs. A 24-unit apartment complex also had to be evacuated.</li></ul> January 11th, 2012 [126] Contrary to the popular assertion by proponents of Anthropogenic Global Warming that species are moving north due to warmer temperatures, this winter, native species in Alaska could not cope with the extreme cold and their wings froze. *Bering Sea storm freezes wings of some sea birds, bald eagles* *Alex DeMarban* Is state of Alaska overhyping danger of moose-vehicle collisions? Fans of the hit cable television show "Deadliest Catch" surely know the dangers posed by freezing sea spray as it crusts over crab boats.

[126] http://www.alaskadispatch.com/article/bering-sea-storm-freezes-wings-some-sea-birds-bald-eagles

But they might not know that birds in Western Alaska fishing ports face the same problem. That's apparently what happened on Tuesday, when a Dutch Harbor storm spat so much sleet that ice-glazed seabirds couldn't take off and bald eagles cracked as they spread their wings for a frigid, labored flight, according to a report from a local agent with the University of Alaska Fairbanks Marine Advisory Program and radio station KUCB.

The professor, Reid Brewer, awoke early Tuesday morning to discover the storm caked the Aleutian Island city in ice. To get into his car, he cracked off the wintry varnish with a sledgehammer. "Cars were frozen like ice cubes," he said.

Later, a city worker picked up a crested auklet he'd found along the roadside in hilly terrain, an odd place for a seabird. They're often on the water or in flight, not standing around, and they usually don't let humans get close.

But the man set the auklet in a box and the bird defrosted as he drove it to Brewer. Brewer was set to release the bird into the bay -- he'd sought advice from the International Bird Rescue in Anchorage and the Alaska Department of Fish and Game -- when he noticed a couple of murres marooned on shore with icy wings.

Later, he noticed bald eagles with frost-covered wings and matted heads having difficulty flying.

"It limited their flight to 10 meters (about 33 feet)," he said. "It looked they were weighed down."

Freezing rain is part of life on the island wedged between the Bering Sea and Pacific Ocean, where sea-brewed tempests whip down mountains and winter temperatures seesaw above freezing.

Brewer worried the seabirds may have been iced-over too long, leaving them unprotected by the oil-secreting glands that typically keep them dry and warm.

But once released in the bay, the birds shot away and disappeared beneath the surface, where they'll hopefully be when the next storm strikes.

December 14th, 2011
Anchorage blown away by extreme weather.

- On the edge of what other city in North America can you get knocked flat by hurricane-force winds in a blizzard roaring up the suburban neighborhood street you've boldly started down in an effort to find out what caused the power outage? And where else in the country would they knock on the door of a neighbor to tell him the high-voltage lines carrying power across the valley have been torn completely off the pole next to his house, leading him to look at you and ask, "You'll do anything for a cheap thrill, won't you?" And then laugh as another gust hit, and his house shook, and the adjacent power line whipped around like it was going to crack? But then they've been through this a few times before.
- Only a week earlier, the hurricane-force winds that rolled across the Anchorage

Region	Highlight

Hillside tore a separate insulator loose from the same power pole and left the line bouncing and swaying in the wind. It was a different line from the one that tore an insulator out of the cross bar this time. That line then hit another, caused a whole lot of sparks, and kicked out a breaker.

- Everyone in Anchorage should own a Coleman lantern or some equivalent, as this is a city vulnerable to power outages either by wind or earthquake. The semi-official report from the National Weather Service, recorded at a home in the neighborhood, was 97 mph. T)he interim director of Alaska Climate Science Center wondered if the big blows might be linked to the warming off the ocean. Storms generate significantly more energy over warm water than cold. .The warm water in the North Pacific was thought to be a significant player in what some called the "Arctic hurricane" that ripped into the Bering Sea and Western Alaska last month. The storm pounded the coast with winds up to 90 mph and left widespread damage. The Governor later declared the region a disaster area. This year has produced more than its share of blasts (odd for a La Nina winter). The latest storm is still too fresh for anyone to get a full damage assessment. It began to die on Sunday night, sometime after the power came back on around supper time.

November 2011

- Alaska's storm listed among most significant events for November - The massive "extra-tropical cyclone" that walloped Bering Sea and Northwestern Alaska communities with high winds and blizzard conditions last month was named one of the eight significant climate events to strike the United States in November. Winds gusted to over 80 mph and the storm surge topped 8 feet, marking the strongest storm to impact the region in decades."
- Alaska's storm began as an intense low pressure system that formed southeast of Japan on Nov. 7 and grew stronger and more intense as it rolled across the North Pacific toward Alaska's Aleutian frontier. As this cyclone whirled into position, the storm's interior pressure dropped 50 millibars in 24 hours to a minimum of 944 millibars — comparable to a Category 1 hurricane. "Waves to 35 feet and 100 mph winds were recorded offshore as the storm approached. Hurricane force winds and blizzard conditions affected coastal Alaska. Storm surges of up to 10 feet affected communities along Alaska's west coast - causing flooding, some structural damage and property loss...
- An ice zone connected to land had not yet developed to lessen the impact of large waves striking the coast." NOAA reported the *state saw the sixth coolest temperatures since 1918*. Overall, 2011 has delivered almost exactly average temperatures to the Far North state — the 43rd coldest of the past 94 years. Despite lots of snow in southern Alaska, November was the 41st driest in the record, with 2011 giving the state the 41st driest January-November since 1918.

Note: The last paragraph from globaldisasterwatch.blogspot.com was an obvious attempt to minimize the December winter storm in the hope that it was a one-off. Events in January obviously proved otherwise. Such minimization and pre-conceptual watering down of cold weather events by co-citing previous warm weather at the same time has now become ubiquitous in the mainstream and pseudoscience media.

Region	Highlight
Australia	

January 12th, 2012 [127]

Cold snap sets new record low temperatures [during the Australian summer]

- The weather bureau says an extreme cold front has broken a series of low temperature records for Canberra, Goulburn and the Snowy Mountains.
- The southern tablelands and Victoria's Alpine region have also been hit by the summer chill.
- A rapidly moving cold front from Antarctica moved though Tasmania, Victoria, New South Wales and the ACT yesterday.
- The icy and changeable weather delivered a low of -4 degrees Celsius and a dusting of snow to the Snowy Mountains.
- Forecaster Sean Carson says the snow is unseasonable, but not rare."In fact, it was only three to four years ago they had a 20cm centimeter fall in January," he said.
- *In Canberra, the mercury dropped to 1.6C, eclipsing the record of 1.8C set in 1956.*
- *Goulburn experienced -0.1C, beating the previous record January low of 1.4C.*
- The front has now moved east over the Tasman Sea.

January 5th, 2012 [128]

With typical Australian humor, the below article tells the truth about the end of drought on the Australian continent and an unusually cool summer, despite the heat wave in and around Perth hyped by proponents of AGW.

Once again, proponents of AGW were caught prematurely crying out global warming wolf back in the Southern Hemisphere Summer of 2009 - 2010 when the fires in the then drought-ridden Australia were puffed up ad-infinitum as an "example" of global warming. Hyperbole such as "loss of 20% of moisture in the last 20 years", "summer climate moving 20 kilometers towards the equator every year, etc.

Drought-breaking La Nina made the continent cooler
BY: GRAHAM LLOYD, ENVIRONMENT EDITOR
From: The Australian January 05, 2012 12:00AM

AUSTRALIA did its best for global cooling in 2011 but it had nothing to do with the federal government's carbon tax.

Rather, back-to-back La Nina weather systems that caused widespread flooding and ended the 10-year drought also pushed temperatures below the 30-year average for the first time since 2001, resulting in the coldest autumn since at least 1950.

As with the economy and this year's start to summer, last year's weather was a two-speed affair.

---

[127] http://www.abc.net.au/news/2012-01-12/cold-snap-hits-south-east/3768810?section=act
[128] http://www.theaustralian.com.au/news/health-science/drought-breaking-la-nina-made-the-continent-cooler/story-e6frg8y6-1226236841072

Region	Highlight
	According to the Bureau of Meteorology's annual climate statement for 2011, cooler temperatures in Sydney, the sub-tropics and tropics offset above-average conditions in Victoria, South Australia, Western Australia and Tasmania.

And for those looking to the figures to disprove climate change, the Bureau of Meteorology says Australia was the only continent to record cooling and the nation's 10-year temperature average trend was still up.

"In 2011, the La Nina and heavy rainfall acted like an evaporative cooler for Australia," said bureau climate change spokesman David Jones.

"The year 2010 was relatively cool in recent historical context and 2011 was cooler again."

Mr. Jones said there was no evidence to link the strong La Nina weather systems with changing global temperatures.

"We have had this regular cycle of La Nina and El Nino," he said. "The strongest El Nino on record was in 1997 and we have seen one of the strongest La Ninas on record in 2010-11."

Mr. Jones said the climate science was not very clear on what would happen with El Nino and La Nina patterns, particularly at this early stage of global warming.

"We have only seen one degree of warming so far but we will see substantially more as we move through the century, but it is probably too early to draw any concrete relationship between hotter temperatures and La Nina," he said.

"One simple thing we can say is we know La Nina are historically cooler for Australia but there is a big difference between variability and climate change."

The BOM climate statement said Australia's mean rainfall total for last year was 699mm, which was 234mm above the long-term average of 465mm, making it the third-wettest year since comparable records began in 1900.

The Australian area averaged mean temperature was 0.14C below the 1961-1990 average of 21.81C. Last year, maximum temperatures averaged 0.25C below normal across the country, while minimum temperatures averaged 0.03C below normal.

"Despite the slightly cooler conditions, the country's 10-year average continues to demonstrate the rising trend in temperatures, with 2002-2011 likely to rank in the top two warmest 10-year periods on record for Australia, at 0.52C above the long-term average," the bureau said.

"If you are interested in determining whether the planet is warming, you look at the global temperature," Mr. Jones said.

"Australia follows the global trend closely, but it can vary." |

Region	Highlight
	Mr. Jones said it was likely that the current La Nina weather pattern would decline this autumn and there would be a rebound in average temperatures.

Commenting on the start to summer, Mr. Jones said "there has been quite a bit of bias in the perception of what summer has been like".

"Sydney and Brisbane have had a gloomy start to summer with a lack of high temperatures.

"The average has been quite low but more remarkable has been the lack of any spikes.

"If you were in Melbourne, Adelaide, Hobart or Western Australia there has been nothing remarkable about summer so far."

December 11th, 2011
- Queensland's *record-breaking cold start to summer* has eased and "almost normal" summer conditions were expected on Saturday.
- The cold snap - following hard on the heels of Brisbane's driest and warmest spring in two years - saw records broken from the city south to the Gold Coast and Stanthorpe and west to St George.
- Much of the state was warming up after almost a week of surprisingly cold conditions, with Brisbane and Amberley hitting 25C, Coolangatta 24C and Toowoomba 20C on Thursday. Coolangatta made 19.7C at 9am, beating the previous low maximum set in 1965 of 20.4C.
- Brisbane had an equal low maximum of 19.9C, the same as 1963

Region	Highlight
Southern Hemisphere	Aside from the unusually blistering conditions in and around Perth, Western Australia, there were no other extraordinary heat waves of note reported from the Southern Hemisphere, although as per the previous, there were multiple instances of unusual cold.

Below are some of the temperatures from key cities across the Southern Hemisphere on December 31st, 2011:
- Cape Town: High: 23C, Low: 16C
- Sydney: High: 23C, Low: 16C
- Sao Paulo: High: 23C; Low: 16C
- Perth: High: 27C; Low: 20C

Nothing to write home about just after the Southern Summer Solstice.

An unseasonable snowstorm hit the US East Coast, with some areas of Massachusetts seeing more than 27in (68cm) of snow. This was the earliest snowstorm on the US East Coast in decades.

http://www.bbc.co.uk/news/world-us-canada-15502380

Snowstorm hits US East Coast killing at least nine

31 October 2011 Last updated at 05:05 GMT

An unseasonable snowstorm has hit the US East Coast, with some areas of Massachusetts seeing more than 27in (68cm) of snow.

The authorities say at least nine people have died in snow-related accidents.

More than three million homes have lost their electricity supply from Maryland to Massachusetts - some residents may be without power for several days.

The snowfall eased on Sunday, as the storm headed north from Maine.

It had worsened as it moved north, with states of emergency declared in New Jersey, Connecticut, Massachusetts and parts of New York.

Communities in western Massachusetts were among the hardest hit.

New York saw unusually early snowfall Nantucket in Massachusetts experienced wind speeds of 69mph (111km/h), a National Weather Service (NWS) statement said.

A number of deaths were reported in the storm:

- Four people were killed in two separate crashes on an icy road in Philadelphia, while falling snow killed an 84-year-old man in Temple, Pennsylvania

- A traffic accident killed one person in Colchester, Connecticut
- Traffic accidents killed a 54-year-old New York woman, and a person in New Jersey
- In Springfield, Massachusetts, a man died when he touched a protective rail surrounding downed power lines

Connecticut Governor Dannel P Malloy cautioned the 750,000 people who were without electricity in his state that the effects of the storm would still be felt after the snowfall stopped.

"If you are without power, you should expect to be without power for a prolonged period of time," CBS News quoted him as saying.

New Jersey Governor Chris Christie's house was one of the 600,000 suffering power cuts in the state.

'Not even Halloween'

West Milford, New Jersey, about 45 miles (70km) north-west of New York, saw 19in of snowfall, and Hillsboro, New Hampshire, saw 21.5in.

"I can't believe it's not even Halloween and it's snowing already," Carole Shepherd of Washington Township in New Jersey told Associated Press after shovelling her driveway.

Biggest snowfall in each state:

- Connecticut: 17in - Bristol
- Delaware: 2in - Newark
- Maine, 7.7in - Grays
- Maryland: 11.5in - Sabillasville
- Massachusetts: 27.8in - Plainfield
- New Hampshire - 21.5in, Hillsboro
- New Jersey: 19in - West Milford
- New York: 17.9in - Millbrook
- Pennsylvania: 16in - Huff Church
- Rhode Island - 3.6in, North Foster
- Vermont: 13in - Wilmington
- Virginia: 9in - Skyland
- West Virginia: 14in - Mount Storm

Source: NWS (National Weather Service)

In New York City, a new record for October snowfall was set when 1.3in fell in Central Park. Only three other snowy October days have been recorded in the park in 135 years of record-keeping.

Most of the Occupy Wall Street protesters in New York's Zuccotti Park saw out the storm.

Nick Lemmin, of Brooklyn, told AP he had "slept pretty well", although fellow protester Adash Daniel headed off after three weeks in the park, saying: "I'm not much good to this movement if I'm shivering."

On Sunday, passengers were stranded for more than seven hours on one JetBlue flight in Hartford, Connecticut.

On Saturday, flights were delayed at Newark airport in New Jersey, which was being lashed by heavy rains and winds.

Amtrak reported massive disruption to train services, including a 13-hour delay for passengers on one train in central Massachusetts.

John LaCorte, a National Weather Service meteorologist in Pennsylvania, told the agency that the last time the state saw a major storm so early was in 1972.

"This is very, very unusual. It has all the look and feel of a classic mid-winter nor'easter," he said.

High pressure over south-eastern Canada had fed cold air south and into moisture from the North Carolina coast.

In New England it is usual for measurable snow to fall in early December.

NWS meteorologist Bill Simpson said temperatures could return to normal by the middle of next week.

"This doesn't mean our winter is going to be terrible. You can't get any correlation from a two-day event," he said.

Bill Simpson was, of course, right to be cautious. The winter turned out quite different to what was expected. The Eastern and Central United States subsequently experienced a mild winter as opposed to the historic cold which refrigerated Eurasia for several weeks.

Unlike the prophesies of climate pseudoscience, such as TIME Magazine with the absurd headline: "The Year That Winter Forgot: Is It Climate Change?", meteorology is still based on science. Meteorologists correctly hesitate to make any predictions beyond a few days. When they do make slightly longer-term forecasts, they are nearly always careful to add a lower probability of accuracy to such guesses. By contrast, one wonders what possesses climate pseudoscience to confidently predict the climate decades down the road, such as Hansen's 1988 prediction of sea level rise inundating a good part of New York by now. Even worse, when Mother Nature fails to behave as predicted, there is hardly ever, any admission of error on the part of the proponents of Anthropogenic Global Warming. What one gets instead are fantastic attempts to rationalize nature to preconception.

June to September 2011

In the heyday of Anthropogenic Global Warming, during the relatively warm years at the turn of the century when spring came early and winters were milder, proponents confidently terrified the population of the world with harebrained prognostications of unbearably hot summers, even earlier springs, vanishing winters, disappearing snow, etc. They even strongly recommended to the people of the Alps and tourists alike to pack up their ski gear, shut down the ski stations, and focus instead on summer sports. In other words, the alarmism was akin to replacing snowboarding with mountain biking in the winter.

Then, much to the chagrin of the global warming doomsayers, the weather inexplicably changed from the end of 2007. Harsh winters, not experienced for decades returned with a vengeance. At first the proponents of AGW tried to rationalize the cold with pseudoscience explanations for one-off exceptions. There will be examples of such rationalizations later in this chapter. Nonetheless, cold winters continued to persist and spring arrived early no more. It took the alarmists more than two years to invent the new story of "climate weirding": now extreme cold weather became a "sign" of a warming planet, even if this was not part of the plot previously. Therefore, when snow persisted in Central Europe in the late spring and summer of 2011, and the Southern Hemisphere, once again had a cold winter, this was definitely "heat" in disguise.

Skiers were still delighted in late spring (June 2011)

http://www.youtube.com/watch?v=-wSrkPx1wEs

"Monday was bright and sunny, about 80 degrees [Fahrenheit]; Tuesday brought rain; and, this is what we woke up to on Wednesday, June 1, 2011, in Wengen, Switzerland."

So much for early spring....

Summer Festival impacted by snow in Verbier in July 2011

Verbier is not a resort at high altitude. Barely 1,500 meters (or 5,000 feet) above sea-level, tourists often use the village as a base to visit the higher mountains that encircle it. Nonetheless, being very prudent as the Swiss often are, Verbier has its annual summer music festival in late July. Still, in 2011, the "warming" planet did not prevent the midsummer revelry from being impacted by highly unusual snow. The outdoors were not really welcoming to tourists caught in summer clothing, although the music show had to go on indoors. The below picture counters a thousand words of global warming rationalizations.

http://www.blipfoto.com/entry/1288682

"We woke to a new layer of snow just above the village. Not the kind of weather we should expect, even in the Alps, in July. The music continues, though, perfect in this magical place. More masterclasses and concerts today"

Incidentally, the climactic snow line, the point above which snow and ice cover the ground throughout the year, is as follows: [129] [130]

Alps (northern slopes)	48°N	2500–2800 meters
Central Alps	47°N	2900–3200 meters
Alps (southern slopes)	46°N	2700–2800 meters

In other words, the snow line is well above Verbier.

Additionally, temperatures even on the plateau of Switzerland, about 300 to 500 meters above sea level were about 30 degrees Fahrenheit below average for a good part of July 2011.

Snow disrupted the Tour de France in July 2011

http://www.cyclingnews.com/news/tour-de-france-to-just-avoid-snow-on-galibier

Tour de France to just avoid snow on Galibier

By: Pierre CarreyPublished: July 19, 2011, 03:29, Updated: July 19, 2011, 03:32Edition:First Edition
Cycling News, Tuesday, July 19, 2011Race:Tour de France

New snowfall expected Tuesday but then, the sun might help the race

On Sunday, around 200 cyclists had to be rescued from the Col du Galibier while on a "Brevet alpin de cyclotourisme." Dressed for summer conditions, they bore the brunt of 3°C temperatures before reaching the top of the pass, which was covered by snow. Cyclingnews understands new falls of 15 to 20cm of snow over 2,500m are expected Tuesday in Galibier, which peaks at 2,645m.

It could be that the Tour de France faces a serious problem as it will climb these roads twice, Thursday to a mountaintop-finish and Friday on the route to Alpe d'Huez.

The race might avoid the snow, however.

"The Tour is pretty lucky," forecaster Yan Giezendanner told Cyclingnews. "Imagine if the snow would fall Thursday and Friday instead of Tuesday and last Sunday: the road would certainly be blocked."

A Météo-France meteorological expert, specializing in mountain climates, Giezendanner says the "sun will melt the snow" before the stage on Thursday. "There might be only some 'névé' on the climb but nothing that will adversely affect the race," he adds, referring to the patches of residual snow which have been partially melted, refrozen and compacted.

Thanks to the warmth of the sun, the temperature at the top of Galibier might be sustainable for the riders, both Thursday and Friday.

---

[129] Flint, R. F. (1957). Glacial and Pleistocene geology. John Wiley & Sons, Inc., New York, xiii+553+555 pp
[130] Wilhelm, F. (1975). Schnee- und Gletscherkunde [Snow- and glaciers study], De Gruyter, Berlin, 414 pp.

In the same area, Tour de France organisers in 1996 were been forced to cancel the main part of stage 9 from Le Monêtier-les-Bains to Sestriere, because of the iced rain in the Col d'Iseran and the 100km/h wind in Galibier. The stage was reduced from 189 to 46 kilometers and only held on the Montgenevre and Sestriere climbs. Bjarne Riis took the yellow jersey that day, taking the first steps of his overall victory.

Mountains are well-known for their capricious weather in summer. "I remember 20cm of snow fell on La Plagne resort 1989 the first of August and hail storms in the Galibier the 2nd of July last year," recalled Laurent Cadars, the manager of the Valloire Council which encapsulates the Galibier climb.

Forecaster Giezendanner confirmed this year's snowfalls are indeed seasonal.

He predicts rain on stage 16 to Gap, Tuesday. "Rainfalls will go with the race almost the whole day. Sunny periods will be rare. The race the bad weather will both head north."

Stage 17 to Pinerolo will be quite warm, with a 30°C temperature expected at the finish.

Thursday, the finish to the top of Galibier will be sunny too and the riders will have a tailwind during the stage. The temperature at the finish is estimated to be about 6°C.

It'll be a bit warmer Friday for the second Galibier ascent, en route to Alpe d'Huez, with 7 to 8°C at the summit of mythical pass and 12 to 13°C at the finish.

"Of course the temperature is differently appreciated with the sun," Giezendanner said. "6°C temperatures are really sustainable if the sky is clear but you'll be frozen if you are in the shade. Fortunately the next days at the Tour won't be cloudy too much."

In 2011, it snowed in the Alps in both July and August and September.

Under objective circumstances, a white summer in Central Europe should have made the headlines in the mainstream and pseudoscience press. Yet there was very little, if no mention of this very significant weather event, despite the publicity associated with summer events such as the Tour de France and the Montreux Jazz Festival. It seems that near-unified silence was considered the best way to attempt to protect the pre-conceptual myth of global warming.

Early snow in the Alps in September 2011

Proponents of Anthropogenic Global Warming selectively puffed-up the dry November of 2011 and the late start of the ski season in the mountains of Central Europe in yet another forlorn bid to restore some of their long diminished credibility. As usual, their hype did not tell the full story. In 2011, the Alps experienced unseen levels of snowfall in September before the dry spell set in until the beginning of December. And we all know what has happened since December.

http://www.australianclimatemadness.com/2011/09/early-snow-in-switzerland-alarmists-blame-global-warming-in-3-2-1.../

http://www.expatica.com/ch/news/swiss-news/record-snowfall-in-switzerland-45-cm-in-st-moritz_176497.html

Early snow in Switzerland. Alarmists blame "global warming" in 3, 2, 1...

Tuesday, 20 September 2011 14:38 pm · 23 comments

RECORD SNOWFALL IN SWITZERLAND - 45CM IN ST MORITZ

Snow fell in the Swiss Alps overnight Sunday to levels unseen for the month of September, Swiss weather agency Meteosuisse reported on Monday.

In the ski resort of St Moritz, in the southeast canton of Grison, a total of 45 centimetres

(nearly 18 inches) of snow was recorded on Monday morning, it said.

The weather agency said the high levels of precipitation were due to a cold front which lowered the snow line to 800 metres (2,600 feet).

Rainfall was also higher than usual, with around 100 liters per square meter measured in the town of Santa Maria, also in Grison, the highest level since records began in 1901, Meteosuisse said.

The snowfall also provoked traffic disturbances in the mountains, with the St Bernard, Flueela and Nufenen passes closed, according to ViaSuisse, which reports on the condition of Swiss roads.

The Gothard, Lukmanier and Oberalp passes are also covered in snow, it added.

The mild summer was not limited to Central Europe. In the Eastern Mediterranean, for example, daytime high temperatures averaged a comfortable 25°C, when just a few years ago in the mid-2000s, they were typically at 35-45°C during the month of August.

Extreme cold in the Southern Hemisphere in August 2011

The Southern Hemisphere often tends to be ignored, except when misused by proponents of Anthropogenic Global Warming to explain away and rationalize a cold spell in the Northern Hemisphere. Therefore, let us check what happened during the Southern Hemisphere winter of 2011.

Australia and New Zealand, for example, experienced severe cold during August. There was even heavy and very unusual snow in tropical Auckland in the North Island - the first snowfall since 1939. And the South Island was literally crippled by the most severe cold spell in fifty years. And, as the below article mentions, "due to the lack of extensive landmasses south of the equator, mid-latitudes of the Southern Hemisphere rarely experience the severe cold and snow familiar to many areas of the Northern Hemisphere. The New Zealand cold wave comes just weeks after unusual snow in South Africa and parts of Chile."

http://www.washingtonpost.com/blogs/capital-weather-gang/post/rare-snowstorm-cold-blast-hit-new-zealand/2011/08/16/gIQAfK4BJJ_blog.html

Rare snowstorm, cold blast hit New Zealand

Posted at 10:15 AM ET, 08/16/2011

By Justin Grieser

An unusual outbreak of cold and snow has gripped many parts of New Zealand after a strong

Antarctic front pushed northwards into normally mild parts of the country.

Blizzards across the lower South Island dumped heavy snow in Christchurch and Queenstown, closing airports, cutting power, and creating treacherous road conditions across the island nation. Southern regions and higher elevations have reported as much as 30-60 cm (1-2 ft) in recent days.

The wintry blast is being described as a once-in-a-lifetime event, bringing snow to the capital city of Wellington and as far north as the Auckland region for the first time since the 1970s.

Snow actually settled on the ground in downtown Auckland reported IOL news. According to the New Zealand Herald, downtown Auckland last witnessed accumulating snow in 1939. The city also recorded its lowest daily high temperature since official records began there in 1961. On Monday, the temperature only rose to 8.2 C (47 F), breaking the previous record low maximum of 8.7 C set in July 1996. Average high/low temperatures for the city are about 59/45 F even during the second-coldest month of the year.

While this year's snow was short-lived (it later mixed with rain showers), many Auckland residents were nonetheless in awe of the unusual precipitation while it lasted. The Herald writes:

The snow caused waves of excitement in Auckland. Kevin Prohl saw a snow flurry as he was driving...and described it as a fairy tale. "Looking at oncoming drivers and seeing their smiles as we were fascinated by this unusual occurrence - it was truly delightful to see, yet all too short."

Due to the lack of extensive landmasses south of the equator, mid-latitudes of the Southern Hemisphere rarely experience the severe cold and snow familiar to many areas of the Northern Hemisphere. The New Zealand cold wave comes just weeks after unusual snow in South Africa and parts of Chile.

Although I commend the Washington Post for publishing this article, much of the mainstream and pseudoscience media remained silent, once again demonstrating typical warming bias. The test is: before you read this book, were you aware that the Southern Hemisphere experienced unusual cold during their winter of 2011?

Australia was not spared from the cold either. Below is an interesting article from the Guardian newspaper. I will explain why the article is interesting, right after you have had a chance to read it.

http://www.guardian.co.uk/news/2011/aug/01/weatherwatch-australia-floods-storms

Weather: Worst winter storms in Perth

guardian.co.uk, Monday 1 August 2011 23.05 BST
Alison Cobb (MeteoGroup)

South-west Australia, including Perth, was hit by one of its worst storms so far this winter on Wednesday last week. A strong cold front swept through the area at about 7.30am, bringing strong winds and heavy rain. A gust of 57mph was recorded at Swanbourne, situated on the coast, with 45mph recorded at Perth airport. However, from the damage caused to over 30 homes in and around Waroona, gusts were estimated to have reached 78mph.

Following more than 30 fatalities, floods and destruction across Luzon in the northern Philippines early last week, tropical storm Nock-Ten continued north-westwards and hit southern China on Friday afternoon. Nock-Ten made landfall in the city of Wenchang, in south China's Hainan province. Dongfang in the west of Hainan recorded 141 mm of rain on Friday, accompanied by winds of up to 63mph. This made Nock-Ten the most powerful storm to hit China so far this year.

With a high pressure system close by, much of eastern Europe had a hot spell last week. This heat instigated thunder-storm development on Thursday, with 51mm of rain falling in Aluksne, Latvia, in the 12 hours up to 9pm. Also on Thursday, the temperature in Moscow reached 33.8C, almost 10 degrees above the average for this time of year. However, there was an end to this dominant high pressure on Friday, as a cold front swept across, bringing in slightly fresher air.

The reader is requested to take note of the final paragraph, where the Guardian unnecessarily introduces the hot weather in Eastern Europe, which is completely irrelevant to the subject of the article – the unusual cold in Perth. So why was the paragraph about Eastern Europe added? This is typical of the pre-conceptual warming bias that has become omnipresent in most of the mainstream and pseudoscience media.

March 2011

Advocates of Anthropogenic Global Warming often like to intimate that, in a warming planet, tropical species are moving towards the poles to formerly colder latitudes. That is the myth. Now, let us look at the reality in March 2011.

Musk oxen, related to sheep, have been hardened to extreme cold conditions well before the first human being ever set foot on this planet. They roamed the tundra and steppes during the

Pleistocene era and are among the most famous *Ice Age survivors* of the far north. Yet some of them failed to survive the cold March of 2011 in this age of the "warming" planet.

http://www.reuters.com/article/2011/03/22/us-alaska-muskoxen-idUSTRE72L87F20110322

Dozens of musk oxen found dead near Bering Strait

ANCHORAGE, Alaska | Tue Mar 22, 2011 7:43pm EDT
By Yereth Rosen

(Reuters) - Thirty-two musk oxen were found dead in the ice along the Bering Strait, apparently killed when they drowned in water that surged ashore during a winter storm, the National Park Service said on Tuesday.

The dead animals were found frozen and entombed in shoreline ice at the Bering Land Bridge National Preserve in northwestern Alaska, the Park Service said. It is suspected that up to 23 more musk oxen may have died in the same event, their bodies buried deeper in the ice, the Park Service said.

The animals, some fitted with radio collars and being tracked as part of a five-year study, were last seen alive in mid-February, said John Quinley, spokesman for the Park Service in Alaska.

They apparently fell victim to a storm later in the month that sent waves up to eight feet higher than normal tide lines, Quinley said.

"The water came up. They didn't leave. And eventually, they all drowned," he said.

There are about 1,000 musk oxen on Alaska's Seward Peninsula, a section of land where the Bering Land Bridge National Preserve is located, Quinley said. The potential loss of 50 or so animals is not considered "a population-changing event," he said.

Scientists plan to return to the site to retrieve teeth and bone samples and to try to determine the dead animals' ages, he said.

There was talk of trying to remove the animals' bodies from the ice in which they are locked, "but it didn't seem very practical to do that," he said.

The Park Service does expect wolves, bears and other animals to scavenge the area, Quinley said. But the agency has warned local villagers the meat is unfit for human consumption and that retrieval of horns from national park lands is not legal.

Musk oxen, related to sheep, roamed the tundra and steppes during the Pleistocene era and are among the most famous Ice Age survivors of the far north.

They are known for their special abilities to withstand the Arctic winters. They have warm, soft hair, called "qiviut," which is shed in the spring and knitted into pricey garments, and a habit of huddling together to protect their young from predators.

There are about 125,000 in the world, mostly in Canada and Greenland, according to state and federal agencies.

Alaska has a few thousand, all of them descendants of animals brought to the territory in 1935 from Greenland by the U.S. Fish and Wildlife Service. Alaska's indigenous population had disappeared by the early 20th century.

The current Alaska population is scattered among remote sites in northern and western Alaska, including the edges of the Bering Strait, and some animals live in farms and at a zoo in Anchorage.

(Editing by Dan Whitcomb and Jerry Norton)

November 2010 to January 2011

Historic cold spell in November 2010

The Northern Hemisphere winter of 2010-2011 started early with a historic cold spell in Europe already in November 2010.

http://www.bbc.co.uk/news/world-europe-11907045

Europe's deadly cold snap maintains grip
4 December 2010 Last updated at 00:44

Northern Europe remains in the grip of heavy snow and freezing temperatures, with more cold weather forecast for the weekend

Dozens of people are reported to have been killed by exposure to the cold or in weather-related accidents.

The snow continues to disrupt transport networks but many airports are resuming a more normal service.

In the Balkans, heavy flooding has forced more than 1,000 people to evacuate their homes.

Temperatures in Poland have fallen to as low as -33C (-27F) in the past few days.

Another 12 people froze to death across Poland on Thursday night, according to police, bringing the total killed there during this cold snap to 30.

Police say many of the victims were homeless people.

Around 150,000 people in the Polish city of Czestochowa were reported to be without heating on Friday.

The cold snap has hit Poland particularly hard Officials in Slovakia said two people had died in snow-related car accidents, while neighbouring Czech Republic also reported several casualties.

Cold weather has also claimed five lives this week in the Baltic state of Lithuania and there have been at least two deaths in Germany.

Siberian air

The snow and ice has meant widespread delays and cancellations to European flights in recent days, with the closure of a number of British airports.

London's Gatwick airport reopened on Friday morning for the first time in two days, but passengers have been warned to expect delays.

While a regular service appears to be resuming in many European airports, arrivals and departure boards are still showing delays and some cancellations.

Eurostar says it expects to operate a significantly reduced service until Sunday, but that no more tickets will be available until Monday.

In other developments

- Denmark mobilised its army to help emergency vehicles make their way through snow in the south-east
- Snowfalls trapped hundreds of motorists in Brittany and Normandy
- In Belgium, 650km (403 miles) of traffic jams were reported, with Flanders hit hard
- Cold air moving down from Siberia has contributed to the wintry conditions in northern Europe.
- Temperatures are an average 5C-10C below average in some major cities.

Extreme Cold To Grip Europe. Forecast -38°C in Switzerland…Will Be Even Colder Later…Pattern Not Seen in 70 Years.

By P Gosselin on 22. November 2010

Computers have been forecasting a wicked cold winter for Europe this year. Looks like it's shaping up to be just that.

Later this week a blast of Arctic air is set to sweep across northern and central Europe, as a huge high pressure zone off in the Atlantic combined with low pressure system Gundula to east over the Baltic pump frigid air over the continent.

One meteorologist says such a weather pattern was last seen 70 years ago.

Wetter.com writes here:

> "Cold air will clash with moist air and, as a result, we will get more and more snow in the days ahead. But that's not all. By and by the cold will tighten its grip, and by Friday temperatures will not get above freezing.
>
> That is still not the end of it. This weekend temperatures nationwide [Germany] will be between 5 below and 13°C below at night. But hang on tight. Next week will be even colder - especially at night, where for example, the temperature in the Black Forest will drop to 26°C below!
>
> And if that is not enough, then drive out to the Swiss mountains. 38°C below is the forecast during the night from Sunday to Monday!"

German tabloid Bild newspaper reports here, quoting meteorologist Dominik Jung of wetter.net:

> "The weekend will be bitter cold. Daytime highs will climb to only -8°C to -2°C . Nighttime temperatures will drop to -10°C. But in areas with clear skies and snow cover, especially in the central mountain regions, temperatures of -25°C will be reached."

Jung says such a weather pattern can stay around a long time before it goes away. He tells Bild newspaper:

> "When cold polar air is really there, than it'll stick around for awhile. At the moment weather models show no end in sight for the coming cold wave. Quite to the contrary, the first weekend of December is going to be even colder, and there's the threat of heavy snowfalls.
>
> We'll be getting temperatures like we would only expect in the dead of winter. A similar weather pattern led to the extreme winter in 1939/1940."

And it's still autumn officially!

Strange how the leaders of the world's most climate-activist countries will be discussing ways to prevent warming in balmy Cancun Mexico just when their respective countries will be struggling in bitter cold. The Gods have a sense of humour.

Enthusiasts of Anthropogenic Global Warming were stunned beyond belief. How could this happen in a warming planet? Did they not predict the end of snow a decade ago? In the typical worst tradition of climate pseudoscience, just as they had tried to do in the harsh winter of 2009 – 2010, the prophets of blistering doom attempted to minimize the event as just a European, not a global phenomenon. They even made use of their recently invented explanation of "climate weirding" in an incongruous attempt to blame the extreme cold on global warming. After all, the only pre-conceptual constant was that the planet was warming. Everything else had to be rationalized to that singular predetermined conclusion in this new world of science twisted into faith.

Two examples are provided below. There were many other such articles flooding the mainstream and pseudoscience media in an obvious attempt to salvage the credibility of Anthropogenic Global Warming.

NASA's Pretext

http://thinkprogress.org/climate/2010/12/12/207188/nasa-explains-how-europe-can-be-so-cold-amidst-the-hottest-november-and-hottest-year-on-record/?mobile=nc

NASA explains how Europe can be so cold amidst the hottest November and hottest year on record

By Joe Romm on Dec 12, 2010 at 10:23 am

How did we get record-breaking November warmth in the middle of a strong La Niña that would normally cool global temperatures (as it did in the fall of 1998, see lower right figure, blue line)? Is the answer the Arctic sea ice death spiral 2010? And is the loss of Arctic sea ice also responsible for the frigid European temperatures?

NASA's James Hansen, Reto Ruedy, Makiko Sato and Ken Lo answer these questions in "2010 " Global Temperature and Europe's Frigid Air," which I repost below with the original maps and charts:

Surface Temperature Anomaly (°C)

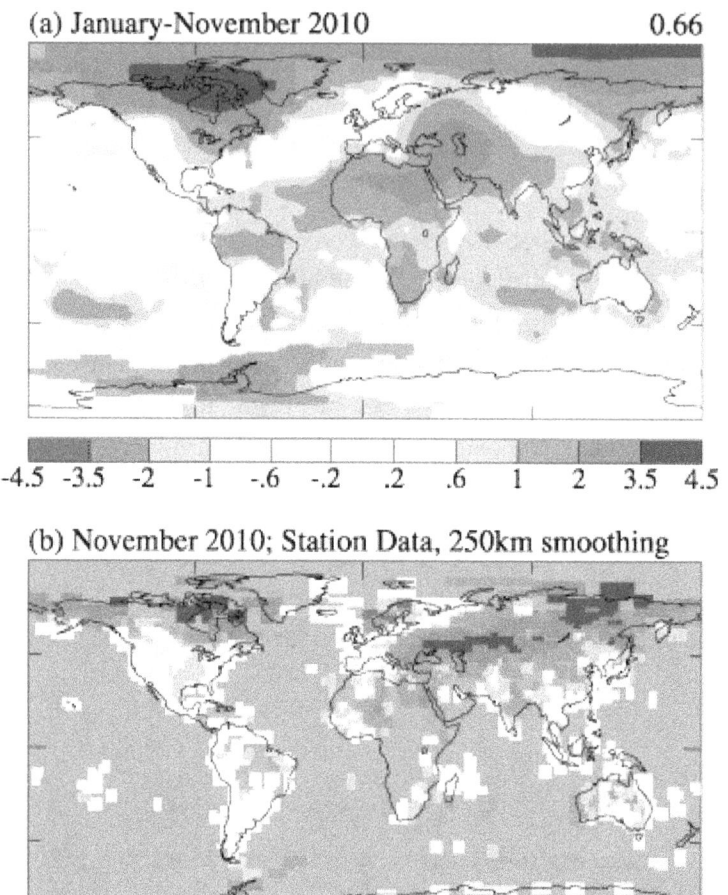

Figure 1: (a) January-November surface air temperature anomaly in GISS analysis, (b) November 2010 anomaly using only data from meteorological stations and Antarctic research stations, with the radius of influence of a station limited to 250 km to better reveal maximum anomalies.

Figure 1(a) [click to enlarge] shows January-November 2010 surface temperature anomalies (relative to 1951-80) in the preliminary Goddard Institute for Space Studies analysis. This is the warmest January-November in the GISS analysis, which covers 131 years. However, it is only a few hundredths of a degree warmer than 2005, so it is possible that the final GISS results for the full year will find 2010 and 2005 to have the same temperature within the margin of error.

As described in an in-press paper at Reviews of Geophysics (see summary PDF) that defines the GISS analysis method, we estimate a two-standard-deviation uncertainty (95 percent confidence interval) of 0.05°C for comparison of global temperatures in nearby recent years. The magnitude of this uncertainty and the small temperature differences among different years is one reason that alternative analyses yield different rankings for the warmest years. However, results for overall global temperature change of the past century are in good agreement among the alternative analyses (by NASA/GISS, NOAA National Climate Data Center, and the joint analysis of the UK Met Office Hadley Centre and the University of East Anglia Climatic Research Unit).

Figure 1(b) shows November 2010 surface temperature anomalies based only on surface air measurements at meteorological stations and Antarctic research stations. In producing this map the radius of influence of a given station is limited to 250 km to allow extreme temperature anomalies to be apparent. Northern Europe had negative anomalies of more than 4°C, while the Hudson Bay region of Canada had monthly mean anomalies greater than +10°C.

The extreme warmth in Northeast Canada is undoubtedly related to the fact that Hudson Bay was practically ice free. In the past, including the GISS base period 1951-1980, Hudson Bay was largely ice-covered in November. The contrast of temperatures at coastal stations in years with and without sea ice cover on the neighboring water body is useful for illustrating the dramatic effect of sea ice on surface air temperature. Sea ice insulates the atmosphere from ocean water warmth, allowing surface air to achieve temperatures much lower than that of the ocean. It is for this reason that some of the largest positive temperature anomalies on the planet occur in the Arctic Ocean as sea ice area has decreased in recent years.

The cold anomaly in Northern Europe in November has continued and strengthened in the first half of December. Combined with the unusual cold winter of 2009-2010 in Northern Hemisphere mid-latitudes, this regional cold spell has caused widespread commentary that global warming has ended. That is hardly the case. On the contrary, globally November 2010 is the warmest November in the GISS record.

Global Land–Ocean Temperature Index

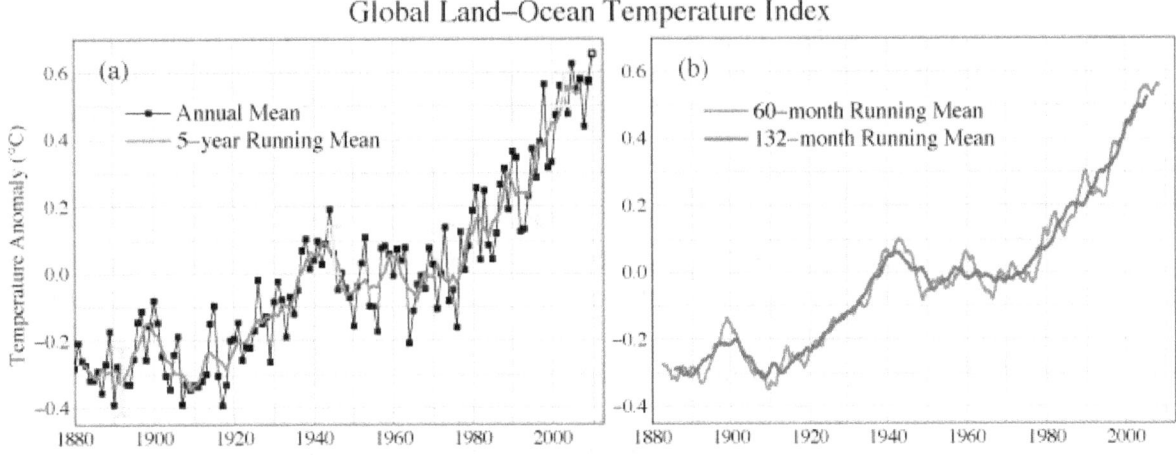

Figure 2(a) illustrates that there is a good chance that 2010 as a whole will be the warmest year in the GISS analysis. Even if the December global temperature anomaly is unusually cool, 2010 will at least be in a statistical tie with 2005 for the warmest year.

Figure 2(b) shows the 60-month (5-year) and 132-month (11-year) running-mean surface air temperature in the GISS analysis. Contrary to frequent assertions that global warming slowed in the past decade, as discussed in our paper in press, global warming has proceeded in the current decade just as fast as in the prior two decades. The warmth of 2010 is especially noteworthy, given the strong La Nina that developed in the second half of 2010. The La Nina, caused by unusually strong easterly equatorial winds, produces the cool anomalies in the tropical Pacific Ocean as cold upwelling deep water along the Peruvian coast is blown westward along the equator.

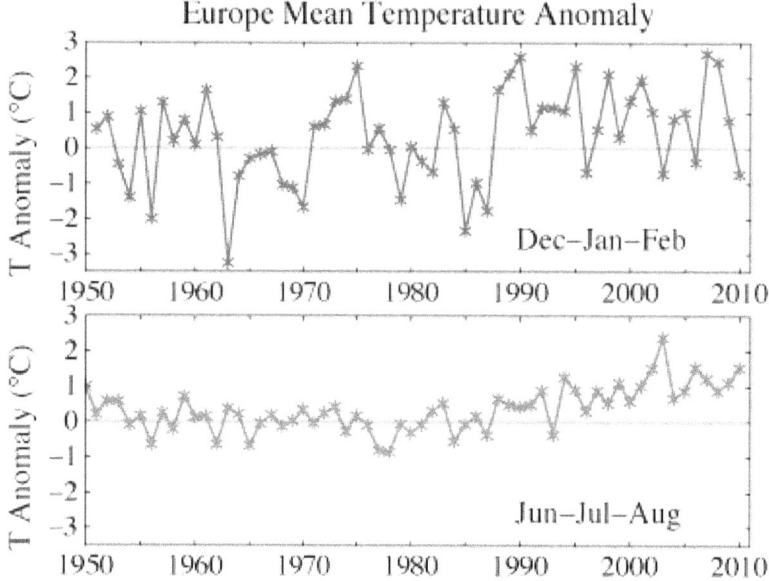

Europe Mean Temperature Anomaly

Back to the cold air in Europe: is it possible that reduced Arctic sea ice is affecting weather patterns? Because Hudson Bay (and Baffin Bay, west of Greenland) are at significantly lower latitudes than most of the Arctic Ocean, global warming may cause them to remain ice free into early winter after the Arctic Ocean has become frozen insulating the atmosphere from the ocean. The fixed location of the Hudson-Baffin heat source could plausibly affect weather patterns, in a deterministic way "" Europe being half a Rossby wavelength downstream, thus producing a cold European anomaly in the trans-Atlantic seesaw. Several ideas about possible effects of the loss of Arctic sea ice on weather patterns are discussed in papers referenced by Overland, Wang and Walsh.

However, we note in our Reviews of Geophysics paper in press that the few years just prior to 2009-2010, with low Arctic sea ice, did not produce cold winters in Europe. The cold winter of 2009-2010 was associated with the most extreme Arctic Oscillation in the period of record. Figure 3, from our paper in press, shows that 7 of the last 10 European winters were warmer than the 1951-1980 average winter, and 10 of the past 10 summers were warmer than climatology. The average warming of European winters is at least as large as the average warming of summers, but it is less noticeable because of the much greater variability in winter.

Finally, we point out in Figure 3 the anomalous summer warmth in 2003 and 2010, summers that were associated with extreme events centered in France and Moscow. If the warming trend that is obvious in that figure continues, as is expected if greenhouse gases continue to increase, such extremes will become common within a few decades.

Reference: Hansen, J., R. Ruedy, Mki. Sato, and K. Lo, 2010: Global surface temperature change. Rev. Geophys., in press, doi:10.1029/2010RG000345.

– James Hansen, Reto Ruedy, Makiko Sato and Ken Lo

Subsequently, NASA's GISS did indeed declare 2010 as the warmest year on record. We have already read previously about the inaccurate predictions of Dr. Hansen as well as examples of the warming bias in his interpretation of data.

John Christy, winner of the 1991 NASA's Exceptional Scientific Achievement Medal (together with Roy Spencer), explains the situation succinctly:

http://www.timesonline.co.uk/tol/news/environment/article7026317.ece

"The temperature records cannot be relied on as indicators of global change," said John Christy, professor of atmospheric science at the University of Alabama in Huntsville, a former lead author on the IPCC.

The doubts of Christy and a number of other researchers focus on the thousands of weather stations around the world, which have been used to collect temperature data over the past 150 years.

These stations, they believe, have been seriously compromised by factors such as urbanization, changes in land use and, in many cases, being moved from site to site.

Christy has published research papers looking at these effects in three different regions: east Africa, and the American states of California and Alabama.

Both Christy and Spencer had to leave NASA, for obvious reasons. In a later chapter, we shall look in more detail at how temperature data published by pseudoscience institutions such as GISS, the CRU, etc., have significant credibility issues, due to the influence of pre-conceptual warming, to put it politely.

Global Warming blamed for the extreme cold ("Climate Weirding")

http://www.reportingclimatescience.com/news-stories/article/global-warming-shares-blame-for-europes-cold-weather-says-climate-scientist.html

Global warming shares blame for Europe's cold

02.12.2010 10:18 Age: 2 yrs
By: Leon Clifford

Global warming may be contributing to the current heavy snow and subzero temperatures across Europe, which has closed major airports, caused hundreds of highway accidents and several deaths. Cold weather in Europe is often associated with a weather system known as the North Atlantic Oscillation (NAO) but there is also a significant impact from current low levels of sea ice in the Barents-Kara Sea, according to a leading climate scientist. Low levels of sea ice in the Barents-Kara Sea are currently close to the record lows seen in the harsh winter of 2005 and 2006....

WARMER OCEANS MELT SEA ICE AND LEAD TO THE EFFECT

Warming oceans, thought by many to be associated with climate change, are contributing to reductions in sea ice in the Arctic area. Computer models suggest that a reduction in sea ice in the eastern Arctic leads to a loss of ocean heat and a consequent warming of the lower atmosphere which can trigger atmospheric circulation anomalies that can in turn lead to an overall cooling of northern continents, according to Petoukhov's research which was published in the Journal of

Geophysical Research in November. This can result in a continental-scale winter cooling reaching, on average, −1.5C colder than it would otherwise have been.

A drastic reduction of sea ice was observed in the Barents-Kara Sea north of Norway and Russia during the cold European winter of 2005 and 2006 and the exposed sea surface lost a lot of warmth to the normally cold and windy arctic atmosphere. Warming of the air over the Barents-Kara Sea seems to be associated with bringing cold winter winds across Europe.

The researchers simulated this situation using a general circulation climate model and incrementally reduced the sea ice cover of the eastern Arctic in the model from 100 per cent cover down to 1 per cent cover in order to analyse the relative sensitivity of wintertime atmospheric circulation. The simulations demonstrated that lower-troposphere heating over the Barents-Kara Sea in the eastern Arctic caused by the sea ice reduction may result in a strong anticyclonic anomaly over the Polar Ocean and anomalous easterly advection over northern continents and a consequent cooling. An abrupt transition between different regimes of the atmospheric circulation in the sub-polar and polar regions may also be very likely.

This correlation between the sea ice reduction and the continental cooling is strong, according to the research. Other explanations linking cold winters and global warming include reduced solar activity and changes in the Gulf Stream are less strongly correlated. However, the NAO could interact with sea-ice decrease, the study concludes and one could amplify the other. This is what may be happening at the moment, suggested Petoukhov.

Above, the reader has read several pages of pseudoscience explanation for the extreme cold in Europe in November 2010. To summarize, proponents of AGW pre-conceptually rationalized the extreme cold as a *European* phenomenon as a result of a weather pattern caused by *planetary warming.* For the sake of argument, let us assume that their explanation was correct. Therefore, to be logical, let us now examine the weather conditions in the rest of the world during December 2010 and January 2011.

Cancun, Mexico

Let us start with Cancun, the site of yet another expensive United Nations conference on "saving the planet" from global warming. So, how was the weather in Cancun during the conference? In one short phrase, the weather was *abnormally cold*.

http://www.freerepublic.com/focus/f-news/2641425/posts

Record Cold at Cancun Climate Confab
New American ^ | December 11, 2010 | W. F. Jasper
Posted on dimanche 12 décembre 2010 14:14:47 by IbJensen

As the United Nations opened its latest conference on global warming, Mother Nature sent snowstorms and freezing temperatures that disrupted travel all across Europe and much of the Northern Hemisphere.

Even Cancun, Mexico's sunny resort city that hosted the confab, was not spared the chill. The UN summit, known as COP16 (the 16th Conference of Parties on global warming), concluded Saturday morning after an all-night marathon session. Cancun may not have experienced blizzards and ice, but it did, nevertheless, get hammered with record low temps for the month of December.

Meteorologist Anthony Watts, who runs the global warming skeptic blog "Watts Up With That?," couldn't resist rubbing it in. In a December 10 posting, entitled "Gore Effect" on Steroids: Six straight days of record low temperatures during COP16 in Cancun Mexico — more coming, Watts wrote:

The irony, it burns. Do you think maybe Gaia is trying to send the U.N. and the delegates a message? One record low was funny, three in a row was hilarious, a new record low for the month of December was ROFL [Rolling On Floor Laughing], but now six straight days of record lows during the U.N. COP16 Global Warming conference? That's galactically inconvenient. The whole month so far has averaged below normal...

How cold did it get? Temps were in the low 50s (Fahrenheit), for the most part, so no one got frostbite. Not bone-chilling, but also not the bikini temperatures and margaritas-on-the-beach experience most snowbirds are expecting when they ditch frigid Oslo, London, and Kalamazoo for Mexico's fabled sunny clime. Watts shows the official low temperatures for the past six days of the conference (December 5-10) and then matches them with the recorded historic lows for the same dates. The conference dates either tied the record lows or were lower by 4 to 5 degrees Fahrenheit.

"It is likely we will see a full week, possibly 8 days of record lows, and another new all time record low for the month of December is possible also," says Anthony Watts.

This recent cold wave in Cancun has already added to the lore of the "Gore Effect," which is defined in the Urban Dictionary as:

The phenomenon that leads to unseasonably cold temperatures, driving rain, hail, or snow whenever Al Gore visits an area to discuss global warming.

Poor Al Gore. He doesn't even have to show up at a global warming event anymore for record cold — often accompanied by paralyzing blizzards — to dampen, or even cancel, the event. The following are a few of the many links to stories from around the world about the Gore Effect provided by Marc Morano at the Climate Depot blog site:

Melbourne temperatures plummet in anticipation of Al Gore's visit next week!

First October snow since 1922 blankets London as global warming bill debated - October 2008

'Gore Effect': Driving snow froze the hopes of organizers of "the biggest global warming protest in history" in Washington

The Gore Effect brings snow to New York City

Al "Mr. Global Warming" Gore didn't show up in Cancun. Perhaps that was by design, since his appearance at last year's Copenhagen summit coincided with the arrival of record cold temperatures and snow storms that forced U.S. House Speaker and her congressional delegation to leave Denmark early — to avoid being snowed in. Even many members of the huge press pool at the Copenhagen summit, almost all of whom were solidly in Al Gore's AGW (anthropogenic global warming) camp, couldn't resist joking about the phenomenon. As I watched the first snow flakes begin to fall last December at Copenhagen's Bella Center, a reporter for Britain's left-wing Guardian commented: "Here comes the snow; Gore's plane must have landed." Similar comments, though considered politically incorrect (if not outright heresy) among the green true believers, were not uncommon in the media center.

Our own video for The New American from the COP15 conference, Gore Effect Impacts Copenhagen Climate provides a glimpse of the deep freeze that Gore's visit was credited with bringing to the event.

As delegates and NGO activists at Cancun insisted that the planet is overheating, millions of people were being deluged with snow and shivering in some of the coldest weather in a century. An extended headline for the U.K.'s Daily Mail on December 9 read:

Now the Army moves in to clear away snow in coldest December for 100 years as fuel runs out at petrol stations in Scotland and East Anglia

British columnist/blogger James Delingpole at The Telegraph aimed this satirical barrage at the global warming alarmists in Cancun — and in the media:

As your boiler breaks down, your pipes freeze, your car won't start, your Ocado delivery fails to arrive, your train is cancelled, your neck is broken after slipping on black ice and you lie in an emergency ward waiting for a doctor to turn up only to learn that they're all off today because of the weather, you might be forgiven for thinking that all this has something to do with global cooling, changes in the Pacific Decadal Oscillation, and the decline in sunspot activity perhaps auguring a new Maunder minimum.

But, of course, you would be wrong, says Delingpole, in his acerbic parody; the "experts" will assure us that all this freezing mayhem is the result of global warming.

Humor aside, how cold did it get in Cancun? Temperatures were in the low 50s (Fahrenheit), for the most part, so nobody got frostbite. Not bone-chilling, but also not the bikini temperatures and margaritas-on-the-beach experience many of the conference attendees were expecting when they ditched frigid Oslo, London, and Kalamazoo for Mexico's fabled sunny clime.

Watts shows the official low temperatures for the six days of the conference (December 5-10) and then matches them with the recorded historic lows for the same dates. The conference dates either tied the record lows or were lower by 4 to 5 degrees Fahrenheit[131].

---

[131] http://wattsupwiththat.com/2010/12/10/gore-effect-on-steroids-six-straight-days-of-record-low-temperatures-during-cop16-in-cancun-mexico/

Below are the temperatures (in Fahrenheit) for conference dates and the record lows:

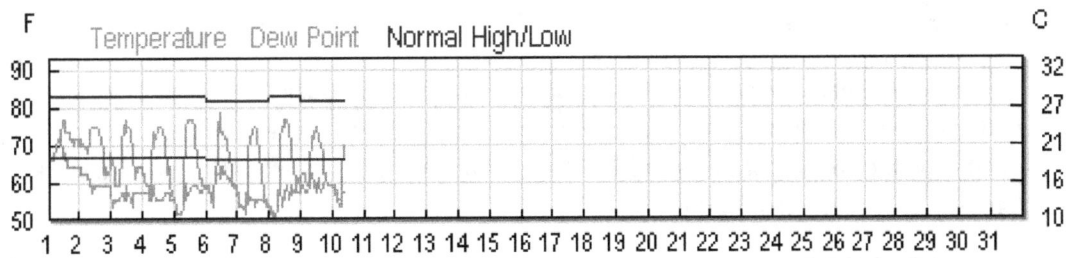

December	Actual Temperature	Record Low
5th	51	51
6th	53	53
7th	53	57
8th	50	50
9th	59	64
10th	55	60

"It is likely we will see a full week, possibly 8 days of record lows, and another new all time record low for the month of December is possible also," Anthony Watts said.

Everyone would agree that Cancun is a quarter of a world away from Europe. I would guess that even global warming alarmists would not try to pretend that Cancun is anywhere near Europe. But you never know.

And it was not just Cancun with unusual cold. According to CNN, some other parts of Mexico had sub-freezing temperatures as low as -6 degrees. People died because they had never in their lifetimes experienced such bitter winter conditions.

United States

Minnesota

On the same day that NASA GISS disseminated misinformation "explaining" the cold weather "in Europe", this is what happened in Minneapolis:

http://sports.yahoo.com/nfl/blog/shutdown_corner/post/Unbelievable-video-of-Metrodome-collapse-from-in?urn=nfl-294816

Unbelievable video of Metrodome collapse from inside stadium
Sun Dec 12 11:53am EST
By Chris Chase

The Metrodome collapsed early Sunday morning after a blizzard dumped 17 inches of snow on Minneapolis. Because of the damage to the stadium, Monday's scheduled New York Giants-Minnesota Vikings game was moved to Detroit's Ford Field.

Cameras were positioned inside the stadium and caught the collapse on tape. To call the video incredible would be an understatement. It looks like the beginning of Armageddon:

The roof over the Metrodome is inflatable, not hard. It's the same sort of bubble you see over swimming pools or tennis courts in the winter, making this collapse almost like a deflated balloon imploding onto the field.

As a result, the damage is not thought to be severe enough to close the stadium for an extended period of time. It will, however, force the Vikings and Giants to play elsewhere on Monday night, continuing the odyssey of this game which began on Saturday when the Giants were stuck in Kansas City because of airport closures.

Heavy snows in Minnesota closed the Minneapolis airport, which left the Giants stranded in the Midwest. The game between the Vikings and Giants, originally scheduled for 1 p.m. ET on Sunday had been moved to Monday night because of the delays. Now the teams will fly to Detroit to play at Ford Field, which will be available after the Lions host the Packers on Sunday. The Giants-Vikings game will kick off at 7:20 p.m. ET. Tickets will be free for anyone who wants to attend. Any ticket holders for the game in Minnesota will receive preferential seating at the 50-yard line if they can make the journey through the snow-ravaged Midwest.

Moving NFL games is rare but has happened occasionally over the past decade. Hurricane Katrina forced the New Orleans Saints to hold a "home game" in New York, while the San Diego Chargers moved a game to Tempe, Ariz., due to California wildfires in 2003.

For once, there was no way for the mainstream and pseudoscience media to cover up this astonishing event, although some did attempt to assign at least part of the blame to the structure rather than the weather. For example, in the above article, you will notice the subtle references to the inflatable roof, Hurricane Katrina and California wildfires.

Minnesota was not the only State in the United States which was unusually cold in the winter of 2010 to 2011. In fact, much of North America suffered one of the severest winters on record. This is especially poignant as the GISS article referenced earlier cited "the Hudson Bay region of Canada had monthly mean anomalies greater than +10°C" to counter the unusual cold in Europe in November 2010.

Florida

In December 2010, Florida had the coldest spell in 169 years. Not that this could have been completely unexpected by Floridians; they had an unusual cold in the previous winter as well.

Let us first look at the weather conditions in Florida in the month of December 2010:

http://www.digitaljournal.com/article/301161

South Florida experiences coldest temperature in 169 years
Dec 7, 2010 in Environment
By Andrew Moran.

Fort Lauderdale - Cold temperatures are spreading across the landscape of the United States. Some states are experiencing temperatures that haven't been felt in more than a century. Southern Florida is one of those states.

At this time of the year, vacationers head down south to the beautiful state of Florida for their sandy beaches, warm temperatures and sunshine. But Floridians woke up to the coldest temperatures in 169 years on Tuesday.

According to the Sun Sentinel, cities like Fort Lauderdale had temperatures in the low-40s (4 degrees Celsius) but it felt more like in the mid-30s (1 degree Celsius) because of the wind chill factor. At 7:24 a.m., Fort Lauderdale broke the 169-year-old record when the temperature dipped to 40 degrees Fahrenheit.

On Monday, orange juice futures skyrocketed to their highest level in nearly 4 years due to concerns over freeze warnings, reports Options Headlines. Another freeze warning was expected until Wednesday.

Fruit damages only occur when temperatures hover around 28 degrees for least 4 hours. Judging by the fruits, there were no indications that temperature reached that point. Parts of the central east coasts did experience, though, temperatures at the 28-degree mark.

""We did not sustain any damage to the trees or the crop," said fruit grower, John Arnold, in an interview with NBC News. "As the winter progresses, it's one cold front after the next and we have to be prepared for the worst. This was like a dry run, so we can test our irrigation measures."

Florida maintains an annual $9 billion fruit industry.

According to the Weather Network, Fort Lauderdale will experience temperatures in the mid-50s (10 degrees Celsius) on Wednesday and go up as the week progresses but it will go back down again by Monday."

As was the case in the previous winter, Florida's citrus industry suffered yet again

http://www.reuters.com/article/2010/12/28/us-florida-citrus-freeze-idUSTRE6BR1J620101228

Cold snap freezes Florida citrus

By Pascal Fletcher

MIAMI | Tue Dec 28, 2010 2:37pm EST

(Reuters) - A second major hard freeze this month iced up oranges and other fruit across Florida's citrus growing regions, causing some fruit damage and raising fears of longer-term impact on groves, growers said on Tuesday.

"We're hearing reports of frozen fruit across all our growing regions ... We definitely had some damage," Andrew Meadows, spokesman for the state's leading growers association, Florida Citrus Mutual, told Reuters.

[132]It was the second significant hard freeze to maul orange and other citrus groves in Florida in two weeks, and farmers fretted over the icy temperatures coming early in the Sunshine State's winter, at a time when trees are heavy with fruit.

Meadows described the reported damage as "all anecdotal at this point", saying it was impossible to immediately quantify the overall impact on Florida's $9 billion citrus industry, which was also hit by freezing weather in January last year.

"I don't think it's catastrophic," he added.

He expected any impact to show up in later U.S. Department of Agriculture revisions of the Florida citrus crop. The Sunshine State yields more than 75 percent of the U.S. orange crop and accounts for about 40 percent of the world's orange juice supply.

With temperatures expected to rise quickly again later this week, there were

also concerns about freeze-hit fruit dropping from trees and rotting, and farmers were stepping up efforts to harvest their damaged fruit and get it to market.

"Growers are scrambling to get the fruit out of the grove into the processing plant," Meadows said.

He added his organization was asking Florida's governor to extend for two more weeks an easing of restrictions on the weight of transport vehicles on the highways hauling harvested produce to market.

Growers in Florida's Highlands, Hardee and Polk counties, among others, reported heavy frost coating the citrus groves.

---

[132] Photo from http://www.usatoday.com/weather/news/extremes/2010-01-11-florida-cold_N.htm

"There's frost everywhere ... it looks like Winter Wonderland out there," said Edward Schwartz, who works at Larry Davis Inc in Wauchula, Hardee County.

In some areas, ice had penetrated the fruit, which can reduce juice yield, growers reported.

"Some guys are telling me they are cutting really solid ice in some fruit ... they are already seeing juice loss, this is going to amplify that," Ray Royce, executive director of the Highlands County Citrus Growers Association in central Florida, told Reuters. Highlands is the second-largest citrus producing county in Florida.

TWO FREEZES IN DECEMBER

Despite the news of the freeze however, orange juice futures ended lower on Tuesday for the first time in seven sessions as profit-taking hit the market.

The key March frozen concentrated orange juice contract sank 6.95 cents, or 4.1 percent, to close at $1.6175 per lb, dealing from $1.615 to $1.664.

The National Weather Service warned of another cold weather snap on Wednesday morning for swathes of central Florida.

Florida growers said they were worried about the recurring incidence of frost coming early in Florida's winter and then striking again, which could inflict more lasting, cumulative damage on the citrus groves.

"We usually don't really get anxious until January, and certainly to have two freezes in December is problematic," Florida Citrus Mutual's Meadows said.

Royce said there was concern the freezes could cause leaf and small twig damage," he added, although he did not see serious "killing" damage to the trees themselves. Leaf and twig damage can affect tree development and the next season's crop.

Typically, citrus can be damaged by four hours or more of temperatures below 28 degrees Fahrenheit (minus 2 Celsius).

Jennifer McNatt of the National Weather Service in the Tampa Bay area said temperatures had fallen below 28 degrees in several citrus growing areas.

In Polk County, Florida's biggest citrus producing county, there was "some isolated damage," said Adam Pate of Statewide Harvesting and Hauling, which harvests for Dundee Citrus.

"The temperatures are supposed to rise and that's always a concern, that fruit will drop prematurely," he added.

Barbara Carlton of the Peace River Valley Citrus Growers Association, which covers several west central counties, said she also expected "some damage" and was worried about the risk of fruit drop if temperatures rise sharply again.

"The fruit will drop, begin to decay, we have to get it in before that happens," she said.

Grower John Arnold of the Showcase of Citrus in southern Lake County said successive frosts and freezes could strip citrus trees of their protective, insulating foliage. "It's like a scab, you keep adding insult to injury ... we still have a longways to go yet this winter," Arnold said.

Florida's Manatees were neither facing extinction from unusual warmth, nor migrating north to non-existent warming in temperate latitudes. Instead, many died from unusually chilly water temperatures at home, fit perhaps for penguins, but not for these beautiful creatures of the Caribbean. 56 Manatees died from the cold in December alone. The cold continued well into January, and, as a consequence, yet another 100 manatees died

http://latimesblogs.latimes.com/unleashed/2010/01/florida-cold-snap-kills-endangered-manatees.html

Florida cold snap kills endangered manatees
January 26, 2010 | 1:46 pm

TALLAHASSEE, Fla. — More than 100 manatees have been found dead in Florida waters since the beginning of the year, mostly victims of a nearly two-week cold snap.

The Florida Fish and Wildlife Conservation Commission says the preliminary cause of death for 77 of the endangered animals is cold stress. They were found from Jan. 1 through Jan. 23.

The Sunshine State saw unseasonably cold weather starting around the first of the year that killed fish and stunned thousands of sea turtles.

Officials say the numbers of dead manatees from the cold is a record for a single year. The previous record, set last year, was 56 deaths from cold stress.

What did NOAA have to say about the coldest December in Florida on record

http://www.srh.noaa.gov/images/mfl/news/2010WxSummary.pdf

2010 South Florida Weather Year in Review
Coldest December on Record Concludes Year of Extremes

December 30th, 2010: Temperature and precipitation extremes marked the weather of 2010 across South Florida. A cool and wet January through March was followed by the hottest summer on record, and then concluded with the coldest December on record for the main climate sites in South Florida (details on the above mentioned periods will be included below) ….

The main culprit behind the cold temperatures in December 2010 was the same one which caused the cold winter of 2009-2010; a strongly negative North Atlantic Oscillation (NAO) and Arctic Oscillation (AO). When these atmospheric oscillations are in the strong negative phase, they essentially "flip" the weather pattern across North America, with upper-level high pressure and relative warmth over Greenland and Northeastern Canada and upper-level low pressure and cold over the eastern Continental United States, including Florida (Figure 1). This pattern forces the jet stream to plunge south from northern Canada into the southeastern U.S., transporting Arctic air masses into Florida.

A pronounced shift in the ENSO (El Niño Southern Oscillation) phase was noted in 2010, from a strong El Niño, or warm, phase to a borderline strong La Niña, or cold, phase. While this may appear at first glance to be a key contributor to the temperature extremes noted across South Florida during 2010, it is believed that it was the strongly negative NAO and AO, not the ENSO phase, which contributed to the cold temperatures in early and late 2010. A strongly phased NAO/AO operating on shorter time scales can override the longer-term ENSO phase.

As mentioned above, South Florida experienced its hottest summer on record in 2010 (with the exception of Naples which recorded its second hottest recorded summer). Despite the record hot summer, average yearly temperatures at the main climate sites will end up around 1 degree below normal, which will be the coolest calendar year since the early and mid 1980s, and among the top 10 on record (except for Miami). At secondary sites Miami Beach and Moore Haven, it was the coolest year on record (please note caveat below table).

Location (beginning of historical record)	Avg 2010 Temp (F)	Departure From Normal (F)	Rank
Miami Int'l (1895)	75.8	-0.9	52nd coolest (Tied)
Fort Lauderdale Int'l (1912)	74.9	- 1.0	7th coolest
Palm Beach Int'l (1888)	74.1	- 1.2	8th coolest (Tied)
Naples Regional (1942)	73.6	- 0.5	6th coolest
Miami Beach (1927) **	74.1	-1.9	Coolest on record
Moore Haven (1918) **	71.2	- 1.9	Coolest on record

** Present Miami Beach and Moore Haven temperature data may not be totally comparable to historical data due to difference in time of daily reports which causes double-reporting of low temperatures.

One cannot help but observe the probable warming bias in the text of the NOAA report:

- The reference to "extreme", implying the new spin of "climate weirding" concocted by Friedman and Sachs, in other words, "extreme" cold now equated to planetary warming, although this was not part of the plot a decade ago, and, ironically, would not be in a year's time when winter "forgot" the Eastern and Central United States.

- The wording around the warm summer in Florida, particularly "despite the record hot summer" appears to be designed to partially anesthetize the reader from the coldest December in the State in 169 years. The admission that temperatures in Florida for the entire year were a degree below normal seems to have been made with a great deal of reluctance.

- Record-cold locations, Miami Beach and Moore Haven were minimized as secondary sites without appropriate explanation and a caveat implying unreliable temperature measurement added.

- The NAO and AO explanation, initially minimized to Europe, was now expanded to include the "eastern Continental United States" as well. No issue; we shall soon examine the weather conditions in the rest of the Northern Hemisphere and discover the truth.

- The El Niño Southern Oscillation (ENSO) has also been a favorite pretext by advocates of AGW, ever since cold winters returned from the end of 2007. The funny thing is that, with all of their comprehensive knowledge of the world's climate that they claim to have, why didn't these "brilliant" pseudoscience organizations predict the shift in the ENSO barely 5 years earlier? Why did these climate "wizards" predict the end of snow instead?

Georgia and the Carolinas

In December 2010, Georgia and the Carolinas experienced the first white Christmas in more than a century. So much for the IPCC prediction at the turn of the century of ice storms replacing snow storms....

http://www.examiner.com/article/atlanta-is-dreaming-of-a-white-christmas-its-first-more-than-a-century

Atlanta is dreaming of a white Christmas: Its first in more than a century
December 23, 2010
Jackie Kass

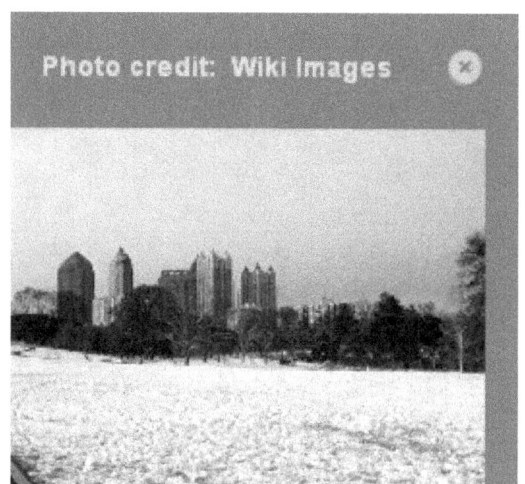

Atlanta is dreaming of a white Christmas, and the possibility of snow on Christmas Day is becoming more and more a reality. In a statement issued early today, the National Weather Service predicted that Atlanta could see 1-3 inches of accumulating snow. The last time Atlanta saw any snow on Christmas Day was in 1993 when a trace fell. The last time there was measurable snow on Christmas Day in Atlanta was more than a century ago in 1882, when .3 inch accumulated.

Other news outlets are also predicting a white Christmas for Atlanta. WSB-TV has the following post on its website: "Cloudy and windy with showers in the morning and snow in the afternoon and evening. Possible accumulations in North Georgia. Temperature in the 30s." The Weather Channel says, "No matter what scenario plays out along the East Coast, confidence is growing that snow will greet parts of the South on Christmas Day through Sunday, December 26th. The same storm that will eventually track along the East Coast will first dive into the South and drop snow and/or sleet over parts of Mississippi, Tennessee, Alabama, Georgia and South Carolina." However, the National Weather Service also adds, "There is still a lot of uncertainty about the storm's projected path. Any deviation to the north or south could mean the difference between rain or snow, as well as total amounts."

The dream did indeed come true. Gone was the nightmare of the near-disappearance of snow even in latitudes much further to the north, as depicted by David Viner, senior "scientist" at the CRU, a decade previously.

Eastern Seaboard

Let us now check what the overall situation was like for the Eastern Seaboard.

http://www.guardian.co.uk/world/2010/dec/27/snow-storm-batters-us-east-coast

Blizzards batter US east coast
Andrew McCorkell
guardian.co.uk, Monday 27 December 2010 13.46 GMT

More than 2,000 flights are cancelled, trains halted and roads littered with abandoned cars

Blizzards whipping across the east coast of America forcing cancellation of more than 2,000 flights and stranding

thousands of rail and bus passengers.

The snow was at its worst this morning, dumping up to 10cm an hour with as much as 140cm in total expected across Rhode Island, Connecticut and eastern Massachusetts.

Forecasters predicted 50mph winds could create deep snow drifts, causing disruption for travellers and commuters returning to work.

In the middle of the traditionally busy post-Christmas period, airlines were not expecting normal service until tomorrow.

More than 1,400 flights out of New York's three main airports were cancelled, while train operator Amtrak cancelled services from New York to Maine after earlier halting trains in Virginia.

Some 2,400 sanitation workers were working 12-hour shifts to clear New York's 9,650km of streets.

Michael Bloomberg, the city's mayor, told a press conference: "I understand that a lot of families need to get home after a weekend away, but please don't get on the roads unless you absolutely have to.

"Our sanitation department has 365 salt spreaders and 1,700 snow plows ready to fight the storm. We also have more than 180,000 tonnes of salt on hand at 30 locations."

The deepest snow – 1.5m – was recorded in Monmouth county, New Jersey, where plows were having difficulty clearing snow because abandoned cars were blocking roads.

Ambulances were unable to reach a passenger bus that stalled on Garden State Parkway.

Emergency officials encouraged businesses to let employees in Rhode Island report to work late, saying road conditions for the morning commute would be treacherous.

Steve Kass, a spokesman for the state emergency management Agency, said: "You don't want to get your employees hurt. The roads are not going to be good, that's for sure."

There has been heavy snow in other parts of the US with a state of emergency called in Maryland, North Carolina, Massachusetts, Maine, New Jersey and Virginia.

South Carolina and Georgia, both southern states, recorded their first white Christmas in more than a century.

Cancelled flights began on Saturday with more expected today, while airlines suspended operations at John F Kennedy International, LaGuardia and Newark International in New Jersey yesterday.

Boston's mayor, Thomas Menino, declared a snow emergency, banning parking on main streets, while the New England Aquarium bubble-wrapped its four 1.5m penguin ice sculptures to protect them from the wind and snow.

The National Weather Service said the blizzards were being caused by a low pressure system off the North Carolina coast which intensified as it moved north-east.

For once, I commend the Guardian Newspaper for not tempering cold reporting with former hot news. Probably by this time, the extent of the epic winter was so brazenly obvious that any mention of warmth would have greatly offended many readers.

In addition, Massachusetts, Maine, Maryland, New Jersey, North Carolina and Virginia all declared emergencies. [133]

New York City had so much snow that even a city accustomed to snowstorms in the winter could not cope.

> http://www.prlog.org/11177656-new-york-snow-storms-chaos-live-weather-cams-across-nyc-and-new-york-state.html
>
> **New York Snow Storms Chaos - Live Weather Cams Across NYC and New York State**
> PRLog (Press Release) - Dec 27, 2010
>
> Snow storms have swept north along the eastern seaboard of the United States, forcing the cancellation of more than 2,000 flights and disrupting rail and road traffic.
>
> These live cameras are monitoring the weather across new York City and State: http://www.myworldwebcams.com/usa/new_york/new_york_webcams.html
>
> The winter storm closed New York airports, stranding thousands of people on a busy post-Christmas travel day.
>
> National rail operator Amtrak suspended passenger services between New York and Boston, where some 30cm (1ft) of snow was expected.

Many plows became stuck in the snow that had accumulated on the streets and several could not get through as many residents had simply abandoned their cars in the middle of roads once it became too hazardous or impossible to move. These cars became stuck as the snow accumulated around them, creating roadblocks for the plows.

In New York City, some 2,400 street cleaners were working 12-hour shifts to clear the snow from the city's 6,000 miles of roads - but inhabitants were advised to stay at home anyway.[134]

Western United States

By the end of December 2010, the mainstream media refreshingly, although only for a brief period, abandoned all aspersions to warming. Nightmares of cold weather dominated the news in all of the major television networks in the United States, Europe, and many other locations in the Northern Hemisphere.

On December 29th, NBC Nightly News summarized the situation in the Western United States as follows: the second winter storm in 2 weeks brought very cold weather and lots of snow to California, Washington State, Nevada, Colorado, Arizona, etc. There was chaos in Washington State as many interstates and roads had too much snow to drive. Lake Tahoe had so much snow, that they could not find tracks of a missing snowboarder.

---

[133] http://www.bbc.co.uk/news/world-us-canada-12081749
[134] http://www.bbc.co.uk/news/world-us-canada-12081749

In fact, by the middle of December 2010, a fortnight before the NOAA report that only referred to "cold over the eastern Continental United States", hundreds of new cold and snow records had already been set all over the United States.

http://wattsupwiththat.com/2010/12/13/hundreds-of-new-cold-and-snow-records-set-in-the-usa/

Hundreds of new cold and snow records set in the USA

Posted on December 13, 2010 by Anthony Watts

While there have been a few high temperature records in the desert southwest and western Oregon, the majority of weather records in the USA this week have been for cold, snowfall, or rainfall. The biggest number of records has to do with the lowest maximum temperature.

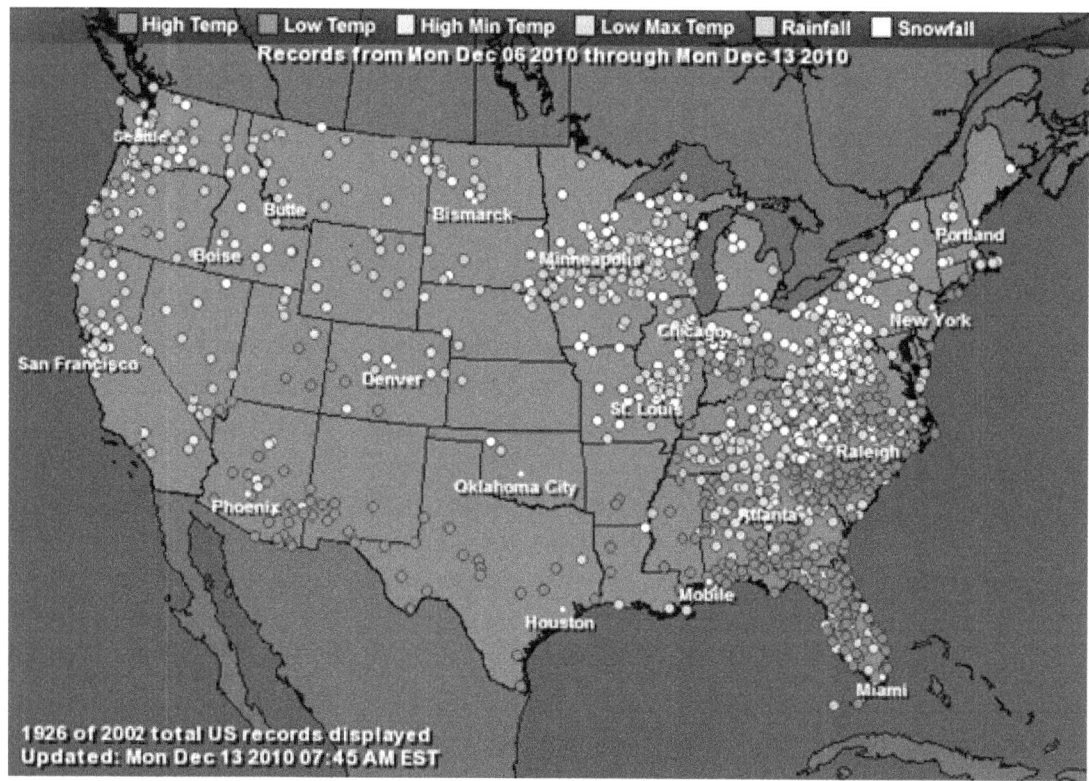

Here's a summary of the weather records:

Record Events for Mon Dec 6, 2010 through Sun Dec 12, 2010

- Total Records: 2002
- Rainfall: 319
- Snowfall: 320
- High Temperatures: 71
- Low Temperatures: 426
- Lowest Max Temperatures: 767
- Highest Min Temperatures: 99

On December 15th, 2010, Brian Williams on NBC Nightly News summarized the situation by stating that 85 million Americans were coddling up and trying to survive in temperatures below freezing day and night. CBS News added that temperatures were below normal in much of the country with some locations being 30 degrees Fahrenheit below average.

The unusual cold continued throughout much of January. By January 12th NBC, CBS and CNN all reported that 49 out of 50 states now had snow, even tropical Hawaii nearly halfway across the Pacific. The Southeast was having one of the snowiest winters ever, many places experiencing snow for the first time since records began. New England was experiencing Arctic temperatures, for example New Hampshire had temperatures down to -30 degrees Centigrade or lower.

January 2010 was declared as the snowiest January in the United States on record.

http://www.masslive.com/news/index.ssf/2011/01/with_more_snow_on_the_way_janu.html
Published: Monday, January 31, 2011, 6:28 PM   Updated: Tuesday, February 01, 2011, 7:25 AM
By Stan Freeman

With a monster winter storm taking aim at a third of the nation Monday, threatening to lay a potentially deadly path of heavy snow and ice from the Rockies to New England, weather trackers here have reported that January has entered the books as the snowiest month ever.

Bradley International Airport in Windsor Locks, Conn., reported 54.8 inches fell in January, and more snow is on the way for the beginning of the new month.

Cities including St. Louis, Kansas City and Milwaukee could be hardest hit by this week's storms, with expected midweek snowfalls of up to 2 feet and drifts piled 5 to 10 feet.

After burying the Midwest, the storm was expected to sweep into the Northeast Tuesday and Wednesday.

Canada is a country used to bitter winters. Nonetheless, by mid-December, hundreds of motorists were stranded in Canada for 24 hours due to extreme cold and snow.

http://www.ctv.ca/CTVNews/TopStories/20101213/eastern-canada-winter-storm-system-101213/

Snow leaves hundreds of drivers stuck on Ont. Highway

Mon. Dec. 13 2010
11:13 PM ET

High winds and snow stranded hundreds of motorists on a stretch of highway in Ontario late Monday night, as fierce winds battered Atlantic Canada.

Ontario Provincial Police estimated that more than 360 vehicles were stuck on Highway 402 near Sarnia, as the surrounding County of Lambton declared a state of emergency.

"Motorists are encouraged to remain in their vehicle and consider pooling vehicle resources until rescuers are able to reach their location," OPP said in a press release. "Motorists are also encouraged to ensure their vehicle exhaust is clear of snow to ensure clean air supply to the vehicle interior.

Meanwhile, winds ripped through Atlantic Canada with gusts measured at 100 kilometers an hour -- 140 km/h in western Cape Breton. They also churned up waves in the ocean, delaying ferries in Nova Scotia and Newfoundland.

The same weather system drenched parts of New Brunswick with heavy rain. Up to 130 millimetres could fall by Tuesday morning, but by Monday night some streets in Fredericton had already been flooded.

The storm is centred on the same cold front that buried parts of the U.S. upper Midwest over the weekend, dumping as much as 50 centimetres of snow on Minnesota and collapsing the roof of the Metrodome in Minneapolis.

Although snowfall associated with the system has eased, the agency said strong northerly winds behind the storm mean squalls and blowing snow.

In Quebec, residents in the south were expecting to see freezing rain, with snow falling in the north.

Temperatures plummeted in Ontario overnight, prompting officials to issue flash-freezing warnings for south-central parts of the province.

"A flash freeze occurs when moisture and some accumulating snow on roads creates icy conditions as the mercury falls quickly below freezing," the weather service explained on its website.

A sharp wind chill made the afternoon temperature in Toronto feel like it was -20 degrees Celsius.

Ontario provincial police were urging people living an hour north of the city to stay off of the roads except in the case of an emergency due to icy conditions and limited visibility.

Earlier in the day, a large swath of southern Ontario was under a weather warning including London, which last week had been punished with more than a metre of snowfall in some areas.

Environment Canada said damaging winds gusting up to 90 km/hr were being felt near the shores of Lake Huron.

Freezing winds coming off the lake were bringing 15 to 20 cm of snow to Georgian Bay and reducing visibility to zero in some areas.

"Dangerous winter driving conditions are likely, due to low to nil visibility in heavy snow, blowing snow, and from accumulating snow. Travellers should adjust plans accordingly," Environment Canada said.

The powerful storm has also cancelled more than 1,600 flights, including several from Canadian airports to New York, Chicago and Boston.

By the end of December, another storm reached Canada's Atlantic coast. Around 27,000 homes in Nova Scotia and 11,000 consumers in the New Brunswick area were left without power, according to reports. Blizzard conditions buffeted New Brunswick, where winds were gusting up to 100km (62mi) per hour. [135]

Greenland

Interestingly, the NOAA report, mentioned earlier, stated: "When these atmospheric oscillations are in the strong negative phase, they essentially "flip" the weather pattern across North America, with upper-level high pressure and relative warmth over Greenland and Northeastern Canada and upper-level low pressure and cold over the eastern Continental United States, including Florida"

---

[135] http://www.bbc.co.uk/news/world-us-canada-12081749

So what was the weather like in Greenland in December 2010?

http://stevengoddard.wordpress.com/2010/12/21/met-office-chief-climate-scientist-has-difficulty-reading-maps/

Met Office Chief Climate Scientist Has Difficulty Reading Maps
Posted on December 21, 2010

This is not a global event; it is very much confined to the UK and Western Europe and if you look over at Greenland, for example, you see that it's exceptionally warm there," she said. "The key message is that global warming continues."

http://www.independent.co.uk/

This is the same misinformation that Hansen has been repeating. Most of Greenland, Alaska and Western Canada are running well below normal temperatures. It is -46F on the Greenland ice sheet headed down to -63F later this week.

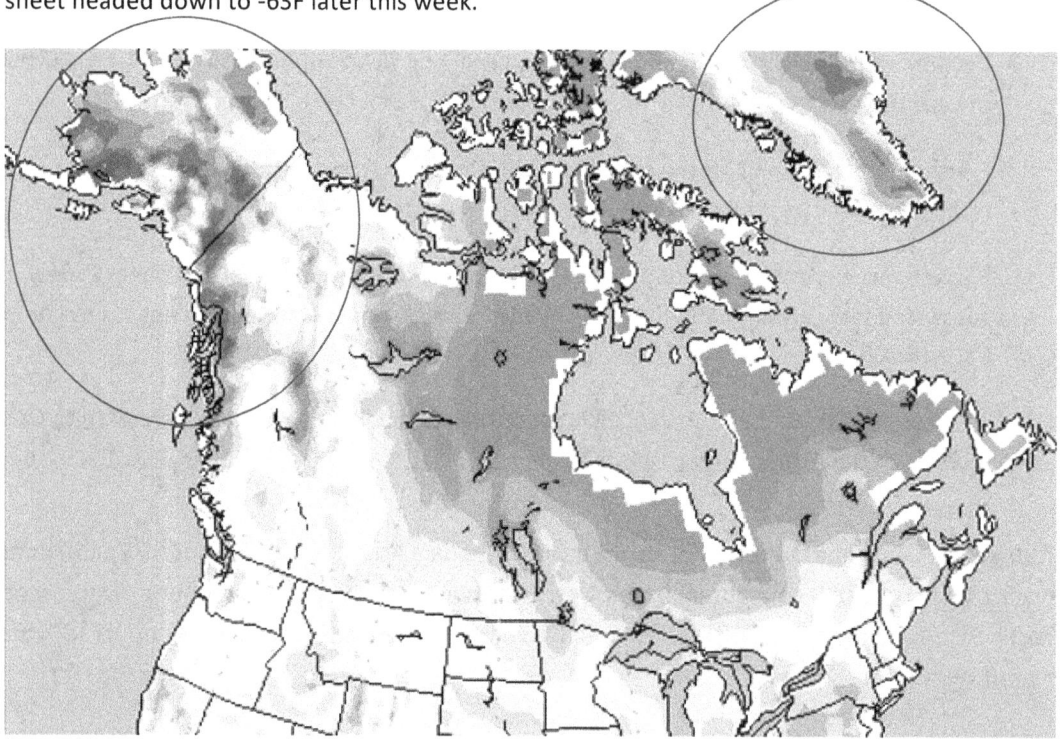

Temperature Anomaly
during the first 7.5-day period from:

Tue, 21 DEC 2010 at 00Z
—to—
Tue, 28 DEC 2010 at 12Z

Note: Temperature anomaly is the difference with the average temperature, i.e. white on the color scale is average, dark blue is -10°C below average, dark red is + 10°C above average

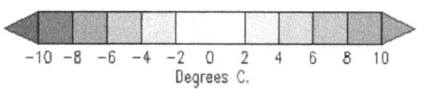

Degrees C.

....

Russia [the other side of the Arctic] is far below normal.

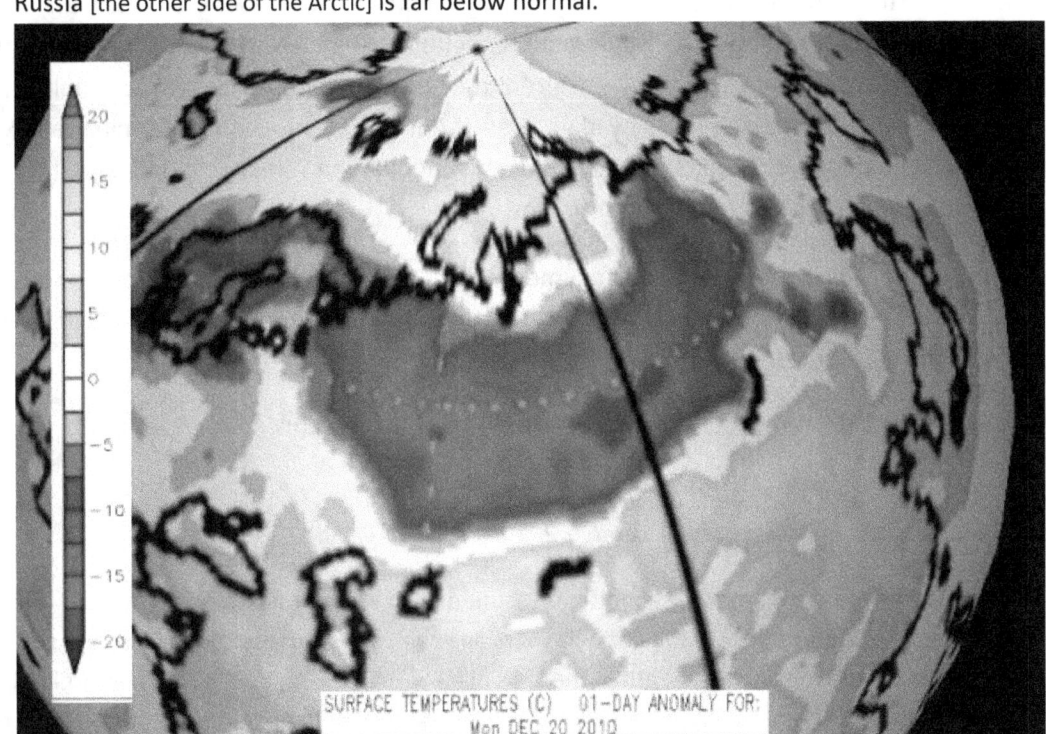

SURFACE TEMPERATURES (C)   01-DAY ANOMALY FOR:
Mon DEC 20 2010

"These people are completely incompetent," says Mr. Goddard. I believe that he is being far too kind.

Not only did the NOAA reports for both Europe at the end of November and the southeastern United States at the end of December, puff up the warmer than average conditions in northeastern Canada and northwestern Greenland to unreservedly imply all of Greenland, but also, even more significantly, a colder than average northern Russia, more than half of the Arctic was not even mentioned. Without a new Climategate leak, it is not likely that the climate pseudoscience community will provide any revisions to their explanation.

It is, therefore, up to each reader to draw his or her own conclusions as to what happened here.

My *personal take* is that there was something far more sinister than simple sloppiness at play here. I believe that local data from northeastern Canada was intentionally misused, while either exaggerating or omitting the rest of the Arctic in order to contrive the semblance of the phenomenon being *global* across the Arctic, when the actual data clearly shows that it was not global. I also believe that this trick was applied in order to concoct, through careful wording, a seemingly plausible rationalization that the extreme cold in much of the Northern Hemisphere in the winter of 2010-2011 was caused by global warming.

Let us return to Europe for a moment. The early start to winter in November did not, by any means, herald a mild December or January. Throughout the winter of 2010 to 2011, Arctic temperatures were prevalent all over Europe. Economic loss from the cold conditions exceeded $100 billion. NASA's Pretexts and the pitiable efforts by advocates of Anthropogenic Global Warming to link the extreme cold to heat ran extremely thin with a shivering population. "Inappropriate" graffiti revealed volumes with regards to public sentiment.

For the purposes of illustration, the book documents the visit of the North Pole to the United Kingdom, although the reader should take note that the situation was very similar in much of the rest of the continent that winter.

December 2010 was the "coldest for a century" in the United Kingdom.

> http://www.metoffice.gov.uk/climate/uk/interesting/dec2010/
> **Snow and low temperatures, December 2010**
>
> Overall, the prolonged freezing conditions resulted in an exceptionally cold December across the UK: the coldest December in the last 100 years and the coldest across central England since 1890. Indeed, this was the coldest month in the UK since February 1986, and in Northern Ireland, the coldest individual month of the last 100 years. Despite a mild first half, the UK also experienced the coldest November since 1993 as a result of the very cold last few days.
>
> From late November to Boxing Day 2010 the UK experienced two spells of severe winter weather with very low temperatures and significant snowfalls.
>
> The first of these spells lasted for two weeks from Thursday 25 November to Thursday 9 December and saw persistent easterly or north-easterly winds bringing bitterly cold air from northern Europe and Siberia, accompanied by snow. Eastern Scotland and north-east England saw the most persistent and heaviest snow, which accumulated to depths of 50 cm or more across the higher ground by the end of the spell. However, lower lying areas were also affected and the snow increasingly spread to other parts of the UK, so that by early December many areas of the UK were under lying snow. Temperatures struggled to rise above freezing during the day and there were very severe frosts at night. Temperatures widely fell below -10 °C on several nights and on occasion below -20 °C in northern Scotland.
>
> This spell of snow and freezing temperatures occurred unusually early in the winter, with the snowfalls judged as the most significant and widespread in late November and early December since late November 1965.
>
> The period from 9 to 15 December saw milder and quieter conditions with a gradual thaw of lying snow. However, a second spell of severe weather began on Thursday 16 December as very cold Arctic air pushed down across the UK from the north. Snow showers affected the north and west on Friday 17 December, while there was heavier snow across southern England and Wales on Saturday 18 December. Further heavy snow affected south-west England on Monday 20 December. The UK remained under bitterly cold Arctic air until Boxing Day, with day time temperatures again failing to rise above freezing and very severe frosts. While there was little further snowfall, lying snow remained until 26 to 27 December.

The second spell of snow and freezing temperatures has been judged the most significant such spell in December since 1981, although late December 2009 to mid-January 2010 (the previous winter) were also broadly comparable to both these spells.

Notwithstanding the resident earth expert George Monbiot, even the global-warming crazed Guardian newspaper had to eat humble pie and report the unusually cold weather. Nonetheless, I give them credit for not pitifully attempting to dilute the cold with some heated fabrication.

http://www.guardian.co.uk/uk/2010/dec/03/snow-travel-chaos-temperature-drop

Snow causes more travel problems as temperatures drop to -20C

Andrew Culf

guardian.co.uk, Friday 3 December 2010 13.46 GMT

The freezing weather, which fell to -20C overnight in parts of Scotland, continues to severely disrupt travel across the country, and has claimed the lives of two pensioners found dead in their gardens in Cumbria.

Lillian Jenkinson, 80, was found dead in her back garden in Workington, on Wednesday, and an elderly man was found dead in his garden in Kirby Stephen on Tuesday, it has emerged.

Forecasters were predicting less snow today and a slow improvement over the weekend, with temperatures climbing above freezing on Saturday, but colder weather could return on Sunday and last into next week, with no sign yet of a big thaw.

"It's going to be really cold into the next 10 days," said Aisling Creevey, a forecaster with Meteogroup, the Press Association's weather division. "The problem is the ground temperature is

lower than the air temperature [which] makes thawing difficult."

Gatwick airport reopened shortly before 7am, bringing some good news for the country's beleaguered transport network, but those intending to fly were warned cancellations and disruption would be inevitable.

Rail companies on busy commuter routes in the south-east were operating emergency timetables, and forecasters warned of widespread icy roads and freezing fog. Police in Kent and Surrey said motorists should not venture out unless their journeys were essential.

Temperatures fell to -20C in Braemar, Aberdeenshire, -15C in Edinburgh, -10C in Glasgow, -7C in Birmingham and -4C in London.

Although it was mainly dry in England and Wales, there was a warning of 4-6cms of snow in Northern Ireland and more light snow showers down the eastern side of Scotland and northern England.

Gatwick, Britain's second-largest airport, which had been closed for 48 hours with 1,400 cancelled flights, reopened this morning after round-the-clock operations to clear 150,000 tonnes of snow.

The airport warned passengers should only turn up for flights after checking with their airline. A spokesman said: "Weather forecasters are predicting freezing fog during most of today, so passengers should expect further disruption, with flights limited and delays and cancellations inevitable. Passengers should also be advised that it is likely to take a few days before flight schedules return to normal."

The airport was operating 10 to 15 flights an hour, compared with the usual 45. Robin Hood airport, in Doncaster, will be closed until 10am on Sunday, but it was hoped that Guernsey, Bournemouth and Southampton airports would reopen around lunchtime today.

The transport secretary, Philip Hammond, has warned train companies to keep passengers better informed of the scale of disruption to services. He said: "What is completely unacceptable is for rail passengers to be kept in the dark about what is happening to their services."

Southern Trains said it was running a revised and reduced service on limited routes, including a half-hourly service on the London to Brighton mainline. Southeastern, which serves Kent and Sussex, said it hoped to run a contingency timetable, but warned there were no services to Tunbridge Wells and Hastings.

Eurostar said it was running a significantly reduced service and warned of 70-minute delays. Northern Rail's services in Yorkshire and Humberside were also severely disrupted.

Motoring organisations said the motorway network was quiet as drivers appeared to heed warnings not to travel. The M20 coastbound in Kent was affected by Operation Stack, where lorries heading for the Channel ports are parked on the carriageway. The M25, M26 and M2 are passable with care.

Chief Superintendent Alasdair Hope, of Kent police, said: "Driving conditions across the county are extremely dangerous and many roads have several inches of snow."

There were fears that fuel supplies were running low in rural areas. The Retail Motor Industry and the Independent Retailers' confederation said some forecourts were out of petrol because tankers were unable to reach remote areas, or to leave the main terminals from two refineries in south Humberside.

The weekend sports programme has also been hit. All of Scotland's Premier League matches are off, plus five matches in England's Championship at Doncaster, Hull, Nottingham Forest, Portsmouth and Sheffield United. Tomorrow's race meetings at Sandown and Wetherby have been cancelled.

A woman who dialled 999 to report the theft of a snowman from outside her home in Chatham, Kent, was labelled "irresponsible" by police. During the call the woman said: "It ain't a nice road, but you don't expect someone to nick your snowman."

http://21stcenturywire.com/2010/12/18/europe-another-record-breaking-winter-what-happened-to-global-warming/

2010: Another record breaking winter. So what happened to global warming?

By Patrick Henningsen

21st Century Wire

Green News

Dec 18, 2010

Britain and Europe have been hit hard for the third straight record-breaking winter season. Labeled by experts as the coldest winter in 100 years and set to blow well into 2011, it is already raising some very interesting questions about the new ideological split we are witnessing throughout society in the much celebrated green debate.

Tonight Britain braces itself for a further 10 inches of snow and more sub-zero temperatures to come- with no let-up, top forecasters have warned. These unusual Arctic conditions are set to last through the Christmas and New Year bank holidays and beyond and as temperatures plummeted to -10c (14f), prompting the UK's Met Office to state that this December 2010 was 'almost certain' to become the coldest since records began in 1910.

Yet another record-breaking winter in Britain, could this be a trend? Go figure. (PHOTO: Patrick Henningsen).

So is it not safe to say that we are witnessing a real, tangible and physical trend here? Unlike the million dollar computer-generated climate model projections produced by the UN's elite circle of research grantees and bursary award-winning climate scientists, this new trend is actually a real one- one we can touch, feel and most importantly... one we can empirically measure.

Indeed, it is Britain who has been hit- yet again, by a siege of blizzards and freezing temperatures. As public transport and utilities face continued disruption in services, major airports are reporting closures as the snow drift continues to pile up. It seems that temperatures will struggle to rising above freezing points for the second straight day and this will sure spell more chaos for the general welfare. For a relatively moderate, low altitude climate zone like the UK, such winter storms can cost lives and create an endless backlog of crisises that municipalities will have difficulty managing.

There is a rather bizarre upside of course. If you count yourself as one of the millions worldwide who find yourself living in constant fear of global warming and climate change, there is one positive reassuring aspect to this now bona fide and well documented global cooling weather trend since 1998. This essentially means that you can now safely get out from under the bed and breathe a sigh of green relief as you look out your window to see everything covered in thick white again. Yes, yes, you are completely and utterly safe from CGI-created scary visions of sea levels rising- as seen in the science fiction "cult" film, Age of Stupid (yes, those are CGI graphics and no, sea levels are not rising), allegedly due to that arbitrary phantom menace... called climate change. But some well-meaning folks seem to have forgotten that the Earth's climate is constantly changing- it always has done. What is the main driver of the Earth's climate; its cloud cover, its weather patterns, its extreme climate cycles? Undoubtedly, it is that big red firey furnace in the sky- the Sun.

Still though, this hasn't stopped thousands of green-washed activists, hippies and guilt-ridden corporate rehab patients in search of a low calorie religious love-in, from chasing their paper tiger into the deepest darkest corner of this endless political forest. A generation lost to indoctrination from up-on-high, high, high up some of the world's leading investment banks and arcane think tanks. And the hippies, well, they are also high.

Lost the plot: What was once a fun green activity for young Euro-hippies has now transformed into a wandering farcical climate circus.

The Club of Rome (official progenitor of the global warming hoax) and the UN's own well-documented programmes of social engineering(Agenda 21) and various departments of political division, all have seen resounding success, particularly between 2004-2008, before the ideological zeitgeist of global warming and its new alter ego, climate change, started heading south for the winter (all the way south to Antarctica, in fact). The inevitable collapse of the Chicago Climate Exchange (CCX) is one recent sign that the whole effort to monetised and control Western lives through this coercive pseudo market is struggling to maintain its illusion of green utilitarianism.

The writing was on the wall this past summer and went unnoticed by most green activists and passive spectators in the great climate debate. Although initial investors and shared holders managed to jump ship with their fortunes in tact, in the end, it was simply "unsustainable".

If you have unwavering faith in men like Al Gore, the UN's own knighted IPCC body of 40-odd climate scientists and the Guardian's resident earth expert George Monbiot, then you have effectively swallowed the blue pill. Here you have a license to switch off your critical thinking faculties because in your mind you can hear the following phrases, successfully implanted there, over and over again. "The debate is over, the science is settled, every scientist agrees". Like a fundamentalist Christian, a Jew, or Muslim, you take the climate scriptures word for word. You are a true believer, one who somehow knows in his heart that there really is a thing called man-made CO2-driven global warming. It simply must be. And there is a ready-made crowd waiting for you at the church, where everyone is singing happily from the same hymn sheet- literally. What once passed for education in the West, was transformed into a top-down waterfall of relentless green propaganda- driven by middle class guilt and a multi-billion dollar gravy train of state-subsidised financial opportunities.

And in the most bizarre turn of green ideology seen yet, more and more liberal-minded zealots are now claiming "that global cooling is what we must expect because of global warming". If you are in any doubt as to the reality of this new claim, just ask any climate change advocate yourself and you will be amazed to hear this new party line stated. Another idea has been successfully implanted into the minds of this faithful flock.

And then you have the skeptics- the demonised, the mavericks, the outcasts (of whom this author is one, and has yet to receive any money from 'big oil' etc) often stoned in public for challenging Herrs Gore, Strong, Hansen, Mann and Jones on certain hack aspects of their sacred computer-modeled science. The doubters knew something wasn't right when Wall Street started its hedging and hyping of the world's most innovative financial instrument yet- carbon emissions. They knew something was off kilter when carbon taxes inevitably became to main thrust of global warming shills and the United Nations. So after a third straight year of frostbite and ice skating down your neighborhood street and into a lamp post, it's gone beyond a joke. You simply have no choice but to swallow the red pill.

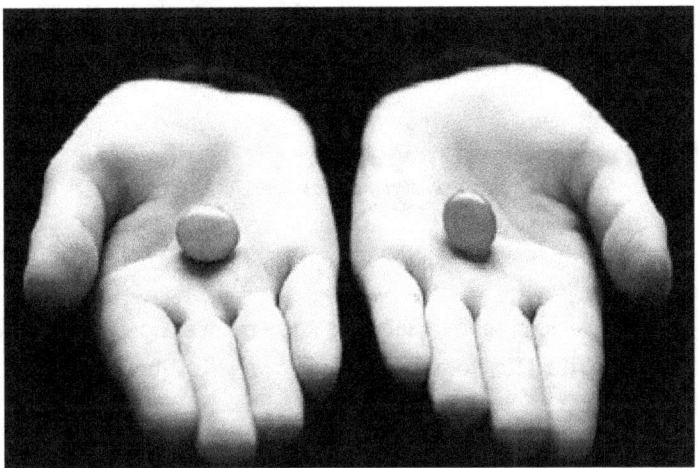

In the end, the climate debate comes down to Alice in Wonderland. Our advice: at least know which pill you have swallowed.

It seems that the only people in denial are the religious followers of the IPCC's new Jonestown Church of climate change... drunk on a delusion that they are, in their own little way, saving the planet from the evil substance known as CO2. It's become a sort of tribal division, where two tribes cannot seem to agree if the Sun orbits the Earth, or the Earth orbits the Sun. Throughout history tribes of people needed mythologies in order to give meaning to their lives. Climate Change is simply the latest mythology for this current epoch. In the 21st century, we thought modern man had surely advanced past this handicap, but alas... old habits die hard.

Alice... are you there Alice?

The Army was obliged to move in to clear away snow in Britain's coldest December for 100 years and fuel ran out of petrol stations in Scotland and East Anglia.

British columnist/blogger James Delingpole at The Telegraph aimed this satirical barrage at the global warming alarmists in Cancun — and in the media:

> "As your boiler breaks down, your pipes freeze, your car won't start, your Ocado delivery fails to arrive, your train is cancelled, your neck is broken after slipping on black ice and you lie in an emergency ward waiting for a doctor to turn up only to learn that they're all off today because of the weather, you might be forgiven for thinking that all this has something to do with global cooling, changes in the Pacific Decadal Oscillation, and the decline in sunspot activity perhaps auguring a new Maunder minimum."

But, of course, you would be wrong, says Delingpole, in his acerbic parody; the "experts" will assure us that all this freezing mayhem is the result of global warming."

Russia

In Russia, Prime Minister Putin had to get personally involved after an ice sheet blanketed the entire city of Moscow, food prices hiked, electricity and running water disappeared, and airports shut down. Freezing temperatures broke electric cables, leaving millions of Muscovites in the dark, including thousands of passengers at Moscow's main airport. The entire city transformed into a "huge ice rink". This was too much even for Muscovites hardened to cold winters.

http://www.rt.com/news/domodedovo-airport-blackout-weather/

Air travel returning to normal following Moscow's big freeze
Published: 26 December, 2010, 11:04
Edited: 27 December, 2010, 22:51

Flights are getting back on schedule in Moscow's main airports, with power returning to one of the capital's worst affected, Domodedovo, following a 12-hour blackout.

However, there are still thousands at Domodedovo airport waiting for their flights, one witness told RIA Novosti. He added that lights in the premises remain dim. There is also a long traffic jam leading to the airport and the square in front of the facility has turned into a huge parking lot.

On Sunday, over a hundred flights to and from the airport were cancelled, while others were diverted, after freezing rain ruptured electricity cables.

All planes heading to Domodedovo were redirected to several other airports, including Sheremetyevo and Vnukovo in Moscow. Meanwhile, Sheremetyevo airport itself had to cancel some sixty flights as the runways had frozen over.

Representative of the federal air traffic agency Rosaviation, Sergey Izvolsky, told RIA Novosti that Domodedovo Airport had to suspend over 100 flights, which were scheduled to carry over 8,000 people.

According to RT correspondent Sara Firth, who was reporting from Domodedovo, the situation at the airport was absolutely chaotic, with crowds of people walking around with their suitcases and not being able to leave.

Domodedovo airport's official website went down several times after being unable to cope with the large number of visitors trying to find out the latest news about flights. The special hotline introduced by airport authorities on Sunday was constantly busy.

The standard response of proponents of Anthropogenic Global Warming to reports of cold weather is that such chill is just "local", "globally it is warmer". Really? We have already confirmed the unusually cold conditions in Europe and North America during the winter of 2010 – 2011. Therefore, let us now study in detail the weather conditions during the same period in what remains of the Northern Hemisphere, in other words, Asia. And we shall not forget the Southern Hemisphere either; in a later section, the Southern Hemisphere winter of 2010 will be described.

Let us commence with Japan. So much snow that accumulated in Japan that Toyota and other Japanese manufacturers were forced to shut down operations

http://www.japantimes.co.jp/text/nn20110118a2.html

Snow snarls but looks to abate

Tuesday, Jan. 18, 2011
Kyodo News

Heavy snow continued to hit areas along the Sea of Japan on Monday as a cold air mass hung over the country, disrupting railway and highway travel and halting factories at Toyota Motor Corp. and other manufacturers.

The Meteorological Agency said the winter atmospheric pattern continued to hover around the archipelago Monday but is expected to gradually recede through to Tuesday.

Snowfall was seen in central Japan as well, forcing Toyota Motor Corp. to idle factories.

In the Chugoku region in the west and on the Sea of Japan coast Monday morning, snow piled as high as 195 cm in Kitahiroshima, Hiroshima Prefecture, 137 cm in Maniwa, Okayama Prefecture, 91 cm in Onan, Shimane Prefecture, and 253 cm in Daisen in neighboring Tottori Prefecture — all record highs in western Japan, the weather agency said.

Nagoya saw 10 cm of snow, while Kyoto got 6 cm, it added.

According to Central Japan Railway Co. (JR Tokai), the Tokaido Shinkansen Line was delayed after bullet trains were slowed by heavy snowfall between Hamamatsu Station in Shizuoka Prefecture and Shin-Osaka Station.

A total of 75 trains ran up to two hours late, affecting some 67,000 people, JR Tokai said.

The Tomei Expressway was also closed between Fukuroi in Shizuoka Prefecture and Toyota in Aichi Prefecture earlier in the day.

Elsewhere, Toyota decided not to open 12 factories in Aichi Prefecture because the snow could disrupt parts deliveries and prevent employees from getting to work. Toyota partly stopped the operation of its Just-In-Time system, which minimizes part stocks and procures additional parts only when necessary.

The snow not only did not abate, but also continued well into the spring. The reader may well remember images of the freezing temperatures and snow in Sendai and other parts of eastern Japan in the aftermath of the massive earthquake and tsunami of March 11th, 2011, complicating the shattered lives of earthquake survivors even further.

Korea

Aside from the cold weather across the northern half of the globe, December 2010 was a very tense time in international geopolitics. In the fall, a South Korean artillery exercise at waters in the south was used as a pretext by the North Koreans to fire around 170 artillery shells and rockets at Yeonpyeong Island in South Korea, hitting both military and civilian targets. The bone-chilling incident occurred on November 23rd, 2010, and was displayed live on international television channels such as CNN. The shelling caused widespread damage on the island, killing four South Koreans and injuring 19. South Korea retaliated by shelling North Korean gun positions. The United Nations declared it to be one of the most serious incidents since the end of the Korean War. By December 18th, former UN ambassador Bill Richardson said tensions had escalated to become "the most serious crisis on the Korean peninsula since the 1953 armistice which ended the Korean War." Full-scale war between North and South Korea, and, by consequence, the United States and China, became a very real possibility.

It was in this exceptionally nerve-racking environment that Richardson traveled to North Korea. The winter weather on the Korean Peninsula during the same period enforced a temporary lull in the confrontation, which gave Richardson just enough time to dissuade the North Koreans from escalating tensions further. In the end, Richardson's remarkable diplomatic skills, together with Mother Nature, succeeded in preventing all-out war in Northeast Asia.

http://www.reuters.com/article/2010/12/17/us-korea-north-richardson-idUSTRE6BG1LN20101217

Bill Richardson says Koreas situation a "tinderbox"
WASHINGTON | Fri Dec 17, 2010 6:53pm EST

(Reuters) - Diplomatic troubleshooter Bill Richardson, visiting Pyongyang to try to ease tensions between North and South Korea, said the situation was "a tinderbox", but urged the North to show restraint and allow contentious military exercises by the South to go ahead.

Richardson said in an interview on CNN he had met with a senior North Korean official and urged "let's cool things down."

"I am urging them extreme restraint ... Let's cool things down. No response. Let the exercises take place," Richardson said.

He said he met Friday with North Korea's top nuclear negotiator and was scheduled to hold talks with the head of the military on Saturday.

"I think I made a little headway," he said in a telephone interview with CNN.

He was not in North Korea as an official U.S. negotiator.

"My sense from the North Koreans is they are trying to find ways to tamp this down. Maybe that will continue today, that's my hope," he said.

"There's enormous potential for miscalculation," he said.

North Korea said earlier in the day it would strike at the South if a live-fire drill by Seoul on a disputed island went ahead, with an even stronger response than last month's shelling of the island that killed four people.

North Korean official news agency KCNA issued the threat as South Korea prepared for firing drills on Yeonpyeong island near a disputed maritime border with the North for the first time since November's exchange of artillery fire.

Richardson suggested the North and South consider holding a summit meeting to discuss ways to avoid an altercation.

"Right now this is a tinderbox. What we need to do right now is not just tamp things down but look at steps that can be taken by the North Koreans, especially such as perhaps allowing the IAEA (International Atomic Energy Agency) to come in and look at the nuclear arsenal."

The United States is working with China, Russia, Japan and South Korea to persuade the reclusive communist state to halt its nuclear weapons programs in exchange for aid and more international recognition.

And while these delicate negotiations were being conducted in Pyongyang, North Korea, South Korea was obliged to delay military drills on Yongpyeong Island due to "poor weather", a decision that may have prevented full-scale war. The South had previously planned to carry out the drills on the island between Saturday, December 18th and Tuesday, December 21st.

China did not escape the severe cold weather during the Northern Hemisphere winter of 2010 – 2011, which proponents of AGW facetiously attempted to hold global warming responsible.

http://www.chinahighlights.com/news/travel-latest/cold-weather-grips-most-of-china.htm

Cold Weather Grips Most of China

Cold weather has persisted in China since the end of December last year, with snowy or rainy weather hitting most parts of China. The temperature in South China hovers around 0° C (32° F), while that of most North China is under 0° C (32° F). The National Meteorological Center (NMC) has issued that most of China will be persistently in grip of cold weather in following three days (from January 5 to January 7).

It is predicted that there will be slight snow or sleety weather in South China, like Guiyang, Changsha, Nanchang, Guilin, Chengdu and Chongqing. In central and north Guizhou Province, and south and west Hunan Province, the freezing rainfall (the rain that quickly turned into ice on the ground which has caused havoc in traffic of Guizhou, Hunan and Chongqing in past three days) continues but will not as severe as before. For North China, the obvious temperature drop happens in west Inner Mogolia, East Jilin Province and east Liaoning Province with the range is about 6° C (43° F), and north wind makes North China wintrier. In north Xinjiang Province and east Inner Mongolia, the highest temperature is still under -20° C (-4° F).

For travelers visiting in China these days, China Highlights has following tips to help them to get the most enjoyment out of their tours.

Pay more attention to the weather forecast and dress accordingly (wear layers to protect you from the high wind, snowy or rainy climate).

Pay particular attention to children and the elderly, since they have more difficulty on regulation body temperature.

Freezing rainfall and snow mantle play adverse impact in traffic, and travelers need to pay more attention to the highway conditions.

The cold weather may influence the scheduled flights or trains, travelers need keep a watchful eye on the traffic information so as not to change the travel accordingly.

One of the pet "theories" of pseudoscience gibberish which global warming alarmists have been broadcasting in recent years is that heat is causing the waters of the Bay of Bengal to expand and rise. In particular, low-lying Bangladesh at the northernmost tip of the bay is being flooded, and as the sea moves inland, a significant section of the 150 million people of the country is being displaced into the northern part of the country and into neighboring India.

To start with, flooding of rivers and cyclones resulting in a temporary rise of the waters of the Bay of Bengal are nothing new in Bangladesh. In addition, catastrophes in the country in recent years may have had more to do with human dam construction rather than human-induced climate change. In 1975, India constructed a dam on the Ganges River 10.3 miles from the Bangladeshi border. Bangladesh alleges that the dam diverts much needed water from Bangladesh and adds a man-made disaster to a country already plagued by natural disasters. The dam has had terrible ecological consequences. In summary, it works this way: in the dry winter months, India blocks the dam, so that there is more water for its citizens in the Indian State of West Bengal. In the wet summer months, India opens the dam, so that excess water reduces the possibility of floods in West Bengal while increasing the likelihood of massive floods next-door in Bangladesh. Tiny Bangladesh, of course, has little leverage with mighty India; therefore this inequitable situation has now lingered on for nearly four decades. Advocates of AGW have naturally misinterpreted this political issue into a fabricated pseudoscience tale of climate change.

So what was the weather like in the winter of 2010-2011? Unusually warm? Not quite. In mid-January, this tropical country experienced its coldest temperature ever on record.

http://bdnews24.com/details.php?id=184206&cid=2.

Lowest temperature in Jessore
Wed, Jan 12th, 2011 1:03 pm BdST

Dhaka, Jan 12 (bdnews24.com)—The country's lowest ever temperature has been recorded in Jessore. The temperature there on Wednesday morning has been recorded as 4.5 degrees Celsius.

According to meteorological department, this was the lowest ever temperature recorded in the country since independence.

A senior meteorologist Abdul Mannan told bdnews24.com that there is a record of temperature dipping as low as 2.8 degree Celsius in 1962.

Due to extreme chilly weather, life in Jessore has almost come to a standstill and the marginal people are facing extreme misery, told Jessore correspondent to bdnews24.com. There was no trace of the sun till 10am due to dense fog.

A cold spell has engulfed the nation for the last few days. Apart from a few areas in Chittagong, the entire country has come under medium to extreme cold wave.

Mannan, on Tuesday morning, said that the intensity of the cold wave increased because Bangladesh was currently in an extended high pressure zone. "The weather may improve in a day or two."

The lowest temperature in Dhaka on Tuesday was recorded as 8.4 degrees Celsius

Bangladeshis are not used to temperatures barely above freezing. For example, central heating does not exist in the country, even for its most prosperous citizens, because, simply put, there has never been a need for such devices hitherto only required in more temperate parts of the world with colder climes.

Just out of interest, when were the hot weather records set in Bangladesh. The below article from the Asian Tribune, while puffing up the heat wave in the country in 2009, at least had the decency to inform the public of when records were set:

http://www.asiantribune.com/node/17080

Bangladesh records highest temperature 42.2 degree Celsius in last 14 years
Mon, 2009-04-27 02:07 — admin
News
M.A Kader-Asian Tribune Correspondent in Bangladesh

Dhaka, 26 April (Asiantribune.com): Bangladesh met office in the last 14 years recorded the highest temperatures a blistering 42.2 degrees Celsius in its Jessore district and 38.7 Celsius in the Dhaka city.

An unrelenting heat wave has been sweeping the country for over a week, and was forecast to continue for at least another few days, a met official said.

The mercury may rise another degree above Sunday's 14-year highs, said the met official.

The highest temperature in 1995 was recorded as 43 degrees in Rajshahi, and 39 degrees in the capital.

Met office sources said the mounting temperature was an impact of global warming. "Bangladesh has been witnessing climate change," he said.

Records of the Department of Environment's Climate Change Cell from 1985-1998, show average May temperatures to have 'risen' by one degree Celcius, and average November (winter) temperatures by 0.5 degrees.

However, data over such a relatively short time is inconclusive. Climate change trends are studied over much longer periods.

On a larger timescale, temperatures have risen 0.5 percent countrywide and 0.7 percent in the capital over the last 50 years, said the met official.

"Unplanned high-rise buildings, high traffic emissions, overuse of air conditioners and lack of greenery are causing the capital's temperatures to be all the more scorching, the official said.

Although the met office recorded highest average temperature at 38.7 degrees in Dhaka, met officials said some of its more densely populated and busy areas including Mohakhali, Gulistan, Farmgate, Motijheel and Tejgaon clocked more than 40 degrees Celsius.

Dhaka's highest ever temperature was recorded at 42.3 degrees Celsius in 1960, while the country's highest was recorded 45.1 degrees in Rajshahi on May 30 in 1972.

....

So, the hottest temperature in Bangladesh was recorded in 1972, four decades ago. And it was very hot in 1960 as well. And one cannot help but wonder about the obvious contradiction between the statement, "temperatures have risen 0.5 percent countrywide", and heat records having withstood the test of time for decades while cold shattered records as recently as 2011. So what has changed over the past 50 years? The answer is nothing except human imagination fabricating another illogical theory cobbled together by pre-conceptual bias compensating for the clear lack of evidence.

Australia

To understand the utter absurdity of pseudoscience claims of advocates of Anthropogenic Global Warming with regards to Australia, one first needs to review what they predicted when bush fires scorched the country during the Southern Hemisphere summer of 2009 – 2010

http://www.reuters.com/article/2010/01/28/us-climate-australia-bushfires-idUSTRE60R15920100128

Australia "faces worse bushfires without CO2 deal"
By Michael Perry
SYDNEY | Wed Jan 27, 2010 10:48pm EST

Australia faces a possible 300 percent increase in extreme bushfires by 2050 unless world leaders can agree to dramatically cut greenhouse gas emissions, a new report said on Thursday.

The report, commissioned by Australia's firefighters and environmental group Greenpeace, said the failure of U.N. climate talks in Copenhagen to agree on a treaty to tackle climate change had left Australia facing future catastrophic bushfire seasons.

The "Future Risk: Battling Australia's Bushfires" report comes only days before the Copenhagen Accord Jan 31. deadline for nations to announce emissions reduction targets.

"Bushfire conditions are clearly changing and there is strong evidence that global warming is making Australia's climate more bushfire-prone," said Jim Casey, secretary of the Fire Brigade Employees Union in Australia's New South Wales state.

"Bushfire seasons are getting longer and fires are becoming more frequent and intense. We have the power to reverse this trend or we can shrug our shoulders, do nothing and play Russian roulette with our lives," Casey said in releasing the report.

Bushfires are a natural phenomenon in Australia, due to the hot, dry climate.

Australia's most deadly bushfires occurred in February 2009 and were blamed on a decade long drought and extreme heatwaves. The "Black Saturday" infernos killed 173 people and destroyed thousands of homes in the southern state of Victoria state.

This Australian summer has again seen extreme bushfires.

THREE SCENARIOS

The bushfire report, based on studies by Australia's peak scientific body the Commonwealth Scientific & Industrial Research Organization (CSIRO), painted three scenarios:

* Under a global climate treaty based on current promises to cut greenhouse gases, Australia's mean temperature would rise by 2 degrees Celsius above 1990 levels by 2050.

This would double the number of severe bushfire days in Australia's most populated southeast corner by 2050. Severe bushfire days would occur once every six months in Sydney.

* Without a legally binding climate treaty the upper forecast temperature rise of 6.4 degrees Celsius globally, by the end of the century, would see Australia experience a 2.8 degree Celsius rise above 1990 levels by 2050.

This is the worst case scenario for Australia which could see up to a 300 percent rise in extreme bushfire days by 2050.

* Under a global treaty with dramatic greenhouse gas cuts, which could see Australia halve its greenhouse emissions by 2050, extreme bushfire danger days would rise by only 8-17 percent.

"Future bushfire danger in Australia will depend heavily on how fast and by how much we act to tackle global warming," said the report.

"The best chance of avoiding a high global warming scenario is through a fair, ambitious and legally binding international treaty to cut emissions," it said.

The firefighters and Greenpeace called on the Australian government to dramatically increase its greenhouse emissions target cuts, but Climate Change Minister Penny Wong on Wednesday announced Australia would stick to its 5 to 25 percent emissions cut range under the non-binding "Copenhagen Accord".

Wong said any decision to opt for a 15 or 25 percent target depended in part on strong steps by India and China to reduce the growth of their greenhouse gas emissions.

"This report shows that unless governments ramp up their targets for cutting greenhouse emissions, we'll be facing more frequent bushfire tragedies on an even greater scale," said Casey. (Reporting by Michael Perry; Editing by Alex Richardson)

It took Mother Nature barely a year to completely extinguish the blazing "300% increase in fires" of global warming alarmists, proving once again that such wild predictions have no basis whatsoever in science.

What a difference a year makes!

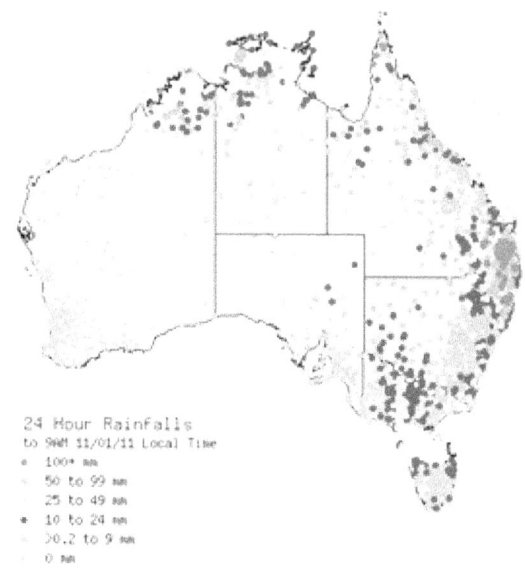

24 Hour Rainfalls
to 9am 11/01/11 Local Time
• 100+ mm
• 50 to 99 mm
  25 to 49 mm
• 10 to 24 mm
  0.2 to 9 mm
  0 mm

http://www.crikey.com.au/2011/01/12/qld-floods-brisbane-ipswich-prepare-for-worst/

'Events don't unfold hour-by-hour, but minute-by-minute': QLD floods

Wednesday, 12 January 2011
by Amber Jamieson

Brisbane and Ipswich are bracing for their worst ever floods, with tens of thousands of homes at risk and fears for citizens' safety, as the toll of dead and missing for the state-wide disaster continues to rise. The banks of the Brisbane River have broken and the flood peak will hit at 4am tomorrow. Already many suburbs are partially submerged, and 20,000 properties are likely to experience complete flooding in Brisbane.

In Ipswich the river is rising, with 1,500 properties already evacuated and over 1,000 people waiting in evacuation centres.

The death toll has risen to twelve people, the latest victims being two men found in Grantham in the Lockyer Valley. The number of missing people has been revised down to 51 people.

We will be back updating this post tomorrow.

Update: 5:45 pm

Most of Brisbane's major roads have now been closed, with the Riverside Expressway now closed. Police are asking for people to avoid driving through Brisbane.

More sad news from ABC 612:

Brisbane's Riverwalk will be demolished: engineers declare it poses a serious safety risk. It is already partially destroyed and submerged.

5:15 pm

The number of missing people has been revised down from 76 to 51 people, although "grave concerns" are held for nine of those missing, announced Neil Roberts, QLD Emergency Services Minister.

But in "a small piece of wonderful news", QLD Police Commissioner Rob Atkinson told of how two men, in two separate incidences in the Lockyer Valley, who were thought to have been swept away by the flood waters have been found safe and sound. "One can only describe it as a miracle based on the circumstances on which they went missing," said Atkinson.

In the latest updates on the floods, the Brisbane River is currently at 4.16 metres. The peak of 5.5 metres is still expected for 4am.

The Bremer River in Ipswich sits steady at 19.4 metres, but current modelling predicts it will rise to 20.5 metres.

It is still possible that the Moggill Ferry may be sunk to avoid further issues, but no resolution has yet been decided. This is only a possibility for The Island Party Barge and the Drift cafe. As Roberts said, "These events unfold not just hour by hour but minute by minute."

Roberts pleaded with residents to only dial 000 in cases of life threatening emergencies, as the phone lines has been clogged. He asked residents to continue to listen to radios and take care on the roads.

The flood crisis is continuing in Chinchilla, where E.coli has been found in the flood waters and water must be boiled before drinking.

Roberts also said things are "touch and go for Goondiwindi."

4:15 pm

Brisbane's inland beach isn't looking so tropical, judging on a photo posted by @rohandwyer:

....

3pm

Jason Whittaker writes: Queensland Premier Anna Bligh has delivered another briefing. As Bligh admitted "We still have a number of families who don't know where their children are."

She acknowledged the town of Chinchilla, north-west of Toowoomba, which is now facing rising floods for the second time. "This is a heartbreaking time for the people of Chinchilla," Bligh said.

The flood peak in Ipswich is expected to be slightly lower than originally

predicted. Water levels are currently around 19.25m, Bligh said, heading to 20.5m in the next couple of hours.

In Brisbane's western suburbs, Gales and Goodna have been "particularly hard hit." Closer to the city, Jindalee, Moggill and Yeronga are most affected.

The Moggill ferry, which runs on a wire across the river, has broken free from one of its guides. Crews have been dispatched to potentially sink or destroy the vessel before it breaks free and becomes a hazard. Other barges in the Brisbane River could also be sunk as a precaution.

In the city, much of the CBD has closed down. Buses to the city have ceased but train services are running where they can on a public holiday-type timetable. Numerous roads are flooded, including the Inner City Bypass. Tolls have been suspended on the Logan Motorway and Gateway Motorway until further notice.

As at 1pm local time, 3,585 people are formally registered in evacuation centres across Brisbane and Ipswich; 400 of those are in Brisbane. The numbers are expected to swell to 2500-3000.

All hospitals are operational and fears that the Wesley Hospital would have to close have been allayed. But all non-urgent surgery in Brisbane hospitals has been delayed.

Meanwhile, Brisbane Mayor Campbell Newman says power has now been cut to some 70,000 homes with more residents affected over the next few hours.

There is no more news on the search and rescue operation in Toowoomba. Police say they are now working with the coroner in the area.

12:45 pm

There are some truly incredibly photos coming out of the floods. From Seven News, the Suncorp Stadium, looking more swimming pool than football oval (there was a minor fire here earlier, but police say it was just a "small fire... in isolated transformer room")

12:00 pm

For a concerning look at the projected modelling for Brisbane's flooding, check out this:

Contrary to myths floating around the internet, the Ipswich water supply has not been cut off, announced the brilliant Queensland Police Media Unit.

But for those needing to travel, there are issues with Brisbane's public transport, with public transport service to be wound down from 1pm, announce TransLink.

Due to Police advice to stay away from the Brisbane CBD, we advise passengers to avoid non-essential travel today and to stay home where possible in the difficult weather conditions.

Bus and train services may be delayed or diverted, and ferry services are cancelled.

Bus and train services are continuing to run where possible, however passengers planning to use public transport should check the status of their intended service.

Check service updates for current bus, train and ferry information.

Disruptions are expected to continue due to bad weather, flooding and king tides.

11:30 am

"We've woken up to a very surreal experience. The sky is blue, we are facing a beautiful Queensland day... but do not take any comfort from the fact that we have blue sky here this morning," warns Queensland Premier Anna Bligh, as she outlined the latest flood news.

The death toll from the floods remains at ten, despite reports to the contrary. Number of missing is now "over 90", said Bligh. However, since the weather is clear today an extensive search and rescue operation will commence in the Lockyer Valley.

The search and rescue operation will be a "very difficult, very urgent and occasionally heartbreaking task", announced PM Julia Gillard, who joined Bligh for the latest press conference in Brisbane.

Already 1,500 properties have been evacuated in Ipswich and 1,200 people are in the 10 evacuation centres there.

Ipswich's flood levels reached 18.9 metres at 7:30am today. In "slightly good news" the peak of 22 metres has been revised to 20.5 metres, said Bligh. It was 20.7 metres in 1974. When the peak comes, 4,000 properties are expected to be affected.

Pontoons and boats are floating down the Brisbane river. The flood levels in Brisbane are at 3.1 metres rising. Bligh warns that while the river is rising slowly this morning, it will be quicker after lunch. It should peak today at 4.5 metres and rise to 5.5 metres tomorrow morning at 4am, where it will then stay until Saturday. In 1974 it peaked at 5.45 metres. The flood levels will be "1974 proportions and slightly higher," advises Bligh.

Bligh also noted that when making those comparisons between the 2011 and the 1974 floods, that people need to remember that those cities are vastly different places, with a significantly higher population, more dense suburban areas and new suburbs, including inner city Brisbane areas.

The numbers of expected flooding in Brisbane are serious. 2,100 streets. 19,700 residences will flood across their entire property. 3,500 commercial properties will experience flooding across their entire property.

In Brisbane 182 people are already registered at the RNA Showgrounds evacuation centre, with a new evacuation centre opening at the QEII stadium on the south side of Brisbane. The Lord Mayor of Brisbane is apparently in talks with churches to see if church halls and church facilities can be used to house those in need of evacuation.

Residents should stay at home and avoid going travelling on the streets. "Do not travel if you do not have to. It is a danger to the rest of the population to have people out travelling unnecessarily", said Bligh. "This incidwent is not a tourist event. This is a deeply serious natural disaster."

Traffic management plans have been put into place and its expected many traffic lights in Brisbane will not function as electricity gets cut off.

It's not just Brisbane and Ipswich battling these floods today. Dalby has rising flood waters and 125 people in an evacuation centre. Chinchilla is expected to receive a flood peak higher than the floods it experienced ten days ago. Condamine was completely evacuated last night. The water at Goondiwindi, a town of 5,00 people, is just half a metre from the levees. Evacuations are also occurring in the tiny town of Texas, population 600. Water in Rockhampton is finally dropping. As is Gympie, but that water will then hit Maryborough. Flooding is also expected in Bundaberg.

As Bligh noted, many of these towns are facing floods for the second time in two weeks and Brisbane and Ipswich residents should "draw inspiration and courage" from them.

"I can say, honestly, I do know what it is like to be worrying about your own family," said an emotional Bligh, as she explained that her own mother is now staying at Bligh's home, since her own house was at risk.

Gillard also announced emergency payments will be made available, with 10,000 payments, worth $17 million already made. Gillard was careful to stress that these are just initial emergency payments and not for rebuilding homes and business. Obviously they will come later.

In more positive news: here is a frog riding a snake in the floods. Will this be the new Sam the Koala?

10:30am

Flood affects journos too, writes Jason Whittaker in today's Media Briefs:

The big guns of breakfast TV are broadcasting through the morning, perched atop Kangaroo Point cliffs overlooking the Brisbane CBD and along the disappearing banks of the Brisbane River. Karl Stefanovic heads Nine's coverage (after broadcasting well into the night yesterday), while Melissa Doyle cut short her holidays to join Larry Emdur (that's the former game show host who Nine now wants to poach to anchor its breakfast news) on an extended Sunrise. ABC1 has flicked the switch to Aunty's 24-hour news station, with Joe O'Brien water-side in Brisbane.

Brisbane's only daily paper The Courier-Mail is headquartered at Bowen Hills, one of the inner-city suburbs expected to face flooding throughout the day. Its online rival, Fairfax's Brisbane Times, has already moved out of its riverside office tower in the CBD and has journalists working in the field, at home and out of stablemate 4BC's Cannon Hill studios. The ABC continues to work out of multiple offices while its new Southbank headquarters is built (the riverside construction site will go under today), with the all-important Local Radio studios in Toowong also in flood's way. The station briefly lost phone services this morning and had to redirect listener calls.

As the 1974 flood comparisions continue, it's interesting to note that Brisbane has over one million more residents today.

10:00 am

The Brisbane River has just broken at Yeronga, with 20,000 properties under threat, reports the ABC. "Hundreds of homes are underwater", said Sky News, who noted that their information came by ABC local radio.

The Brisbane City Council are looking for more volunteers:

Brisbane City Council have asked if anyone would like to lend a hand filling sandbags.

People should preferably wear work boots and high visibility clothing. If volunteers do not have work boots then enclosed shoes must be worn. Light coloured clothing is also acceptable is you do not have high visibility.

9:30 am

A "senior emergency official" says that 30 people are believed to be dead, according to unconfirmed reports in The Australian and also Sky News. Premier Anna Bligh announced yesterday the suspected number of deaths was at least double the current toll.

Ipswich is in immediate danger, with parts of the city already submerged. A flood peak of 21.5 metres is expected to hit the Bremer River in Ipswich at 11am today, a rise of five metres in five hours.

Yesterday afternoon the main street was dry, today it is underwater.

As Ipswich copes with the rising flood waters, reports of looting in flood affected areas has infuriated locals. "If I find anybody looting in our city they will be used as flood markers," declared Ipswich Mayor Paul Pisasale.

Up in Brisbane, a staggering 90,000 Brisbane properties are expected to be flood affected in the coming days, as the Wivenhoe Dam reached 190% capacity last night. The peak for Brisbane is expected early tomorrow morning.

Electricity would be cut to 100,000 properties in Brisbane, warned Energex.

This is the second time in two weeks that Condamine, Warwick, Chinchilla and Dalby communities are facing floods

And again, a year later in the summer of 2011 – 2012, no sign of an increase in bushfires! Not only did the decade-long drought in Australia officially end, but most of Australia had an unusually cool summer. Naturally, global warming alarmists hyped Perth and surroundings, the only region in this vast country with unusual heat this summer.

So, did advocates of AGW offer any apology or explanation for their completely erroneous prediction in 2010? By now, the reader should expect worse. Not only did the pseudoscience charlatans not admit to mistakes made only a year earlier, but also they readily exhibited the intellectual dishonesty to accuse global warming for the floods. Naturally, their prediction of a 300% increase in bush fires over the next half century was quickly forgotten.

I am only comforted by the fact that the aerospace and defense industries are not based on such sloppy "science" or lack thereof. If those areas were subject to the same flimsiness as climate pseudoscience, practically every airplane would have been falling from the sky, and it is more than likely that weapons fired unintentionally would already have resulted in nuclear holocaust.

Still, the visibly false predictions of climate pseudoscience did not prevent Australian Prime Minister, Julia Gillard, from pushing through the Carbon Tax in July 2011, helping to diminish her popularity to the lowest of any Australian leader since World War II. She did so, in full contradiction to the pledge she made during the 2010 election not to impose a Carbon Tax. Such is the power of the Carbon lobby these days.

Fortunately, the people of Australia are, by no means, unintelligent, and have retained their ability to think for themselves. A year later, Julia Gillard is under significant pressure to abandon the misguided carbon tax.

http://www.3aw.com.au/blogs/blog-with-derryn-hinch/carbon-tax-will-julia-gillard-blink/20120503-1y1d2.html

Carbon Tax: Will Julia Gillard blink?

Posted by: Derryn Hinch | 3 May, 2012 - 5:29 PM

It was almost two years ago now that Julia Gillard in the final desperate throes of the 2010 Federal election made the now infamous promise that has haunted the legitimacy of her prime ministership since.

Tony Abbott has reminded us of it almost daily. I'm told his office secretary even has it as her ring tone.

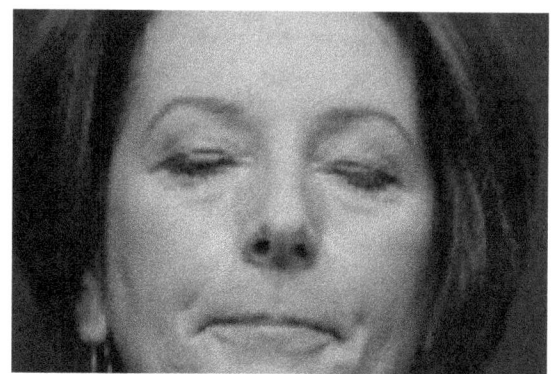

"There will be no carbon tax under a government that I lead," Julia Gillard said.

It has been played and replayed since. It has rarely been out of the news. It has been blamed as the pivotal point for the electorate's loss of faith in her; Lack of trust in her. The start of the offensive Ju-Liar campaign.

And it has lasted longer and been more bitter than any reaction to Paul Keating's broken L.A.W. promise on tax cuts and John Howard's cynical core and non-core promises.

That Carbon Tax is now law. It will become effective from 1 July - less than two months away.

It's back in the news because this week former New South Wales Premier, Kristina Keneally, advised the PM to either reduce it or push it away.

I have said before, if this government has even the proverbial snowflake's chance in hell, Julia Gillard should go on national television and make a speech - if anybody's still listening. She should say she passionately still believes in the Carbon Tax, that it will benefit this country.

She believes the announced compensation will soften the blow for many Australians – for those 'working families'. But she should add: "The rest of the world is not listening. Those that are bringing in a carbon tax are up to $20 a tonne less than us. And, regrettably, because of the current global financial situation, I am withdrawing this legislation until after the next election."

She could say, I guess, that she is cutting it from $23 to $10 but the new Leader of the Greens, Christine Milne, has already said she won't buy that.

So on the premise that the tax will come in in seven weeks' time how will it affect you? Do you believe the compensation will be enough? Are you in that tax bracket where you are going to be clobbered with no rebates at all?

And what about the expected upward spiral in electricity bills, production costs, even – some alarmists say – massive increases in the cost of refrigeration.

By the time of the next election it will have kicked in. Will it be so savage that Tony Abbott won't need any other issue to get the keys to The Lodge?

The big question is: Will Julia Gillard blink? If she doesn't, is her fate sealed? Or won't they even let her lead them to the polls.

Footnote: The Herald Sun today has a big spread on possible replacements for Julia Gillard under the headline 'Two Up for the Big Switch'. The story features Simon Crean and Greg Combet as 'the men most likely'. Yeah, right.

Crean has been there and failed before. And Combet is Gillard's most ardent supporter of the Carbon Tax.

June to August 2010

While the mainstream and pseudoscience media were busy cherry picking the heat wave in Russia, the real weather news was down under. The Southern Hemisphere was experiencing record cold, but the chances are that the reader may never have heard of it before.

South Africa

Epic cold in South Africa resulted in the coldest Soccer World Cup on record. FIFA may have banked on the usually mild winter in the sub-tropical southern tip of Africa and/or global warming, but, as always, Mother Nature had the last word.

http://www.bild.de/news/bild-english/news/freezing-winter-games-in-south-africa-12974394.bild.html

The freezing Winter Games in South Africa

South Africa – the name conjures up images of safari, heat, beaches, sunsets and fine wine.

But what are we seeing at the World Cup?

Brazil's samba stars wearing gloves and roll-neck jerseys under their shirts, coach Dunga in a thick knitted jumper, substitutes in winter hats and anoraks, freezing fans in fleeces and raincoats in the stands.

It's the South Africa Winter Games!

Temperatures of -1C were seen during Brazil's 2-1 victory over North Korea – making it the coldest World Cup ever, for now at least. Captain Lucio, used to the cold after his time with Bayern Munich, said: "It really was freezing tonight."

And coach Dunga, a former Stuttgart player, added: "The cold makes it hard to be creative."

Snow on Table Mountain and chilly rain in Cape Town – Denmark's training session was cancelled because of a flooded training pitch! The first snow for decades fell near to Japan's World Cup base between Cape Town and Port Elizabeth.

Most hotels in Cape Town have no heating. BILD reporter Henning Feindt even had a hot water bottle placed under his bedcovers by the hotel management!

And his colleague Mario Volpe is grateful that he came well prepared for the conditions, saying: "I'm happy that I packed my long johns."

The cold weather has clearly thrown strikers off the scent. A meagre return of 25 goals in the first 16 games is the worst in World Cup history at just 1.56 per match.

Meteorologist Elke Brouwers' forecast for Johannesburg, Pretoria and Bloemfontein, where games are due to take place, will send shivers down the backs of the World Cup stars: "We are expecting a new cold front with frost."

Temperatures of -4 degrees were recorded in Pretoria, which is hosting the German national team. Daytime temperatures, however, will reach 20 degrees, and the German players are capable of dealing with the cold.

Jogi Löw's assistant coach Hansi Flick said: "We have recommended that players should always wrap up warmly and blow-dry their hair after taking showers because of the huge temperature fluctuations."

It seems to have worked so far – the Germans are the only team whose scent for goals hasn't frozen over...

Australia

Australia was no exception to the deep freeze in the Southern Hemisphere

http://globalfreeze.wordpress.com/2010/07/06/record-cold-in-australia/

Record cold in Australia

Adelaide, Alice Springs, Brisbane, Melbourne, Perth, Sydney

Desperate times as Alice shivers on coldest day

http://www.abc.net.au

July 6, 2010

The previous coldest day was in August 1966 when the maximum temperature reached just 7 degrees Celsius.

Adelaide: Record looms, and shivers it's cold

http://www.adelaidenow.com.au

July 05, 2010

The Bureau of Meteorology's early forecast is for an overnight low of just 3C tonight and if correct it will equal the record run of six nights below 5C set in 1982.

Sydney's coldest June morning for more than 60 years

http://www.smh.com.au

June 30, 2010

Sydney's week of cold weather continues, with the city recording its coldest June morning since 1949 when temperatures dived to 4.3 degrees.

Melbourne faces longest cold snap in 14 years

http://www.theage.com.au

July 2, 2010

If you thought it was colder than normal, you're right: Melbourne is within reach of its longest cold snap in 14 years.

Perth on track to beat winter chill record

http://www.watoday.com.au

June 30, 2010

The recent run of cold night weather in Perth could soon become the longest spell on record as the city shivers through its coldest June in 15 years.

Cloud blankets Queensland, Brisbane coldest in two years

http://www.weatherzone.com.au

July 2, 2010

Brisbane only reached a maximum of 16 degrees, four below the July average and the coldest day in almost two whole years. Longreach, near the heart of Queensland, topped at just 14, the second coldest July day in 17 years

South America

And finally South America where the winter was so severe that even the mainstream media could not fully ignore it. Here is CNN:

http://articles.cnn.com/2010-07-19/world/latin.america.weather_1_amazon-region-cold-front-coldest-weather?_s=PM:WORLD

Cold temperatures cause death, damage in South America
July 19, 2010|By the CNN Wire Staff

An intense cold front in southern Latin America continues to blanket the region, causing deaths, school and highway closures, and other woes.

A total of 18 people have died in Bolivia as a direct or indirect consequence of low temperatures, the Peruvian state-run Andina news agency reported. The deaths were spread out throughout the country.

On Monday, Bolivian officials said temperatures in the major city of Santa Cruz de la Sierra would reach 3 C (37 F), the lowest in 29 years, and in other regions the mercury dropped below freezing, Andina reported.

As a precaution, Bolivian authorities canceled school from Monday to Wednesday, the official Bolivian news agency ABI reported.

Police in Paraguay reported eight deaths from hypothermia and two from carbon monoxide poisoning from the use of heating devices. The government opened shelters for the poor, who are picked up at night by military trucks.

Paraguayan authorities also estimated that 1,000 cattle died because of the cold.

In Uruguay, local media reported two weather-related fatalities.

The cold front hit the region on Saturday and was responsible for eight deaths in Argentina over the weekend.

An area of low pressure in the southern hemisphere jet stream pushed deeper north allowing for cold Antarctic air to pool over Chile and Argentina. Below-normal temperatures are expected over the next 48 hours across the region.

Argentina reported Monday that nine of its provinces were feeling temperatures below freezing.

The intense cold will remain in the area at least through Tuesday, Argentina's official news agency, Telam, reported.

Similarly, in Peru, the country's southern Amazon region was experiencing the coldest weather in three years, Andina reported, citing the National Meteorological and Hydrological Service.

In the city of Puerto Maldonado, the temperatures fell to 9 C (48 F). In the Amazon region, the usual lows are in the 20s C (high-60s F).

The cold was also affecting farmers in the Peruvian city of Arequipa, in the Andes Mountains. With temperatures falling there to -17 C (1 F), the cold was too much for the region's Alpaca herds.

Pregnant Alpacas were losing their babies, and young Alpacas were dying, Andina reported. Some 10 percent of the region's 40,000 Alpacas were affected, the news agency reported.

In particular, I found the plight of Alpacas interesting; once again throwing cold water on the myth constantly propagated by global warming alarmists that tropical species are moving to more temperate latitudes due to rising temperatures. On the contrary, evidence appears to demonstrate the exact opposite. Unusual cold in recent years did indeed make life unbearable for manatees, musk oxen, Alpacas and hardened Mongolian livestock (described in a later section), to name just a few species cited in this book, *in their own latitudes*. One wonders if they are considering migrating towards the Equator instead.

Below is more analysis of the bitter winter in South America

http://wattsupwiththat.com/2010/07/20/cold-snap-freezes-south-america-beaches-whitened-some-areas-experience-snow-for-the-first-time-in-living-memory/
Posted on July 20, 2010 by Anthony Watts

From the "weather is not climate" department, more chilling news from the southern hemisphere.

Guest post By Alexandre Aguiar
MetSul Weather Center via ICECAP [136]

A brutal and historical cold snap has so far caused 80 deaths in South America, according to international news agencies. Temperatures have been much below normal for over a week in vast areas of the continent. In Chile, the Aysen region was affected early last week by the worst snowstorm in 30 years. The snow accumulation reached 5 feet in Balmaceda and the Army was called to rescue people trapped by the snow.

In Argentina, the snow in the region of Mendoza, famous for its winery, was described by local meteorologists as the heaviest in a decade. The temperature in the morning of July 16th was the lowest in the city of Buenos Aires since 1991: -1.5C. The cold snap caused a record demand for energy and Argentina had to import electricity from Brazil. Many industries in Argentina were shut down due to gas shortage.

It snowed in nearly all the provinces of Argentina, an extremely rare event. It snowed even in the western part of the province of Buenos Aires and Southern Santa Fe, in cities at sea level.

The most famous beach of Argentina, Mar del Plata, was whitened by the snow in the morning of July 15th, a scene only seen in recent memory in 1991, 2004 and 2007. See image on the right:

The snow was heavy even in Northern Argentina. In Santiago Del Estero, according to media reports, some areas experienced snow for the first time in living memory. In the province of Tucuman, some town saw snow for the first time since 1921 (Gaceta de Tucuman newspaper).

In Uruguay, there were widespread reports of sleet and even snow mixed with rain in towns in the Southern and Eastern part of the country, even in the capital Montevideo. At least two deaths have been blamed in Uruguay on the low temperatures. Hospitals were packed with patients with respiratory illness.

---

[136] http://icecap.us/

In Paraguay, at least nine people died due to the cold weather in only 3 days. Cattle were very affected and one thousand animals died of hypothermia. In Bolivia, dozes of people died in consequence of the very low temperatures. In some areas of the nation the cold period was described as the worst in 15 years. It even snowed in the Chaco of Bolivia, one of warmest areas of South America, where the local population never saw snow before. Classes were suspended in Bolivia for three days to prevent more cold related deaths (El Nacional newspaper from Bolivia).

Southern Brazil was also very affected by the cold air eruption from the Southern Pole. Last week the temperature dropped to -7.8C in the city of Urupema, Santa Catarina. In Rio Grande do Sul, in the hills of the state, temperature felt to -4.9C in the city of Cambara. In the state of Paran�, the low was -6C. Only the nights were freezing, but the afternoons were very cold. In some days, temperature failed to reach 5C in many towns, the first time in a decade. Flurries observed in towns of Rio Grande do Sul, Santa Catarina and Parana and sleet was also reported in Western Santa Catarina.

The most striking scenes came from the top of Morro da Igreja, a 1800 meters elevation in the state of Santa Catarina. The area recorded snow and freezing rain. As anyone can imagine, freezing rain is extremely rare in Southern Brazil. The event was witnessed and photographed by weather observers from MetSul Marcelo Albieri and Caio Souza.

On July 14th, in the afternoon hours, temperatures in the hills of Rio Grande do Sul state in Southern Brazil were lower than in Marambio, the main polar base of Argentina in Antarctica. In Central Brazil, in the tropics, the long streak of cold days was considered extremely rare. It was so cold that thousand of animals died in this region of Brazil known for its cattle, just South of the Amazon basin.

Maybe the most notable fact took place in North South America. The cold reached Amazon and temperatures felt to as low as 7C in towns in the Amazon Forest in the states of Acre and Rondonia. Temperature even felt in Roraima, where the state capital Boa Vista record 20C (normal lows are 25C) and the wind were blowing from the South.

Boa Vista is located at 2 degrees North of latitude, so the influence of the Antarctic cold blast crossed the Equator line and reached towns in the Northern Hemisphere. It would be the same of a cold snap from the Arctic crossing the entire North America continent, the Caribbean and reaching North Brazil in cities at 2 degrees South of latitude as Santarem, a bizarre situation.

And where was most of the mainstream and pseudoscience media? Hyping the heat in Moscow….

http://pgosselin.wordpress.com/2010/07/20/south-america-cold-kills-175-wheres-the-media/
South America Cold Kills 175 – Where's The Media?

When the temperature sinks far below normal and people freeze to death, you hardly hear a peep from the media anymore.

But as soon as the temperature rises a few degrees over normal for a day or more, the media explodes with headlines of "HEAT WAVE!" and "SCORCHING HEAT!".

Thanks to the internet and a few media outlets, few and far in between, the inconvenient news of a devastating, protracted cold snap gripping much of south America is coming out. Read news24.com/World/News.

Here are some grisly statistics:

1. 175 dead in 6 countries

2. 112 people had died of hypothermia and flu in Peru.

3. 16 people froze to death and 11 died of carbon monoxide poisoning due to faulty heaters.

4. Thousands of cattle also froze to death on their pastures in Paraguay and Brazil.

5. In Bolivia, 18 people died, in Paraguay five and two each in Chile and Uruguay. Nine people died of the cold in southern Brazil.

How cold is it? Take a look at this post I put up three days ago. *return-to-1970s-cold*[137]/, showing the 7 day outlook.

You'd think when they're not too busy covering Zsa Zsa Gabor, the media would cover this.

In Germany, it seems the major news magazines are finding space and time to report the cold – along with the reports of the warm weather we're having here, which will end in about 36 hours.

But Germany's ARD Teklevsion has NOTHING. Same goes for ZDF television. Hmmm.

I couldn't find much at the major US news services either. CBS – nothing. At CNN there's a report that's a day old with 17 deaths. Nothing at the right wing Fox either. Maybe others are better at finding big news.

---

[137] http://pgosselin.wordpress.com/2010/07/17/will-2010-stay-hot-and-joe-bastardi-return-to-1970s-cold/

While the epic cold in the Southern Hemisphere went mainly unreported, much of the mainstream and pseudoscience media went on overdrive reporting the heat in Western Russia and unusual flooding in Pakistan, and concocting anthropogenic cause with no basis in science. Reporting heat, while deliberately concealing cold, is equivalent to exaggeration and deception. The attempts at linking the heat wave in parts of the Northern Hemisphere to human factors were so egregious that even some pro-AGW institutions such as NOAA and *the New Scientist* felt it necessary to repudiate such brazen propaganda. Still, the flooding of the air, web and print waves with pseudoscience chicanery did have some degree of success. To this day, there are many people around the world who were effectively brainwashed and still believe that the heat waves of 2010 were caused by Anthropogenic Global Warming. To this day, there are many people around the world who are completely unaware of the bitter cold in the Southern Hemisphere during the same period.

Unfortunately, in the perennially corrupt world of climate pseudoscience, the executors of fraud have unlimited strikes. The same pattern of exaggerating warmth and playing down cold was, and continues to be repeated in "The Year that Winter Forgot" as described in earlier sections.

To start with, here is TIME Magazine leading the global warming charge during the Northern Hemisphere summer of 2010:

http://www.time.com/time/world/article/0,8599,2008081,00.html

Will Russia's Heat Wave End Its Global-Warming Doubts?

By Simon Shuster / Moscow Monday, Aug. 02, 2010

"….there may turn out to be a bright side to Russia's devastating weather: one of the nations most responsible for driving climate change may finally start trying to do something about it."

And here is grist.org / thinkprogress.org engaging in blatant deceit by reporting that climate "experts" agreed that global warming caused the Russian heat wave

http://grist.org/article/2010-08-12-climate-experts-agree-global-warming-caused-russian-heat-wave/

Climate experts agree: Global warming caused Russian heat wave

By Brad Johnson

As Russia chokes from a heat wave of unprecedented ferocity, President Dmitry Medvedev has strengthened his call for the world's leaders to take action to fight global warming pollution. The scientific community has warned for decades that burning coal and oil without limit would intensify heat waves, droughts, and floods. Now that the planet is at its hottest in recorded history, freak climate disasters are arriving with increasing frequency. Some scientists are now stating the obvious: Russia's heat wave simply would not have happened without the influence of fossil fuel pollution on our atmosphere. University of Texas climate scientist Michael Tobis is "hazarding a guess" that "the Russian heat wave of 2010 is the first disaster unequivocally attributable to anthropogenic climate change":

But right now I feel like hazarding a guess. As far as I understand, nothing like this has happened before in Moscow … The formerly remarkable heat wave of 2001, then, is "the sort of thing we'll see more of" with global warming. But it may turn out reasonable, in the end, to say "the Russian heat wave of 2010 is the first disaster unequivocally attributable to anthropogenic climate change."

Meteorologist Rob Carver, the Research and Development Scientist for Weather Underground, agrees. Using a statistical analysis of historical temperature records, Dr. Carver estimates that the likelihood of Moscow's 100-degree record on July 29 is on the order of once per 1,000 years, or even less than once every 15,000 years — in other words, a vanishingly small probability.

However, those tiny odds are based on the assumption that the long-term climate is stable, an assumption that is no longer true.

Like Dr. Tobis, Carver believes that manmade global warming has fundamentally altered weather patterns to produce the killer Russian heat wave. "Without contributions from anthropogenic climate change," Carver said in an email interview with the Wonk Room, "I don't think this event would have reached such extremes or even happened at all":

I agree with Michael Tobis's take at Only In It For the Gold that something systematic has changed to alter the global circulation, and you'll need a coupled atmosphere/ocean global model to understand what's going on. My hunch is that a warming Arctic combined with sea-surface-temperature teleconnections altered the global circulation such that a blocking ridge formed over western Russia leading to the unprecedented drought/heat wave conditions. Without contributions from anthropogenic climate change, I don't think this event would have reached such extremes or even happened at all. (You may quote me on that.)

Just as the Russian heat wave is fueled by global warming, so is the rest of the world's killer climate. World-renowned climatologist Kevin Trenberth explained in an interview with Wired's Brandon Keim that the Eurasian heat wave is part of the even larger circulation pattern that has produced the catastrophic southeast Asia monsoon:

The two things are connected on a very large scale, through what we call an overturning or monsoonal circulation. There is a monsoon where upwards motion is being fed by the very moist air that's going onshore, and there are exceptionally heavy rains. That drives rising air. That air has to come down somewhere. Some of it comes down over the north.

Dr. Rob Carver's analysis of the statistical likelihood of the Moscow heatwave:

Now, let's take a look at July 29, when it cracked 100 F at Moscow Shermetyevo. By our records, the reported high was 100 degrees F and the normal high for that day is 68 F. According to GDAS, the maximum temperature for Moscow on July 29 was 96 degrees F and the normal temperature according to CFSR is 69 F. Using the techniques of Hart and Grumm (2001), the climatological anomaly for maximum temperature is 7 degrees F.

So, using the GDAS and CFSR data, the normalized anomaly of maximum temperature was +3.1. That's near a recurrence interval of once per thousand years which matches the quotes I've heard from Russian met agencies. Now, if we assume the climatological anomaly derived from CFSR data is the same of the observations, the normalized anomaly jumps to +4.5, which translates into "less than once every 15,788".

That however, is a tricky assumption to make. We know that the climatic properties of CFSR and GDAS data have to have some correspondence with what's actually happens in the atmosphere, otherwise weather models wouldn't work. What becomes difficult to quantify (in the time constraints of writing for the public) is how the statistics of the climate properties line up between observations and reanalysis. And at these extremes, it doesn't take much change in the average and standard deviation of a property to dramatically change how unusual an event is. Another possible source of error is the assumption that the climatology of CFSR is the climatology of the operational GDAS. Which is not a slam-dunk since NCEP shifted to a higher-resolution model on July 28. Now, I don't have any information to say the post July 28 GDAS data has different climatological characteristics, but it's a possibility. Another big assumption I make is that daily maximum temperatures follow a Gaussian (normal) distribution and that from 30 years of CFSR data, I can adequately characterize such a distribution.

Brad Johnson is the editor for ThinkProgress Green at the Center for American Progress Action Fund. Brad holds a bachelor's degree in math and physics from Amherst College and master's degree in geosciences from the Massachusetts Institute of Technology. He is the co-author of Technomanifestos, a history of the Information Revolution, and the founder of HillHeat.com, which covers climate policy in our nation's capital.

Even the usually pro-global warming media was obliged to repudiate such baseless claims of anthropogenic cause by saying clearly that the heat wave was caused by "the frozen jet stream"

http://www.newscientist.com/article/mg20727730.101-frozen-jet-stream-leads-to-flood-fire-and-famine.html

Frozen jet stream links Pakistan floods, Russian fires

Raging wildfires in western Russia have reportedly doubled average daily death rates in Moscow. Diluvial rains over northern Pakistan are surging south – the UN reports that 6 million have been affected by the resulting floods.

It now seems that these two apparently disconnected events have a common cause. They are linked to the heatwave that killed more than 60 in Japan, and the end of the warm spell in western Europe. The unusual weather in the US and Canada last month also has a similar cause.

According to meteorologists monitoring the atmosphere above the northern hemisphere, unusual holding patterns in the jet stream are to blame. As a result, weather systems sat still. Temperatures rocketed and rainfall reached extremes.

Renowned for its influence on European and Asian weather, the jet stream flows between 7 and 12 kilometres above ground. In its basic form it is a current of fast-moving air that bobs north and south as it rushes around the globe from west to east. Its wave-like shape is caused by Rossby waves – powerful spinning wind currents that push the jet stream alternately north and south like a giant game of pinball.

In recent weeks, meteorologists have noticed a change in the jet stream's normal pattern. Its waves normally shift east, dragging weather systems along with it. But in mid-July they ground to a halt, says Mike Blackburn of the University of Reading, UK (see diagram). There was a similar pattern over the US in late June.

Stationary patterns in the jet stream are called "blocking events". They are the consequence of strong Rossby waves, which push westward against the flow of the jet stream. They are normally overpowered by the jet stream's eastward flow, but they can match it if they get strong enough. When this happens, the jet stream's meanders hold steady, says Blackburn, creating the perfect conditions for extreme weather.

A static jet stream freezes in place the weather systems that sit inside the peaks and troughs of its meanders. Warm air to the south of the jet stream gets sucked north into the "peaks". The "troughs" on the other hand, draw in cold, low-pressure air from the north. Normally, these systems are constantly on the move – but not during a blocking event.

And so it was that Pakistan fell victim to torrents of rain. The blocking event coincided with the summer monsoon, bringing down additional rain on the mountains to the north of the country. It was the final straw for the Indus's congested river bed (see "Thirst for Indus water upped flood risk").

Similarly, as the static jet stream snaked north over Russia, it pulled in a constant stream of hot air from Africa. The resulting heatwave is responsible for extensive drought and nearly 800 wildfires at the latest count. The same effect is probably responsible for the heatwave in Japan, which killed over 60 people in late July. At the same time, the blocking event put an end to unusually warm weather in Western Europe.

Blocking events are not the preserve of Europe and Asia. Back in June, a similar pattern developed over the US, allowing a high-pressure system to sit over the eastern seaboard and push up the mercury. Meanwhile, the Midwest was bombarded by air from the north, with chilly effects. Instead of moving on in a matter of days, "the pattern persisted for more than a week", says Deke Arndt of the US National Climatic Data Center in North Carolina.

So what is the root cause of all of this? Meteorologists are unsure. Climate change models predict that rising greenhouse gas concentrations in the atmosphere will drive up the number of extreme heat events. Whether this is because greenhouse gas concentrations are linked to blocking events or because of some other mechanism entirely is impossible to say. Gerald Meehl of the National Center for Atmospheric Research in Boulder, Colorado – who has done much of this modelling himself – points out that the resolution in climate models is too low to reproduce atmospheric patterns like blocking events. So they cannot say anything about whether or not their frequency will change.

There is some tentative evidence that the sun may be involved. Earlier this year astrophysicist Mike Lockwood of the University of Reading, UK, showed that winter blocking events were more likely to happen over Europe when solar activity is low – triggering freezing winters (New Scientist, 17 April, p 6).

Now he says he has evidence from 350 years of historical records to show that low solar activity is also associated with summer blocking events (Environmental Research Letters, in press). "There's enough evidence to suspect that the jet stream behaviour is being modulated by the sun," he says.

Blackburn says that blocking events have been unusually common over the last three years, for instance, causing severe floods in the UK and heatwaves in eastern Europe in 2007. Solar activity has been low throughout.

And even NOAA, while maintaining their usual statement of "strong evidence for a warming planet", published a report on August 13th, 2010 that the heat wave was not due to Global Warming.

> http://www.esrl.noaa.gov/psd/csi/moscow2010/

> Greenhouse gas forcing fails to explain the 2010 heat wave over western Russia. ... the current blocking event is intrinsic to the natural variability of summer climate in this region, a region which has a climatological vulnerability to blocking and associated heat waves (e.g., 1960, 1972, 1988). ...

> Our assessment indicates that, owing to the mainly natural cause for this heat wave, it is very unlikely that a similar event will recur next summer or in the immediate future (next decade).

For once, NOAA was right. Such heat did not return to Western Russia in 2011.

Finally, one other act of concealment by the mainstream and pseudoscience media. The heat wave was concentrated in western Russia. Siberia experienced much cooler than normal temperatures.

The following figure shows temperature anomalies for July 20-27 relative to the average for the same dates 2000-2008:

> http://earthobservatory.nasa.gov/IOTD/view.php?id=45069

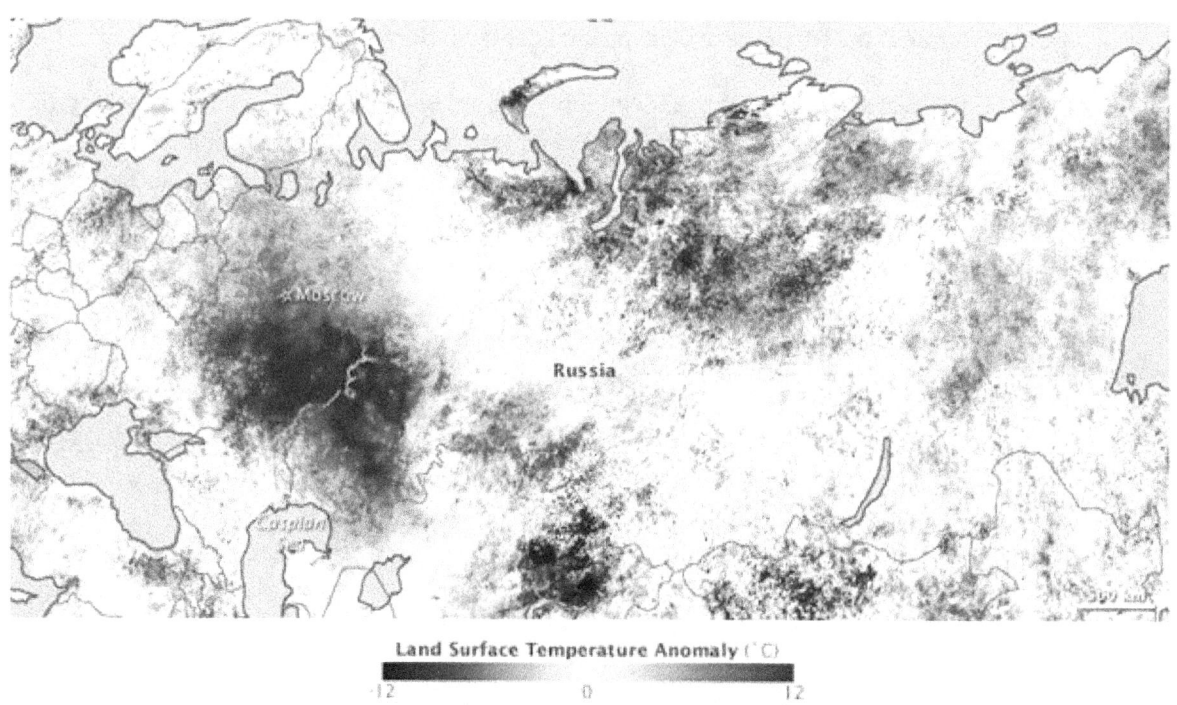

Land Surface Temperature Anomaly (°C)
-12   0   12

Mongolia

Mongolia, right in the center of the continent of Asia, becomes one of the coldest places on the planet in the winter. But the chill of 2009 – 2010 was so intense that much of the livestock, usually hardened to bitter conditions, perished.

Mongolia suffered "the worst winter most people here can remember.... It is supposed to be spring in Mongolia, but you would not know it."

http://news.bbc.co.uk/1/hi/world/asia-pacific/8592408.stm

Struggling to survive Mongolia's freezing winter

Page last updated at 08:58 GMT, Monday, 29 March 2010 09:58 UK

In Mongolia, what aid agencies are calling a slowly unfolding disaster is underway as extreme cold continues to devastate nomadic herder communities. As the BBC's Chris Hogg reports, about 10% of the country's livestock has perished and thousands of families have lost everything.

A prolonged drought has been followed by a bitterly cold winter

It takes about a day to drive from the capital, Ulan Bator, to the worst-affected part of the country, an area called Uvurkhangai where almost a million animals have died.

It is supposed to be spring in Mongolia, but you would not know it.

On the side of a hill in a wind so cold it bites, the Galsaikhan family are feeding their animals.

Some of the sheep and goats are so weak they collapse before they can reach the feeding tins.

The herders' youngest son, just five years old, picks up those that have fallen to the ground and finds them a space at the feeding tins so they can get at the food.

Usually by now tufts of grass would be poking through the snow for the livestock to graze, but not this year.

'Lost everything'

Mongolia is suffering the worst winter most people here can remember.

Last summer a drought made life for the herders hard enough, but weeks of unbearably cold weather with temperatures dropping as low as minus 40C have wrought real damage.

The government says more than four-and-a-half million livestock have died.

Mongolians call this a dzud - a prolonged freeze after a summer drought that destroys the grazing areas.

The Galsaikhan family have lost 800 of their 900 animals.

Chumedtseren, the mother, says the mornings are the worst. "Every day when we wake up we have the same fear. How many have died overnight?"

She says sometimes she and her husband are frightened to go to check.

"If we lose all our animals we'll have lost everything," she says.

The frozen carcasses of the animals lie where they have dropped, several of them in the pen where the others seek shelter from the wind.

No regrets

The reader is requested to note that the pseudoscience community never subsequently explained or expressed regret on how wrong their pretext of cold limited to Europe turned out to be. 2010 continued and continues to be officially the "warmest" year on record. One guesses that global warming enthusiasts hoped that one-sided propaganda, together with short-term public memory would put the cold winter of 2010-2011 behind everyone. Naturally, the mainstream and pseudoscience media also made very little mention of the record-cold Southern Hemisphere winter in the middle of 2010, which had preceded the remarkable cold in the Northern Hemisphere; more about this in a later section.

Ironically, a year later, the relatively mild winter of 2011-2012 in the Eastern and Central United States was grossly overvalued by Time Magazine as winter having forgotten the world. As demonstrated earlier, this statement was not only inaccurate, but was in direct contradiction to the statement by AGW advocates that extreme cold weather is a sign of planetary warming. No matter what the weather, the one unshakeable constant of pseudoscience practitioners is that the globe is getting hotter. Maybe ironic was the wrong word to use – such brazen insincerity is to be expected from minds congested by preconception.

It is interesting to observe that climate pseudoscience organizations rarely correct past reports which are subsequently demonstrated to be inaccurate. They probably hope that nobody will notice, and in any case, are too busy crafting new heaps of misinformation.

To dispel myths that the cold spells of later years were some sort of exotic one-off flukes in a warming planet, this section briefly examines the winter of 2009 – 2010 as well.

Florida

Sadly, extreme cold temperatures in winter are no longer unusual in the Sunshine State. Abnormal freezes in the 2011 and 2012 were preceded by record cold temperatures during the transition from 2009 to 2010.

Here is USA Today on January 11th, 2010

http://www.usatoday.com/weather/news/extremes/2010-01-11-florida-cold_N.htm

Frosty Florida sets record low temperatures

MIAMI (AP) — Record low temperatures chilled Florida from top to bottom Monday, endangering fruit and vegetable crops and taxing the power grid of a state unaccustomed to the cold.
The National Weather Service reported 36 degrees at the Miami airport, beating an 82-year-old record of 37 degrees. It dipped to 42 degrees in Key West, one degree off the record and the second-coldest reading since 1873.

"I even had ice on my car this morning, which was an unbelievable sight for Miami," said Dan Gregoria, meteorologist with the National Weather Service.

It was 14 degrees Monday morning in Tallahassee, breaking the record of 15 set in 1982. Record-tying lows of 29 were observed in Orlando, and Tampa's 25-degree weather beat its old record of 27.

South Florida is usually around 68 degrees [Fahrenheit] this time of year.

A runner bundles up as she exercises on the boardwalk on Monday in Miami Beach. The National Weather Service recorded 36 degrees in Miami Monday morning, which beat the 82-year-old record of 37 degrees.

By Joe Raedle/Getty Images

Cold records were set all over Florida. Below is a list of several new record lows as temperatures in the Sunshine State fell well below freezing into the 20s and even teens [Fahrenheit][138]:

City	New Cold Record January 2010 (°F)	Old Cold Record (°F)	Colder than Previous Cold Record by:	Year of previous Cold Record	No. of Years since previous Cold Record
Tallahassee	14°	15°	1°	1982	28
Brooksville	15°	26°	11°	1942	68
Gainseville	17°	20°	3°	1959	51
Tampa	25°	27°	2°	1982	28
Sarsota	28°	29°	1°	1959	51
Orlando	29°	29°	0°	1982	28
Fort Myers	31°	32°	1°	1959	51
St. Petersburg	33°	34°	1°	1977	33
West Palm Beach	33°	34°	1°	1927	83
Miami	36°	37°	1°	1927	83
Key West	42°	48°	6°	1970	40
Average			2.2°		50

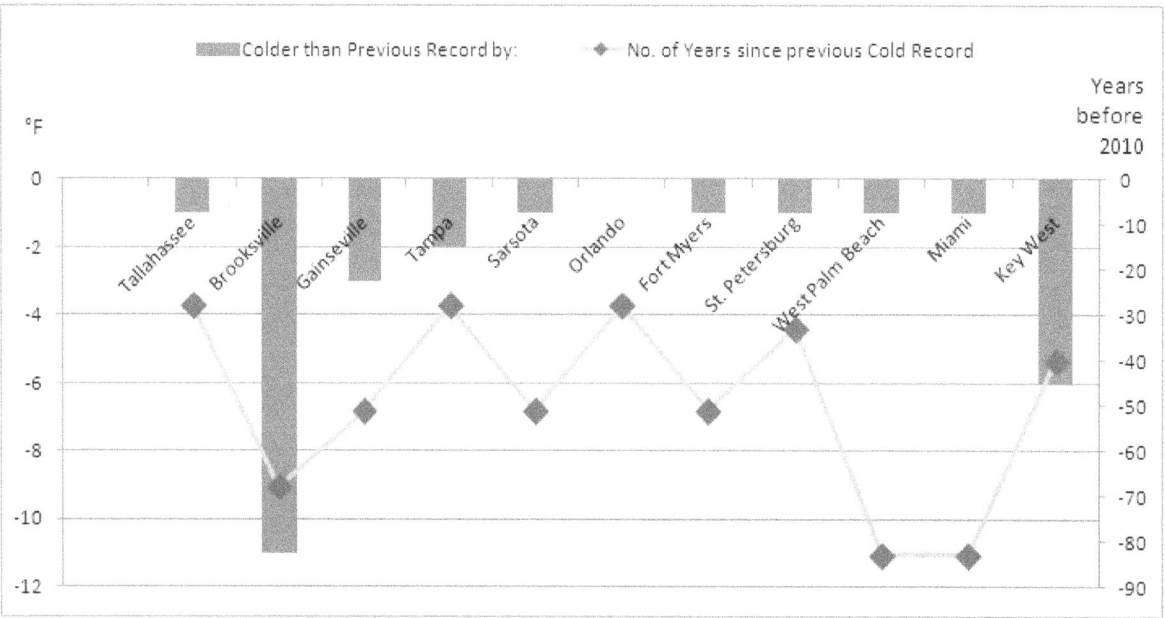

In summary, Florida had experienced temperatures 2.2°F colder than the previous record, which, on average, was set a half-century previously. Strange for a warming planet, isn't it?

Remarkably, even in the far south - the Florida Keys, temperatures plummeted to the low 40s, barely above freezing. "Climate weirding" - cold is warm – at play again, correct?

---

[138] http://www.weather.com/outlook/weather-news/news/articles/florida-record-cold-temperatures_2010-01-06

Typically when one cites cold weather to proponents of Anthropogenic Global Warming, their usual retort is to accuse the messenger of cherry picking. The modus operandi is actually quite simple: unusually cold weather is cherry picked weather and/or a sign of warmth in disguise; however, unusually hot weather is always climate – sure signs of planetary warming.

So the cold in Florida in the winter of 2010 was just a fluke, right? One could not be more wrong. Through December 2009 and January 2010, a dramatic cold spell swept across Eurasia, England and parts of North America. A spectacular image capturing this phenomenon was a satellite photo revealing the whole of Britain covered in snow.[139]

Global Warming Alarmists, stunned by the cold, used the same pretext as they would in the following year and the year after. Here's skepticalscience.com: "The cold snap is due to a strong phase of the Arctic Oscillation. This is causing cool temperatures at mid-latitudes (e.g. - Eurasia and North America) and warming in Polar Regions (Greenland and Arctic Ocean). The warm and cool regions roughly balance each other out with little impact on global temperature." [140] Don't worry, we shall not go into the details of the Arctic Oscillation explanation yet again; we already did that in great depth when we examined the cold winter of 2010-2011.

Some other "experts" added El Nino into the mix of "explanations". [141]

The addiction of pre-conceptualism never ceases. A warming planet is Rome, and all roads naturally lead to Rome, even if they have to be dug up, renovated and/or reconstructed altogether – the exact opposite of the open-minded practice of good science.

[139] http://news.bbc.co.uk/2/hi/uk/8447023.stm
[140] http://www.skepticalscience.com/December-2009-record-cold-spells.htm
[141] https://sites.google.com/site/whythe2009winterissocold/

And here is how Europe looked like in the winter of 2009 – 2010[142].

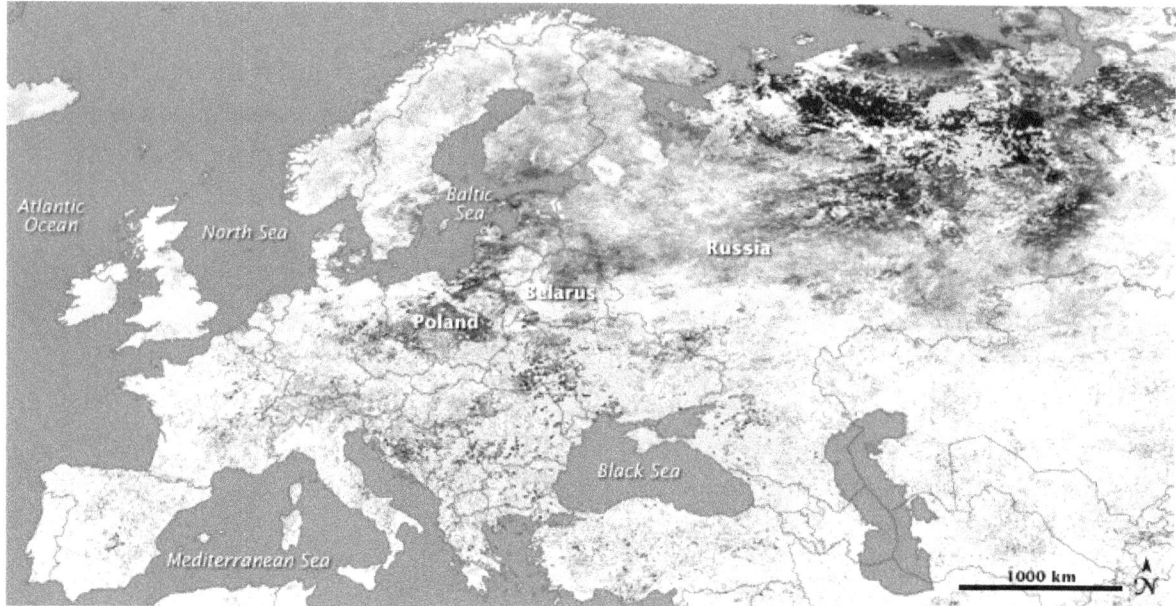

Instead of repeating the rebuttal of the flimsy and panicked "cold is warm" explanations of global warming alarmists, let us do something a little more interesting. Let us examine briefly the previous winter of 2008 – 2009. Here's what happened:

http://www.dailymail.co.uk/news/article-1118244/Americans-suffer-record-cold-temperatures-plunge-40C.html

Americans suffer record cold as temperatures plunge to -40C
By Julian Gavaghan
UPDATED: 09:50 GMT, 16 January 2009

Americans were today shivering as bitter arctic winds caused temperatures to plunge to record-breaking levels in many parts of the vast country.

There are even fears that crowds planning to watch Barack Obama's presidential inauguration next week could suffer hypothermia and frostbite in sub-zero conditions.

This winter has been one of the toughest in decades with temperatures today reaching as low as -38C in large areas of the Midwest and -40C in the coldest place.

But even on the east coast – where conditions are typically milder than the fridgid hinterlands – the icy blast was being felt.

New York endured a -14C chill today and further south Washington – which hosts Mr Obama's inauguration on Tuesday – plunged to 11 degrees below zero.

---

[142] http://earthobservatory.nasa.gov/IOTD/view.php?id=42067

Some places have recorded record lows – with the temperature in Flint, Michigan dipping to an incredible -28C.

Bone-chilling winds – known as the Alberta Clipper after a Canadian province – have also led to record snowfall

Dave Mangin braves sub-zero temperatures as he walks along the shore of Lake Michigan, in Milwaukee

Chicago has suffered the most consecutive days of snowfall since records began in 1884 after enduring nine days of blizzards.

But it's not just the northern part of the nation feeling the freeze.

By tomorrow, the Gulf Coast states will face a hard freeze with temperatures around -6C in places like Tulsa, Oklahoma, which is on the same latitude as Tunisia.

But there are parts of the U.S. that aren't freezing and fighting Jack Frost.

The west coast is seeing some summerlike beach weather.

San Fernando, California, clocked a 32C high and Southern California temperatures will remain warm thanks to a lingering high pressure system.

The coldest place in the continental United States has been International Falls, Minnesota on the Canadian border.

Overnight, the city had temperatures approaching -40C with chills making it feel like -60C.

'It's kind of raw. It really cuts your skin,' said Belmer Cole who maintains an outdoor high school ice rink in the city.

The colder temperatures make the ice dry and brittle, which makes skating tougher.

'When the ice gets this cold, and the temperature this cold, the puck freezes to the point where if it hits a goal post, sometimes the puck would break into pieces,' said International Falls High School hockey coach John Prettyman.

But residents of International Falls are used to bitter weather. It's a way of life for them, especially the ice fishermen.

'I'm still standing here. Parts aren't falling off yet. There's really no bad weather, just bad clothes,' said Woody Woods.

 A bird flies through steam hovering over the Menomonee River, Milwaukee: Temperatures in the region are below zero

Brr-ruff: Even dogs are suffering from the Arctic blast.

But chilly temperatures are potentially deadly for those without heat. The cold can be more harmful than even the hottest summer days.

'The problem is that cold hurts you much more rapidly and more uniformly. You can put a number of folks in the desert, many of them will do fine,' said Wallace Carter, of New York Presbyterian Hospital.

'You can take 10 people, put them out in sub-zero weather and they will all start to have serious side effects and will ultimately die.

'Most people think that they are well prepared and in fact they're not well prepared. People underestimate how incredibly insidious and violent cold temperature is.'

All of the above did not stop Hansen's GISS from catastrophically declaring 2009 as the "second warmest" year on record.

http://www.giss.nasa.gov/research/news/20100121/

"2009 was tied for the second warmest year in the modern record, a new NASA analysis of global surface temperature shows. The analysis, conducted by the Goddard Institute for Space Studies (GISS) in New York City, also shows that in the Southern Hemisphere, 2009 was the warmest year since modern records began in 1880."

After all, it was possible to declare anything by applying the latest in "hide the decline" and other temperature manipulation techniques. There will be more about this later. [143]

Education

The difference now was that, unlike the situation during previous decades of real increases in temperature, global warming alarmism was now fast losing credibility in the general population. As usual, the masters of fraud had underestimated the basic intelligence of ordinary human beings. So a massive campaign was launched in the media and elsewhere to "educate" them.

In the wake of the Climategate scandal in late 2009, "re-educating" an already skeptical adult population would indeed be difficult. Belief in Anthropogenic Global Warming has dipped significantly in Britain and even in traditionally AGW-friendly Germany.

http://www.nytimes.com/2010/05/25/science/earth/25climate.html

Climate Fears Turn to Doubts among Britons
By ELISABETH ROSENTHAL
Published: May 24, 2010

Nowhere has this shift in public opinion been more striking than in Britain, where climate change was until this year such a popular priority that in 2008 Parliament enshrined targets for emissions cuts as national law. But since then, the country has evolved into a home base for a thriving group of climate skeptics who have dominated news reports in recent months, apparently convincing many that the threat of warming is vastly exaggerated.

A survey in February by the BBC found that only 26 percent of Britons believed that "climate change is happening and is now established as largely manmade," down from 41 percent in November 2009. A poll conducted for the German magazine Der Spiegel found that 42 percent of Germans feared global warming, down from 62 percent four years earlier.

And London's Science Museum recently announced that a permanent exhibit scheduled to open later this year would be called the Climate Science Gallery - not the Climate Change Gallery as had previously been planned.

---

[143] This book only selectively describes some of the cold weather events in the past few years. For a more comprehensive list, the reader can refer to the following sites: http://www.iceagenow.com/ and http://iceagenow.info/

"Before, I thought, `Oh my God, this climate change problem is just dreadful,' " said Jillian Leddra, 50, a musician who was shopping in London on a recent lunch hour. "But now I have my doubts, and I'm wondering if it's been overhyped."

Perhaps sensing that climate is now a political nonstarter, David Cameron, Britain's new Conservative prime minister, was "strangely muted" on the issue in a recent pre-election debate, as The Daily Telegraph put it, though it had previously been one of his passions.

And a poll in January of the personal priorities of 141 Conservative Party candidates deemed capable of victory in the recent election found that "reducing Britain's carbon footprint" was the least important of the 19 issues presented to them

And, in the United States, here's a recent Rasmussen Poll from August 3, 2011

http://www.rasmussenreports.com/public_content/politics/current_events/environment_energy/69_say_it_s_likely_scientists_have_falsified_global_warming_research

69% Say It's Likely Scientists Have Falsified Global Warming Research
Wednesday, August 03, 2011

The debate over global warming has intensified in recent weeks after a new NASA study was interpreted by skeptics to reveal that global warming is not man-made. While a majority of Americans nationwide continue to acknowledge significant disagreement about global warming in the scientific community, most go even further to say some scientists falsify data to support their own beliefs.

The latest Rasmussen Reports national telephone survey of American Adults shows that 69% say it's at least somewhat likely that some scientists have falsified research data in order to support their own theories and beliefs, including 40% who say this is Very Likely. Twenty-two percent (22%) don't think it's likely some scientists have falsified global warming data, including just six percent (6%) say it's Not At All Likely. Another 10% are undecided. (To see survey question wording, click here .)

The number of adults who say it is likely scientists have falsified data is up 10 points from December 2009 .

Therefore, aside from making the most hay[144] out of the relatively mild winter in the continental United States and the drought that followed, proponents of AGW have widened their focus to include a bid to start indoctrinating kids in kindergarten.

http://www.washingtontimes.com/news/2012/jun/5/teaching-global-warming-in-kindergarten/

Teaching global warming in kindergarten

By Joy Pullmann

Tuesday, June 5, 2012

The Public Broadcast Service recently reported that increasing numbers of educators are teaching about the controversy over climate change. This has the scientific establishment doubling down on efforts to feed children their mantra: There is no debate.

There is no man behind the curtain, Dorothy. The "consensus" has spoken.

Except not among the hoi polloi. Eighty-two percent of science teachers report they have faced skepticism about climate change from students, according to the most recent poll from the National Science Teachers Association. Fifty-four percent report encountering the same skepticism from parents.

Those students and parents are not alone. Growing numbers of U.S. citizens are taking a critical look at the outlandish, government-laden "solutions" climate-change alarmists promote. Majorities of Americans in Rasmussen polls consistently disagree that human activity has caused global warming, and over the past five years of Pew Forum polls, fewer have been willing to say solid evidence shows it's a serious problem. In Pew's most recent poll on the subject, global warming slid to be U.S. voters' last priority.

In response to this turn of events, alarmists are engaging in a renewed public-relations campaign, most prominently including the May 11 release of draft science standards for elementary school students intended to apply nationwide.

The Common Core state education standards list what math and language-arts information and skills children should master in each grade. Forty-five states adopted those standards in 2009 and 2010 under heavy incentives from the Obama administration.

The Common Core next expands into science standards, which 26 states have committed to helping develop and implement. The draft standards integrate global warming and other overplayed worries about human impacts on the planet, starting in kindergarten.

This early-grade tendentiousness will create a foundation for ideas that build on long links of suppositions: catastrophic, man-made global warming; the evils of fossil fuels ("explain differences between renewable and nonrenewable sources of energy" in fourth grade) and the need for the Environmental Protection Agency and other government agencies to strangle human liberties in order to "save the planet."

Fifth-graders are to examine Earth's temperature increases and believe they destroy penguin habitat and erode coral reefs. Middle schoolers will have to accept that "human activities have significantly altered the biosphere, geosphere, hydrosphere and atmosphere" and agree

---

[144] With the connivance of the perennially biased mainstream and pseudoscience media

continuous monitoring is necessary to undergird "social policies and regulations that can reduce these impacts." They also must acknowledge the "disciplinary core idea" that "human activities, such as the release of greenhouse gases from burning fossil fuels, are major factors in the current rise in Earth's mean surface temperature ('global warming')." High schoolers will have to understand that "though the magnitudes of human impacts are greater than they have ever been, so too are human abilities to model, predict and manage current and future impacts."

There's lots of trust here in big government and phony climate models but not much interest in how all this affects normal folks. That's OK - we have too many of them, the curriculum teaches. High schoolers will be told our rising global population makes "land for agriculture or drinkable water ... scarcer and more valued."

Note the implication that kids who do not accept some very specific and politically motivated scientific claims are not properly educated. This contradicts the Common Core's requirement that students pose and respond to "questions that probe reasoning and evidence; ensure a hearing for a full range of positions on a topic or issue; clarify, verify or challenge ideas and conclusions; and promote divergent and creative perspectives."

Sure, that's exactly what a single set of nationwide science standards created by a technocratic, politicized science establishment will promote. Close that curtain, kid.

The European Union and Global Warming

Take a moment to picture the European Union (EU) in 2011. What images come to mind?
- Debt crisis
- 800 billion Euro bailout fund
- Greek default → massive economic contraction and increase in unemployment
- Spain and Italy with unsustainable debt → bond interest rates at 7% or higher
- 25% unemployment in Span / 50% of Spanish youth without jobs
- Bailouts by Northern Europe (principally Germany) of Greece, Italy, Portugal, etc.
- Immense austerity imposed by the troika of the International Monetary Fund (IMF), European Central Bank (ECB) and EU on bankrupt nations receiving bailout money, such as Ireland, Portugal and Greece
- Self-imposed austerity in other indebted countries such as Spain, Italy and even the United Kingdom hoping to avoid bankruptcy
- etc.

So the politicians of the nearly defunct European Union committing to giving away $30 billion a year to developing nations to "combat" climate change would seem absurd, correct? But this is exactly what happened at the climate summit in Durban, South Africa at the end of 2011.

http://www.greenbusinessguide.co.za/eu-nations-get-cold-feet-over-climate-change-fund/

EU NATIONS GET COLD FEET OVER CLIMATE CHANGE FUND
International News

09 May 2012

BRUSSELS (Reuters) – EU nations have yet to come up with a plan on how to fill a multi-billion euro fund to help tackle climate change, even as the region's executive body hosts talks with countries likely to bear the brunt of extreme weather.

The European Union recommitted to providing 7.2 billion euros ($9.4 billion) for the fund over 2010-12, according to draft conclusions seen by Reuters ahead of a meeting of EU finance ministers next week.

But after that, how much cash will flow is unclear as the text, drafted against the backdrop of acute economic crisis in the euro zone, states the need to "scale up climate finance from 2013 to 2020", but does not specify how.

The Green Climate Fund aims to channel up to $100 billion globally per year by 2020 to help developing countries deal with the impact of climate change.

Its design was agreed at international climate talks in Durban last year.

Europe's Climate Commissioner Connie Hedegaard is fighting to build on the fragile international agreement in Durban.

On Monday and Tuesday, together with the Danish EU presidency, she is holding informal discussions in Brussels with members of what some call the "coalition of ambition", ahead of U.N. talks in Bonn later this month.

The coalition is a union of the EU, the Alliance of Small Island States and the Least Developed Countries. At the U.N. talks in Durban, it played a lead role in forging agreement on keeping alive the Kyoto process, the only global framework on addressing global warming.

Hedegaard told reporters the talks in Brussels were "informally testing different ideas".

"We are all in agreement: no back-tracking, no less ambition. What binds us is this idea we will push for ambition."

Representing the low-lying Marshall Islands, Tony deBrum, minister in assistance to the president, said the EU's leadership had thrown "a life-line" to parts of the world most vulnerable to rising sea levels.

"For us, this is a matter of the survival of a people," he said. "A culture, a language, a way of life."

'STUBBORN'

Non-governmental organization Oxfam said "intransigence" from some EU member states was putting the coalition at risk as they are arguing against firm commitments to finance after 2012.

"At a critical moment in the fight against climate change, Europe looks to be sitting back rather than stepping up," Lies Craeynest, Oxfam's EU climate change expert, said.

Debate within the EU has also focused on how much of the bloc's $30 billion share of the $100 billion should come from the private sector, which would reduce the need for public funds.

The draft conclusions ahead of the May 15 EU ministerial meeting noted "further efforts are required to clarify the concept of private financing and its contribution to the $100 billion".

Representing the Danish EU presidency Martin Lidegaard, minister for climate, energy and building, told reporters there were gaps to fill in terms of pledges to curb greenhouse gas emissions and promises of finance.

"There are some areas where only public funds will do," Lidegaard said, but added public finance could also help to leverage private funds.

"Between public and private is definitely an area we have to explore."

Some money could come from levies on shipping and aviation, although these sources are contentious.

Since January, all airlines using EU airports have been required to offset their carbon emissions using the EU's Emissions Trading Scheme (ETS), prompting an outcry from some non-EU nations.

The level of protest has intensified effort at the U.N.'s International Civil Aviation Organization (ICAO) to come up with a global plan for tackling airline emissions. If achieved, that would justify modifying the EU scheme, the Commission has said.

Experience suggests it would be rash to hope for too much from ICAO. The European Commission said the EU was driven to bringing all airlines into its scheme because more than a decade of talks at ICAO had produced no result.

The Commission is also working with the International Maritime Organization (IMO) to find a solution for shipping emissions.

The draft conclusions called on "the EU and its member states to further engage effectively in negotiations in ICAO and IMO to support carbon pricing schemes".

It is no wonder that European politicians have kept the Durban commitment very quiet. One cannot even begin to conceive an act more irresponsible from the leaders of a region in economic stagnation and teetering on the verge of bankruptcy. Bragging about giving away $30 billion annually may not have gone down very well with a population suffering from high unemployment, austerity, increases in the retirement age, massive cuts in government benefits and spending, etc. Wouldn't that money have been better spent to first help distressed people at home? My guess is that most Europeans are not even aware of this suicidal act of "generosity".

In summary, the following parody summarizes the shifting narrative of Anthropogenic Global Warming.

Cold is absence of heat but to proponents of AGW, cold is evidence of warming.

Day is night.

Up is down.

Left is right.

Blue is orange.

Dry is wet.

And "extreme" is anything the observer has not witnessed before in his or her lifetime.

To be convinced by AGW, one either needs to blindly trust what one hears in the media or one has to believe all of the above.

****

To be continued in Part II of the AGW Saga

Key Themes in Part II:

How temperature data has been tampered to concoct the appearance of continued planetary warming
Details of the Climategate revelations and their implications
A list of key scientists who disagree with AGW (so much for consensus....)
The facts and the hype about glacier and polar ice cap melts
The facts and the hype about sea level rise and disappearing islands and coastlines
More about the techniques used to spin and distort in the mainstream media
Negative economic impacts resulting directly from carbon mania
Al Gore (one cannot, of course exclude him in a dissertation on AGW)